Fourth Edition

Modern Elementary Algebra

Vivian Shaw Groza

SAUNDERS COLLEGE PUBLISHING
Philadelphia New York Chicago San Francisco
Montreal Toronto London Sydney Tokyo
Mexico City Rio de Janeiro Madrid

Address orders to:
383 Madison Avenue
New York, NY 10017

Address editorial correspondence to:
West Washington Square
Philadelphia, PA 19105

Text Typeface: Helvetica
Compositor: York Graphic Services, Inc.
Acquisitions Editor: Leslie Hawke
Project Editor: Sally Kusch
Copyeditor: Bonnie Boehme
Managing Editor & Art Director: Richard L. Moore
Design Assistant: Virginia A. Bollard
Text Design: Caliber Design Planning, Inc.
Cover Design: Larry Didona
New Text Artwork: Vantage Art
Production Manager: Tim Frelick
Assistant Production Manager: Maureen Read

Cover credit: Cover photograph by John Drooyan © 1982

Library of Congress Cataloging in Publication Data
Groza, Vivian Shaw.
Modern elementary algebra.
Rev. ed. of: Modern elementary algebra for college students. 3rd ed. c1977.
Includes index.
1. Algebra. I. Title.
QA152.2.G76 1983 512.9 82-60629
ISBN 0-03-042651-0

Photo Credits

Cameo of François Viète on page 26 courtesy of Brown Brothers.
Cameo of Girolamo Cardano on page 55 courtesy of The New York Public Library, Astor, Lenox, and Tilden Foundations.
Cameo of René Descartes on page 272 and cameo of Pythagoras on page 302 courtesy of David Eugene Smith Collection, Columbia University Libraries.

Modern Elementary Algebra, Fourth Edition ISBN 0-03-042651-0

Library of Congress catalog card number 82-60629.

345 016 98765432

CBS COLLEGE PUBLISHING
Saunders College Publishing
Holt, Rinehart and Winston
The Dryden Press

Preface

For this fourth edition, both the exercise sets and the exposition have been carefully revised.

Each chapter now begins with a list of objectives, typical problems related to the objectives, and text locations of the topic involved. The set of problems in this list could be used as a pretest for the chapter.

The exercise sets still contain a generous supply of technique problems and applications. The major improvements are as follows:

1. The odd-numbered problems are now completely parallel with the even-numbered problems.
2. The problems in each set now begin with easy problems and progress to those of greater difficulty.
3. Verbal problems have been updated, especially with regard to recent financial changes.

In keeping with several recommendations, the answers to all the problems now appear in the back of the book.

Several changes have been made in the Contents.

Root extraction has been removed from Chapter 1. This topic is covered in Chapter 6 and is not needed in the development of the first five chapters. Many beginning students have experienced difficulty with this concept at this time.

Quite a few alterations were made in Chapter 3, Operations on Polynomials. Each special product and its related factoring form are now placed in one section. Product of Sum and Difference and Factoring Difference of Squares are in Section 3.5; Product of Related Binomials and Factoring Simple Trinomials in Section 3.6; Perfect Square Trinomials in Section 3.7; and General Trinomials in Section 3.8.

The FOIL method has been included as a technique for multiplying binomials. It has proved to have great value in promoting student understanding.

The special forms, Sum and Difference of Cubes, have been removed from Chapter 3. It was felt that these topics are more appropriate for a second course in algebra.

In Chapter 4, Fractions, multiplication and division have been placed before addition and subtraction. This is a more natural order, which facilitates the development and understanding of the more difficult topic of addition.

Simplification of quotients involving radicals has been added to Chapter 6, Quadratic Equations. This permits a change in the technique for solving $ax^2 + bx + c = 0$ for $a \neq 1$ by completing the square (that is, "divide by *a*" instead of "multiply by 4*a*"). The same method is then used to derive the quadratic formula.

There is an Instructor's Manual to accompany the text, containing four tests for each chapter, with answers. The manual also has multiple-choice questions, with the questions identified by chapter number. Comments about objectives, possible omissions, possible instructor functions, and common errors are included in the hope of providing a useful service. The Manual also contains six final examinations.

Acknowledgements

I would like to thank the following people for their many helpful and constructive suggestions: James Kelley, Samford University; Henry Cataldo, Los Angeles City College; James Smith, New Mexico State University; Dorothy Thames, Lamar University; Stewart M. Venit, California State University-Los Angeles; Mark Glucksman, El Camino College; Terry Slaubaugh, Blue Ridge Community College; Ellen Church, Golden West College; and Edward Hoff, Glendale College.

Vivian Shaw Groza

Contents

1 Basic Concepts

Objectives

OBJECTIVES	TYPICAL PROBLEMS	LOCATIONS
Read and do the operations of addition, subtraction, multiplication, division, squaring, and cubing.	1–8. Do each indicated operation. 1. $8 \cdot 5$ 2. $9 + 8$ 3. $20 - 4$ 4. 5^2 5. $3(9)$ 6. 9^3 7. $\frac{39}{3}$ 8. $(4)(25)$	*Section 1.1* *Example 1* *Exercises 1–40*
Translate word expressions about numbers into symbols.	9–14. Translate into symbols. 9. The sum of x and 6 10. The product of 8 and y 11. The quotient when 5 is divided by n 12. The difference when 4 is subtracted from x 13. The square of c 14. The cube of 8	*Section 1.1* *Example 2* *Exercises 41–80*
List special subsets of the set of natural numbers.	15–20. List each set. 15. The multiples of 6 between 10 and 50 16. The factors of 110 17. The even factors of 110 18. The odd factors of 110 19. The prime factors of 110 20. The composite factors of 110	*Section 1.2* *Examples 1–8* *Exercises 1–48*

Do operations involving grouping symbols.

21–24. Do the indicated operations.

21. $25 - (7 - 3)^2$

22. $80 - 2 \cdot 5^2 - (20 - 5)$

23. $\dfrac{6^2 - 6 \cdot 4}{3 \cdot 6 - 3 \cdot 4}$

24. $8 + 2\{9 - 3[7 - (8 - 3)]\}$

Section 1.3
Examples 1–4
Exercises 1–60

Evaluate algebraic expressions for a given value of one or more variables.

25–26. Evaluate.

25. $x^2 - (x - 3)(x + 2)$ for $x = 6$

26. $C = \dfrac{35(A + B)}{A^2 - B^2}$ for $A = 7$ and $B = 2$

Section 1.4
Examples 1–12
Exercises 1–66

State the meaning of a negative number for a given meaning of the corresponding positive number.

27. State the meaning of -6 if $+6$ means a gain of 6 pounds.

Section 1.5
Examples, Table 1.2
Exercises 21–40

Name the coordinate of a point on a horizontal number line.

28–30. Name the coordinate of the point on the number line indicated by the dot.

28. 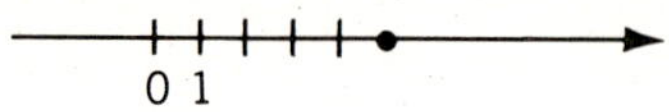

29. 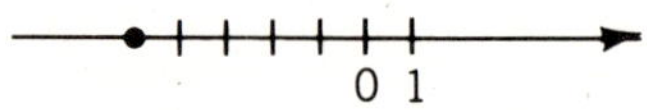

30. 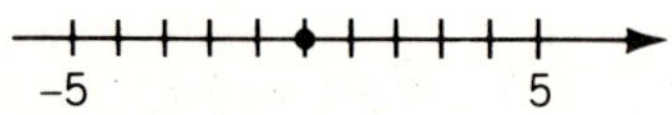

Section 1.5
Examples, Figures 1.4, 1.6
Exercises 41–50

Name the coordinate of a point on a vertical number line.

31–33. Name the coordinate of the point on the number line indicated by the dot.

31. 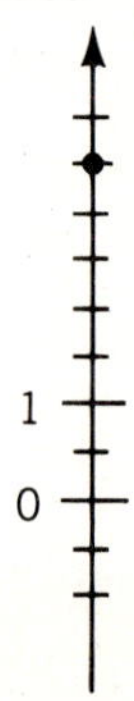

32. 

33. 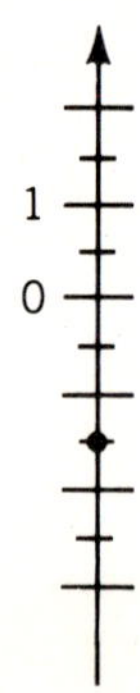

Section 1.5
Examples, Figure 1.5
Exercises 51–60

3

Draw the graph of a rational number on a horizontal number line.

34. Graph each number on a horizontal number line.

a. $2\frac{1}{3}$ b. -2

Section 1.5
Examples, Figures 1.4, 1.6
Exercises 61–70

Draw the graph of a number on a vertical number line.

35. Graph each number on a vertical number line.

a. $1\frac{3}{8}$ b. $-2\frac{7}{8}$

Section 1.5
Examples, Figure 1.5
Exercises 71–80

Simplify expressions involving absolute value symbols.

36–37. Simplify.
36. $|-5| + |-3|$
37. $-(|-9| - |-5|)$

Section 1.6
Examples 1–3
Exercises 1–36

Insert one of the relation symbols, $<$, $>$, or $=$, between two numbers to form a true statement.

38–40. Insert $<$, $>$, or $=$ between each pair of numbers to form a true statement.
38. 5, 8
39. -5, -8
40. $|-5|$, 5

Section 1.6
Examples 4, 5
Exercises 37–62

Graph order relations on a number line.

41–43. Graph each set of numbers on a horizontal number line.
41. $x \leq 4$
42. $x > -3$
43. $-4 \leq x < 3$

Section 1.6
Examples 6–8
Exercises 63–80

Add signed numbers.

44–48. Do each addition.
44. $(-6) + (-9)$
45. $2 + (-8)$
46. $8 + (-3)$
47. $-(-10 + 3)$
48. $-15 + 17 + (-13)$

Section 1.7
Examples 1–7
Exercises 1–96

Subtract signed numbers.

49–52. Do each subtraction.
49. $24 - 50$
50. $-12 - 8$
51. $-9 - (-2)$
52. $7 - (-10)$

Section 1.8
Examples 1–4
Exercises 1–100

Multiply signed numbers.

53–56. Do each multiplication.
53. $8(-6)$
54. $-8(-6)$
55. $2(-4)^2$
56. $(-3)(-4)(-5)$

Section 1.9
Examples 1–5
Exercises 1–70

Divide signed numbers.	57–62. Do each operation, if possible. 57. $\frac{72}{-9}$ 58. $\frac{-32}{4}$ 59. $\frac{-65}{-5}$ 60. $\frac{8-8}{8}$ 61. $\frac{2-6}{6-2}$ 62. $\frac{2-6}{6-6}$	*Section 1.10* *Examples 1–3* *Exercises 1–80*
Evaluate expressions involving signed numbers.	63. Evaluate $b^2 - 4ac$ for $a = 2$, $b = -5$, $c = -3$.	*Section 1.11* *Examples 1–3* *Exercises 1–60*

Answers to Typical Problems

1. 40
2. 17
3. 16
4. 25
5. 27
6. 729
7. 13
8. 100
9. $x + 6$
10. $8y$
11. $\frac{5}{n}$
12. $x - 4$
13. c^2
14. $8^3 = 512$
15. {12, 18, 24, 30, 36, 42, 48}
16. {1, 2, 5, 10, 11, 22, 55, 110}
17. {2, 10, 22, 110}
18. {1, 5, 11, 55}
19. {2, 5, 11}
20. {10, 22, 55, 110}
21. 9
22. 15
23. 2
24. 14
25. 12
26. 7
27. A loss of 6 pounds
28. 5
29. -5
30. 0
31. $3\frac{1}{2}$
32. 0
33. $-1\frac{1}{2}$
34.

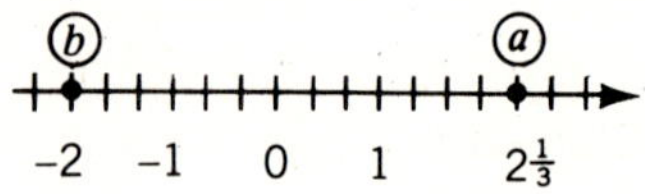

35.

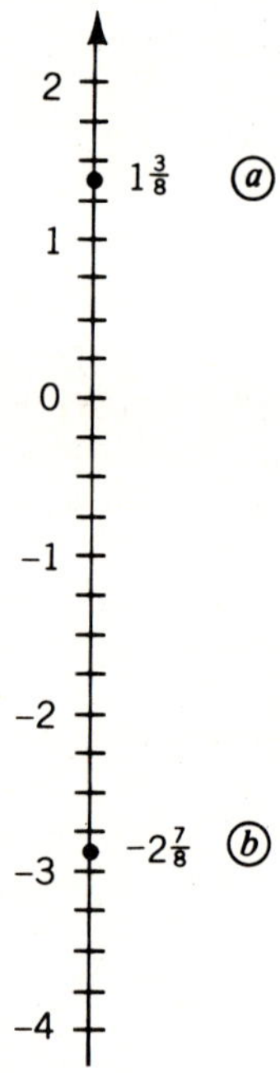

36. 8
37. -4
38. $5 < 8$
39. $-5 > -8$
40. $|-5| = 5$

41. 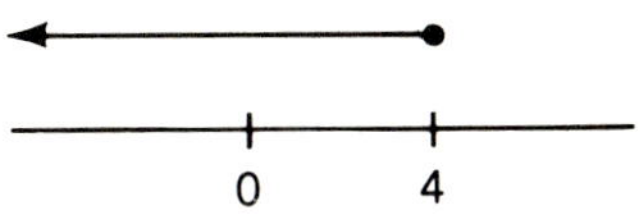

42. 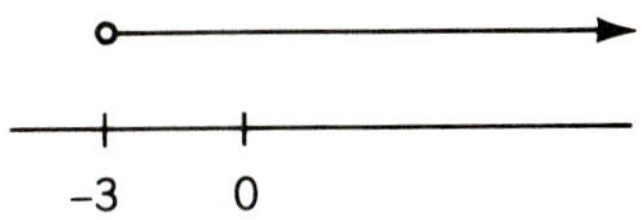

43. 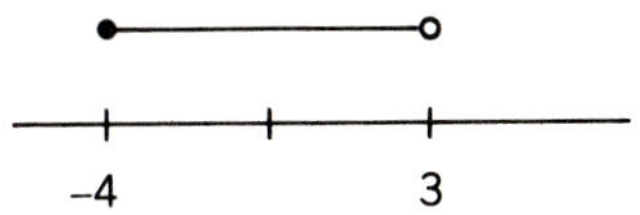

44. −15
45. −6
46. 5
47. 7
48. −11
49. −26
50. −20
51. −7
52. 17
53. −48
54. 48
55. 32
56. −60
57. −8
58. −8
59. 13
60. 0
61. −1
62. Undefined
63. 49

Elementary algebra is a study of **numbers** and their properties.

Number problems have played an important role in people's existence for thousands of years. At first these problems were written in words. Gradually, the words were replaced by abbreviations of the words, and finally the words and abbreviations were replaced by **symbols.**

The Word Algebra

The word *algebra* first appeared around A.D. 825 in the work *Hisâb al-jabr w'al-muqâbalah* by the Arabic scholar al-Khowârizmi. Literally this title means "science of restoration and opposition." A more common meaning of *al-jabr* was introduced into Spain by the Moors: An *algebrista* was a person who reset or "restored" broken bones.

The red-and-white-striped pole in front of barber shops goes back to the days when barbers also practiced medicine; in Spain barbers also had these words above the door: *Algebrista y Sangrador* ("bonesetter and blood-letter").

The use of symbols reduced the amount of writing that was necessary and enabled the problem to be seen and understood more rapidly.

For example, contrast the verbal account of the financial transaction stated below with its symbolic representation, which follows.

VERBAL ACCOUNT

A dealer sold a car for two thousand three hundred ninety-five dollars. The car cost him one thousand nine hundred eighty dollars. His profit was four hundred fifteen dollars.

SYMBOLIC ACCOUNT

$$\begin{array}{ll} \$2395 & S \\ \underline{\$1980} & \underline{C} \\ \$\ \ 415 & P \end{array}$$

The letter *S*, written at the right of \$2395, indicates that this amount is the selling price of the car. Similarly, the letter *C* is used to indicate the cost, and the letter *P* is used to indicate the profit.

It is easy to see that, in general, the profit is found by subtracting the cost from the selling price. In symbols, one writes

$$P = S - C$$

In summary, symbols are used in algebra for the following reasons:

1. To economize the amount of writing that is necessary.
2. To promote clarity and understanding of the problem.
2. To facilitate the discovery and generalization of number properties.

Early Names for Algebra

Algebra was not always called algebra. Since the unknown quantity was referred to as the *thing*, which in Italian is *cosa*, we find algebra called the *regola de la cosa* by some Italian writers, *die coss* by German writers, and *cossike arte* by the English.

In the fifteenth and sixteenth centuries, Italians referred to algebra as the *greater* art and to arithmetic as the *lesser* art. Cardano (1501–1576) called his book on algebra *Ars magna;* other Italians used *l'arte maggiore*. The Spaniard Juan Diez used *l'arte mayor* in his book *Sumario compendioso,* published in 1556, the first mathematical work published in the New World.

The name *algebra* was established through the success of a book by the Swiss mathematician Leonhard Euler. His *Vollstaendige Anleitung zur Algebra* ("Complete Introduction to Algebra"), published in 1770, was so popular that it served as the model of algebra texts for many years.

1.1 Numbers, Numerals, Operations

Numbers and Numerals

The abstract concept of number was developed to answer the question "How many?" The first numbers to be invented were the **counting numbers:** one, two, three, four, five, and so on.

To keep records of numbers, people invented symbols or numerals to serve as the names of numbers.

Numerals are symbols that name numbers according to a specified system, such as the Hindu-Arabic system of numeration. The Hindu-Arabic system of numeration is the modern decimal system in use today. The symbols 1, 2, 3, 4, 5, 6, 7, 8, and 9, which name the first nine counting numbers, and the zero symbol, 0, are called the **digits.** The numerals that name the other counting numbers are formed from these ten digits by using the principles of a positional numeral system with base ten.

For example, 346 is the *numeral name* of a number whose *word name* is three hundred forty-six.

Letters of the alphabet are also used as the names of numbers. A letter is used to designate a number whose numeral name is unspecified.

For example, in the formula $P = S - C$, the letters P, S, and C are used to designate numbers that represent the profit, the selling price, and the cost, respectively.

As another example, $y = x + 5$ means that y represents a number that is five more than the number represented by x.

A **constant** is a letter that names exactly one number. Letters at the beginning of the alphabet, such as a, b, or c, are usually used to name a constant. The Greek letter π (read "pi") is used to designate the constant ratio of the circumference C of any circle to its diameter d; that is, $\pi = \frac{C}{d}$.

A **variable** (also called an unknown) is a letter that names a number belonging to a set consisting of more than one member.

A **set** is a well-defined collection of objects called *elements* or *members* of the set.

Any letter can be used to name a variable, but usually the letters at the end of the alphabet, such as x, y, and z, are used, following the example of the French mathematician René Descartes, who perfected much of the symbolism of algebra.

Operations

Elementary algebra is concerned with six operations that are performed on numbers:

1. Addition
2. Subtraction
3. Multiplication
4. Division
5. Raising to a power
6. Root extraction

All these operations are indicated by symbols. The set of numbers on which the operations are performed is always designated. In this section, all numbers will be restricted to belong to the set of **whole numbers**—that is, the number zero and the natural numbers (counting numbers).

Furthermore, the operations discussed in this section will be restricted to addition; subtraction; multiplication; division; and two special cases of raising to a power, namely, squaring and cubing.

Addition: $x + y$

The result of adding two numbers is called the **sum** of the two numbers. The numbers that are added are called **terms.**

In symbols, $x + y$ indicates the sum of the terms x and y and is read "x plus y" or "the sum of x and y."

For example, the sum of x and 5 is $x + 5$.

Subtraction: $x - y$

The result of subtracting one number from another is called the **difference** or **remainder.**

In arithmetic, the name **subtrahend** is given to the number that is subtracted from the other number, called the **minuend.**

In algebra, both the subtrahend and the minuend are called **terms.**

In symbols, $x - y$ indicates the difference between the terms x and y, when y is subtracted from x. Thus $x - y$ is read "x minus y" or "y subtracted from x."

For example, the difference when 5 is subtracted from x is $x - 5$.

Multiplication: xy, $3x$, $3 \cdot 4$, $3(4)$, $(3)(4)$

The result of multiplying two numbers is called the **product.** The numbers that are being multiplied are called **factors.**

In arithmetic, multiplication is indicated by the "cross" symbol, $\times$, such as 3×4. However, because this symbol can be confused with the letter x used to name an unspecified number, the $\times$ symbol is not used in algebra to indicate multiplication.

In algebra, multiplication is indicated in three different ways. If two numerals are used to name the factors, then the dot symbol, $\cdot$, is used, as in $3 \cdot 4$, which means the product of 3 and 4, or 12. The dot is written at half the vertical height of the numerals so that it will not be confused with the decimal point.

If two letters or a numeral and a letter are used to name the factors, then multiplication is indicated by juxtaposition—that is, the symbols for the numbers are written side by side. For example, $3x$ means the product of 3 and x, and xy means the product of x and y. Twice x is written $2x$.

Multiplication is also indicated by writing one factor next to the other factor with either factor or both enclosed in parentheses. For example, 3(4) and (3)(4) mean the product of 3 and 4.

Division: $\frac{x}{y}$

The result of dividing one number by another is called the **quotient.** The name **dividend** is given to the number that is being divided by the other number, called the **divisor.**

In arithmetic, division is indicated by the symbol $\div$. In algebra, division is indicated by using the fractional notation, $\frac{x}{y}$. The expression $\frac{x}{y}$ is read "x divided by y" or, more informally, "x over y."

For example, the quotient when x is divided by 5 is written $\frac{x}{5}$. This can also be read "x over five."

Squaring: x^2

Multiplying a number by itself is called **squaring.** The result of squaring is called the **square** of the number. The numeral 2, called an **exponent,** is written to the upper right of another number symbol, x, called the **base.** The exponent 2 indicates that the base x is to be used twice as a factor.

For example, 3^2 (read "three square") means $3 \cdot 3$, or 9. Similarly, 5^2 means $5 \cdot 5$, or 25. In general, x^2 means xx.

Cubing: x^3

The product obtained by using a number as a factor three times is called the **cube** of the number. The numeral 3, the exponent, is written to the upper right of another number symbol, x, the base. The exponent 3 indicates that the base x is to be used three times as a factor.

For example, 5^3 (read "five cube") means $5 \cdot 5 \cdot 5$, or 125. In general, x^3 means xxx.

Squaring and cubing are special cases of the more general operation called **raising to a power,** written as x^n, where n is a counting number. The exponent n indicates that the base x is to be used n times as a factor.

TABLE 1.1 **Summary of Operation Symbols**

Operation	Numerals	Numeral and Letter	Two Letters
Addition	$3 + 4$	$x + 4$	$x + y$
Subtraction	$7 - 3$	$x - 4$	$x - y$
Multiplication	$3 \cdot 4$ or $3(4)$ or $(3)(4)$	$3x$	xy
Division	$\frac{12}{3}$	$\frac{x}{3}$	$\frac{x}{y}$
Squaring	3^2	x^2	(Not Applicable)
Cubing	5^3	x^3	

Table 1.1 summarizes operation symbols. Numerical tables are also useful for finding squares, cubes, square roots, and cube roots of some natural numbers. An extensive table is found on the inside cover of this book.

Exercises 1.1

1–40. Do each indicated operation. See Example 1.

Example 1 Do each indicated operation:

a. $6 \cdot 4$ b. $20 - 8$ c. $\frac{15}{3}$ d. $9 + 7$ e. 6^2

f. 2^3 g. $5(3)$ h. $(30)(40)$

SOLUTION

a. $6 \cdot 4 = 24$ (Multiplication)
b. $20 - 8 = 12$ (Subtraction)
c. $\frac{15}{3} = 5$ (Division)
d. $9 + 7 = 16$ (Addition)
e. $6^2 = 6 \cdot 6 = 36$ (Squaring)
f. $2^3 = 2 \cdot 2 \cdot 2 = 8$ (Cubing)
g. $5(3) = 15$ (Multiplication)
h. $(30)(40) = 1200$ (Multiplication)

1. $4 \cdot 6$
2. $(8)(9)$
3. $\frac{12}{6}$
4. $\frac{25}{5}$
5. $7 + 8$
6. $6 + 7$
7. $12 - 4$
8. $14 - 3$
9. 5^2
10. 3^2
11. 4^3
12. 2^3

13. $9 + 0$

14. $0 + 4$

15. 7^2

16. 8^2

17. 9^3

18. 4^3

19. $23 - 0$

20. $56 - 0$

21. $\frac{6}{1}$

22. $\frac{1}{1}$

23. $(10)(20)$

24. $35 + 65$

25. $47 + 53$

26. $7 \cdot 200$

27. $0 \cdot 9$

28. $\frac{5}{5}$

29. $\frac{200}{10}$

30. $8(0)$

31. $\frac{200}{200}$

32. $\frac{1000}{40}$

33. $100 - 0$

34. $40 - 0$

35. $100 - 6$

36. 3^3

37. 6^3

38. $1000 - 25$

39. $8(125)$

40. $9 \cdot 81$

41–80. Translate into algebraic expressions. See Example 2.

Example 2 Translate each of the following into algebraic expressions:

	SOLUTION
The product of 3 and y	$3y$
The quotient when t is divided by 3	$\frac{t}{3}$
The cube of a	a^3
The sum of 4 and z	$4 + z$
The square of b	b^2
The difference when c is subtracted from 8	$8 - c$

41. The sum of x and 8

42. The sum of y and 5

43. The product of 8 and x

44. The product of 4 and t

45. The square of n

46. The square of y

47. The cube of t

48. The cube of c

49. The difference when 4 is subtracted from y

50. The difference when 1 is subtracted from x

51. The quotient when x is divided by 6

52. The quotient when y is divided by 5

53. The product of s and t
54. The product of y and z
55. c squared
56. y squared
57. z cubed
58. n cubed
59. x plus y
60. a plus b
61. x times y
62. a times b
63. The difference when x is subtracted from y
64. The difference when a is subtracted from b
65. The quotient when x is divided by y
66. The quotient when a is divided by b
67. The difference when x is subtracted from 7
68. The difference when x is subtracted from 10
69. The quotient when 5 is divided by y
70. The quotient when 10 is divided by x
71. The sum of the square of x and the square of y
72. The sum of the square of a and the square of b
73. The difference when the square of t is subtracted from the square of 5
74. The difference when the square of 6 is subtracted from the square of n
75. The quotient when the sum of a and b is divided by the product of a and b
76. The quotient when the product of x and y is divided by the sum of x and y
77. The product of 5 and the cube of n
78. The product of 7 and the cube of k
79. The difference when the square of 4 is subtracted from the product of 4 and y
80. The difference when the product of 6 and x is subtracted from the square of 6

1.2 Special Sets of Numbers

Particular sets may be defined by two general methods, the *listing* method and the *description* method.

In the **listing method,** the set is defined by listing or stating the names of its members enclosed by braces and separated by commas.

For example, the set of **digits** may be defined by the listing method as follows:

$\{0, 1, 2, 3, 4, 5, 6, 7, 8, 9\}$

As another example, the set of **natural numbers** (another name for the set of counting numbers) may be defined as follows:

$\{1, 2, 3, 4, 5, 6, 7, 8, 9, 10, \ldots\}$

where the three dots indicate that this list continues in this manner without ending.

Note that zero is *not* a natural number.

The set of **whole numbers** is the set consisting of the number zero and the natural numbers.

The set of natural numbers and the set of whole numbers are infinite sets, whereas the set of digits is a finite set.

In the **description method,** the set is defined by "describing" the set or stating a property possessed by each member of the set. For example,

S is the set of digits exactly divisible by 3

Using the listing method,

$S = \{0, 3, 6, 9\}$

The **empty set,** also called the **null set,** is the set that has no members. It is denoted symbolically by $\varnothing$ or by $\{\quad\}$.

For example, the set of digits that are greater than 17 is the empty set.

The empty set is considered a finite set.

A **universal set,** U, is a set to which the elements of all other sets in a particular discussion must belong.

For example, let $U = \{1, 2, 3, 4, 5, 6, 7, 8, 9\}$. Now if A is the set of numbers that are exactly divisible by 3, then $A = \{3, 6, 9\}$. The number 12 cannot be a member of A because 12 does not belong to the universal set, U, that was selected.

In a particular discussion, the universal set that is selected is always stated or understood.

In this section our universal set will be the set of natural numbers. In other words, the word *number* shall mean *natural number.*

Definition

We say that the number r is a **multiple** of the number s, or that r is **divisible** by s, if r can be obtained by multiplying s by some natural number.

Example 1 List the set of multiples of 6 that are less than 50.

SOLUTION The multiples of 6 are obtained by multiplying 6 by each of the natural numbers. Since $6(1) = 6$, $6(2) = 12$, $6(3) = 18$, $6(4) = 24$, $6(5) = 30$, $6(6) = 36$, $6(7) = 42$, and $6(8) = 48$, the set of multiples of 6 less than 50 is

$$\{6, 12, 18, 24, 30, 36, 42, 48\}$$

Definition

We say that the number r is a **factor** of s, or that r is a **divisor** of s, if r can be obtained by dividing s by some natural number.

Example 2 List the factors of 15.

SOLUTION The factors of 15 are obtained by dividing 15 by natural numbers. When the quotient and the divisor are natural numbers, both these numbers are factors. Since

$$\frac{15}{1} = 15$$

1 and 15 are factors, and since

$$\frac{15}{3} = 5$$

3 and 5 are factors. These are the only possibilities. Thus the factors of 15 are the set

$$\{1, 3, 5, 15\}$$

Definition

A natural number n is **even** if and only if 2 is a factor of n.

Example 3 List the set of even numbers less than 25.

SOLUTION $\{2, 4, 6, 8, 10, 12, 14, 16, 18, 20, 22, 24\}$

Definition

A natural number n is **odd** if and only if it is not even.

Example 4 List the set of odd numbers less than 20.

SOLUTION $\{1, 3, 5, 7, 9, 11, 13, 15, 17, 19\}$

Definition
A natural number p is a **prime** if and only if p has exactly two factors, namely p and 1.

The set of the first ten primes is $\{2, 3, 5, 7, 11, 13, 17, 19, 23, 29\}$. Note that 1 is not a prime, since it has only one factor.

Definition
A natural number n is a **composite** if and only if $n \neq 1$ and n is not a prime.

The set of the first ten composite numbers is $\{4, 6, 8, 9, 10, 12, 14, 15, 16, 18\}$.

Example 5 List the factors of 70 that are prime numbers.

SOLUTION First find the factors of 70:

$$1 \cdot 70 = 70, \quad 2 \cdot 35 = 70,$$
$$5 \cdot 14 = 70, \quad 7 \cdot 10 = 70$$

The factors of 70 are $\{1, 2, 5, 7, 10, 14, 35, 70\}$. Select from these the numbers that are prime. The prime factors of 70 are the set

$$\{2, 5, 7\}$$

Example 6 List the factors of 70 that are composite.

SOLUTION We first find the factors of 70 (see Example 5). We reject 1, because 1 is not composite, and we reject all prime factors. Those remaining are the composites. Therefore the set of composite factors of 70 is

$$\{10, 14, 35, 70\}$$

Example 7 Write 70 as the product of prime numbers.

SOLUTION $70 = (2)(5)(7)$

Example 8 Write 2200 as the product of prime factors.

SOLUTION Divide 2200 by prime numbers beginning with 2, then trying 3, then 5, then 7, then 11, and so forth.

$$\begin{array}{rr} 2 & \underline{)2200} \\ 2 & \underline{)1100} \\ 2 & \underline{)550} \\ 5 & \underline{)275} \\ 5 & \underline{)55} \\ 11 & \underline{)11} \\ & \underline{)\ 1} \end{array}$$

Thus, $2200 = 2^3 \cdot 5^2 \cdot 11$

Exercises 1.2

In Exercises 1–38, list each of the following sets.

1. The even numbers between 17 and 25
2. The even numbers between 35 and 45
3. The odd numbers between 56 and 63
4. The odd numbers between 70 and 80
5. The multiples of 3 between 29 and 50
6. The multiples of 4 between 29 and 50
7. The multiples of 10 that are less than 100
8. The multiples of 12 that are less than 100
9. The factors of 28
10. The factors of 24
11. The factors of 30
12. The factors of 26
13. The divisors of 50
14. The divisors of 18
15. The prime numbers that are less than 20
16. The prime numbers between 30 and 50
17. The prime numbers between 50 and 80
18. The prime numbers between 80 and 100
19. The composite numbers between 19 and 27
20. The composite numbers between 71 and 83
21. The factors of 28. Which are prime numbers?
22. The factors of 24. Which are prime numbers?
23. The prime factors of 30
24. The prime factors of 26
25. The prime factors of 50
26. The prime factors of 18
27. The even factors of 36
28. The odd factors of 42
29. The odd factors of 16
30. The even factors of 35
31. The composite factors of 19
32. The composite factors of 48
33. The common multiples of 3 and 5 that are less than 50. (A common multiple is a multiple of *both* 3 and 5.)
34. The common multiples of 4 and 6 that are less than 50
35. The common multiples of 18 and 12 that are less than 200
36. The common multiples of 52 and 65 that are less than 300
37. The common divisors of 12 and 18. (A common divisor is a divisor of *both* 12 and 18.)
38. The common divisors of 30 and 45

Write each number in Exercises 39–48 as the product of prime numbers.

39. 15	40. 14	41. 18	42. 40	43. 66
44. 52	45. 78	46. 60	47. 98	48. 75

1.3 Grouping Symbols

After an operation has been performed on two numbers, an operation can be performed again on the resulting number and a third number. This process can be continued as often as desired.

In order to indicate which operation was done first, which second, which third, and so on, grouping symbols are used. The grouping symbols that are most frequently used are

> parentheses, ( ), braces, { }, brackets, [ ], and the bar (also called vinculum), ——

For example, to indicate that 3 is to be subtracted from the difference 12 − 5, any one of the following symbolic expressions can be used:

$$(12 - 5) - 3$$
$$\{12 - 5\} - 3$$
$$[12 - 5] - 3$$

If the operations are done as indicated, then first one obtains that 12 − 5 is 7 and then that 7 − 3 is 4. Therefore (12 − 5) − 3 results in the number 4.

If the parentheses or other grouping symbols were not used, the order in which the operations are to be done would not be clear, and in some cases it would be possible to obtain as a result a different number from the one intended.

For example, consider 12 − 5 − 3. If 3 is first subtracted from 5, then the difference is 2. When 2 is subtracted from 12, the result is 10. Thus the expression 12 − 5 − 3 is ambiguous—that is, more than one number can be obtained as a final result, depending on how the expression is interpreted.

Grouping symbols are used to avoid ambiguity. Thus (12 − 5) − 3 results in 4 only and 12 − (5 − 3) results in 10 only, since it is understood that the operation within the parentheses is to be done first.

Several kinds of grouping symbols are used to make an algebraic expression that involves more than two operations easier to read. For example, one way to indicate that the sum of 3 and the difference $8 - 2$ is to be subtracted from 12 is as follows:

$$12 - \{3 + (8 - 2)\}$$

This means that the operation within the innermost grouping symbols, the subtraction of 2 from 8, is to be done first. Next, the result 6 is to be added to 3. Finally, the sum of 6 and 3, or 9, is to be subtracted from 12, giving a final result of 3.

This combination of operations could also have been written, using parentheses only, as follows:

$$12 - (3 + (8 - 2))$$

Note that the use of two kinds of grouping symbols causes the algebraic expression to be read more clearly and more rapidly.

By adopting certain conventions, some grouping symbols can be omitted from certain expressions. This yields an expression that is simpler to read, and the intended meaning is still clear. The following convention is adopted:

Convention

1. The operations within the innermost set of grouping symbols are done first, following the procedure in step 2. (Recall that the bar used in division is a grouping symbol.)
2. Unless grouping symbols indicate otherwise, the operations are done in the following order:

   First, squaring and/or cubing, as read from left to right.
   Second, multiplication and/or division, as read from left to right.
   Third, addition and/or subtraction, as read from left to right.

Thus $5 + (2^3)$ can be written more simply as $5 + 2^3$, since the cubing must be done first, in agreement with the above convention. Thus $5 + 2^3$ means $5 + 8$, or 13, whereas $(5 + 2)^3$ means 7^3, or 343.

Similarly, $5 \cdot (2^3)$ can be written as $5 \cdot 2^3$, with $5 \cdot 8$, or 40, as the intended final result, since the cubing must be done before the multiplication.

If one wants to indicate that 5 is to be multiplied by 2 and that the resulting product is to be cubed, grouping symbols must be used as follows: $(5 \cdot 2)^3$, which means 10^3, or 1000.

The number indicated by $5 \cdot 2 + 7$ is the sum of 10 and 7, or 17, since the convention requires that the multiplication be done before the addition. Thus $5 \cdot 2 + 7$ is another name for 17.

To indicate that 2 is to be added to 7 and that the resulting sum is to be multiplied by 5, one writes $5(2 + 7)$, which names the number $5 \cdot 9$, or 45.

The product $5(2 + 7)$ could also have been written as $5 \cdot (2 + 7)$. However, since the numerals 5 and $(2 + 7)$ are written in juxtaposition (side by side), the operation of multiplication is clearly indicated and the dot is not necessary. Greater simplicity and clarity are achieved by writing only the symbols that are necessary.

Example 1 Do the indicated operations: $13 - (5 - 2)^2$.

SOLUTION

1. Simplify the expression inside parentheses: $5 - 2 = 3$. Thus

$$13 - 3^2$$

2. Square: $3^2 = 9$. Thus

$$13 - 9$$

3. Perform subtraction: $13 - 9 = 4$. Therefore

$$\begin{aligned} &13 - (5 - 2)^2 \\ &= 13 - 3^2 \\ &= 13 - 9 = 4 \end{aligned}$$

Example 2 Do the indicated operations:

$$25 - \{3 + [10 - (2 + 3)]\}$$

SOLUTION

1. Do the operation within the parentheses and obtain

$$25 - \{3 + [10 - 5]\}$$

2. Do the operation within the brackets and obtain

$$25 - \{3 + 5\}$$

3. Do the operation within the braces and obtain

$$25 - 8$$

4. Subtract 8 from 25 and obtain 17 as the final result.

Example 3 Simplify $5 + 2 \cdot 3^2 - (12 - 9)$.

SOLUTION

$$\begin{aligned} &5 + 2 \cdot 3^2 - (12 - 9) \\ &= 5 + 2 \cdot 3^2 - 3 && \text{(Do inside parentheses)} \\ &= 5 + 2 \cdot 9 - 3 && \text{(Square first)} \\ &= 5 + 18 - 3 && \text{(Multiply before adding or subtracting)} \\ &= 20 && \text{(Add and subtract from left to right)} \end{aligned}$$

Example 4 Simplify $\dfrac{8^2 + 8 \cdot 6}{2 \cdot 8 + 2 \cdot 6}$

SOLUTION The bar used in division is a grouping symbol, so the operations in the numerator and in the denominator must be done before the division:

$$\frac{8^2 + 8 \cdot 6}{2 \cdot 8 + 2 \cdot 6} = \frac{64 + 8 \cdot 6}{2 \cdot 8 + 2 \cdot 6} \quad \text{(Squaring first)}$$

$$= \frac{64 + 48}{16 + 12} \quad \text{(Multiply)}$$

$$= \frac{112}{28} \quad \text{(Add)}$$

$$= 4 \quad \text{(Divide)}$$

Exercises 1.3

Do the indicated operations.

1. $3 + (5 + 7)$
2. $4 + (9 + 8)$
3. $(3 + 5) + 7$
4. $(4 + 9) + 8$
5. $22 - (10 - 3)$
6. $(50 - 20) - 10$
7. $(22 - 10) - 3$
8. $50 - (20 - 10)$
9. $3(4 \cdot 5)$
10. $2(6 \cdot 7)$
11. $(3 \cdot 4)(3 \cdot 5)$
12. $(2 \cdot 6)(2 \cdot 7)$
13. $(9 - 3)(6 - 4)$
14. $(15 - 2)(12 - 8)$
15. $9 - 3(6 - 4)$
16. $15 - 2(12 - 8)$
17. $5 \cdot 4^2$
18. $8 \cdot 5^2$
19. $(5 \cdot 4)^2$
20. $(8 \cdot 5)^2$
21. $8 + 2^3$
22. $7 + 3^3$
23. $(8 + 2)^3$
24. $(7 + 3)^3$
25. $7 \cdot 2 + 3$
26. $8 \cdot 4 + 5$
27. $7(2 + 3)$
28. $8(4 + 5)$
29. $16 - 8 - 3$
30. $34 - 21 + 2$
31. $25 - 12 + 7$
32. $80 - 45 + 5$
33. $6 + 3 \cdot 4^2$
34. $80 - 2 \cdot 5^2$
35. $30 - (2 \cdot 5 - 2)$
36. $25 - (7 + 3 \cdot 4)$
37. $8 - 2(5 - [2 + 1])$
38. $9 + 3(6 - [3 - 1])$
39. $(8 - 2)(5 - [2 + 1])$
40. $(9 + 3)(6 - [3 - 1])$

41. $40 - \{8 + [12 - (4 + 5)]\}$
42. $20 - \{15 - [8 + (7 - 5)]\}$
43. $6^2 - 2(3 \cdot 6) + 3^2$
44. $5 \cdot 4^2 - 7 \cdot 8 + 4 \cdot 2$
45. $2\{[5^2 - 3(4 + 2)] + 1\}$
46. $7 - [6 - \{5^2 - 4(2 + 3)\}]$
47. $2[(7^2 - 3^2) - (4^2 - 3^2)]$
48. $5\{(9 + 4)^2 - (9 - 4)^2\}$
49. $\dfrac{5^2 - 4^2}{5 - 4}$
50. $\dfrac{2^3 + 3^3}{2 + 3}$
51. $\dfrac{5^2}{5} - \dfrac{4^2}{4}$
52. $\dfrac{2^3}{2} + \dfrac{3^3}{3}$
53. $\dfrac{8 \cdot 7 + 8^2}{2 \cdot 7 + 2 \cdot 8}$
54. $\dfrac{20^2 - 4 \cdot 20}{4 \cdot 20 - 4^2}$
55. $\dfrac{30 - 2(5 - 2)}{(30 - 25) - 2}$
56. $\dfrac{20 - [10 - 2(7 - 3)]}{20 - [10 + 2(7 - 3)]}$
57. $(9 + [5 - 2])(9 - [5 - 2])$
58. $[4^2 + (8 + 6)][4^2 - (8 + 6)]$
59. $4^2 + 2\{15 - 2[3^2 - (3 - 1)^2]\}$
60. $7^2 - (6^2 - [5^2 - \{4^2 - (3^2 - 2^2)\}])$

1.4 Equality, Substitution, Formulas

The Equality Relation

An equation is a statement that has the form $A = B$, which is read "A equals B." The symbol $=$ indicates that A and B are names of the same number.

As examples,

$$12 + 3 = 15$$

means that $12 + 3$ (the sum of 12 and 3) names the same number as the numeral 15, and

$$12 \cdot 3 = 36$$

means that $12 \cdot 3$ (the product of 12 and 3) names the same number as the numeral 36.

It is often convenient to use the symbol $\neq$, which means "does not equal." For example, $3 \neq 5$ is read "three does not equal five" and means that the numerals 3 and 5 do not name the same number.

Symmetric and Reflexive Axioms

An equation does not have to be a true statement. As examples,

$6 + 4 = 10$ is a true statement
$6 + 4 = 15$ is a false statement

Some statements are true just because of the form they have. Two cases involving the equality relation are so important that they have special names. These are the **symmetric** and **reflexive axioms** of equality. (An axiom is a statement that is accepted without proof.)

Reflexive Axiom	*Symmetric Axiom*
$A = A$	If $A = B$, then $B = A$

EXAMPLES OF THE REFLEXIVE AXIOM:

$$3 = 3$$
$$x = x$$
$$2x + 1 = 2x + 1$$

EXAMPLES OF THE SYMMETRIC AXIOM:

If $6 + 4 = 10$, then $10 = 6 + 4$
If $10 = x + 3$, then $x + 3 = 10$

The reflexive axiom means that we call an equation true whenever both sides of the equation are identical.

The symmetric axiom means that we can exchange the sides of an equation without changing the truth or falsity of the equation.

Substitution

Since the relation $r = s$ is understood to mean that the variables r and s are names of the same number, it seems reasonable that one name can replace the other in a given statement without changing the truth or falsity of the statement. Furthermore, if one name replaces the other in an algebraic expression, the number that is named by the algebraic expression remains the same. This property is called the **substitution axiom** for the equal relation.

The Substitution Axiom

If $r = s$, then r may replace s or s may replace r in an algebraic expression without changing the number that is being named or in an algebraic statement without changing the truth or falsity of the statement.

For example, consider the statement $y = x + 5$.

Now if $x = 3$, then 3 can replace x in this statement, and one obtains $y = 3 + 5$. This substitution process may be repeated by using the fact that $3 + 5 = 8$. Replacing $3 + 5$ by its equal, 8, in the statement $y = 3 + 5$, one obtains $y = 8$.

The substitution axiom may be used to *evaluate* algebraic expressions. To evaluate an algebraic expression means to replace each letter by the numeral to which it is equal and then to perform the operations that are indicated, replacing the names of the numbers by simpler names that do not involve the operation symbols.

Example 1 Evaluate $3(x + 4) - \frac{x}{2}$ for $x = 6$.

SOLUTION

1. Rewrite the given expression:

$$3(x + 4) - \frac{x}{2}$$

2. Remove the letter and hold its place with open parentheses:

$$3((\quad) + 4) - \frac{(\quad)}{2}$$

3. Insert the numeral to which the letter is equal within the parentheses:

$$3((6) + 4) - \frac{(6)}{2}$$

4. Do the indicated operations:

$$3(10) - 3$$
$$30 - 3 = 27$$

Example 2 Evaluate $\dfrac{x^2 + 2xy - 3y^2}{x + 3y}$ for $x = 8$ and $y = 4$.

SOLUTION

1. Rewrite the given expression:

$$\frac{x^2 + 2xy - 3y^2}{x + 3y}$$

2. Replace one letter, say x, by open parentheses:

$$\frac{(\quad)^2 + 2(\quad)y - 3y^2}{(\quad) + 3y}$$

3. Insert the value for x:

$$\frac{(8)^2 + 2(8)y - 3y^2}{(8) + 3y}$$

4. Replace the other letter, y, by open parentheses:

$$\frac{(8)^2 + 2(8)(\quad) - 3(\quad)^2}{(8) + 3(\quad)}$$

5. Insert the value for y:

$$\frac{(8)^2 + 2(8)(4) - 3(4)^2}{(8) + 3(4)}$$

6. Evaluate (do the operations):

$$\frac{64 + 2(32) - 3(16)}{8 + 12}$$

$$= \frac{64 + 64 - 48}{20}$$

$$= \frac{128 - 48}{20}$$

$$= \frac{80}{20}$$

$$= 4$$

Formulas

Many problems in mathematics, science, business, and other areas are most easily solved by using formulas.

A **formula** is a symbolic statement indicating what operations are to be performed on certain numbers with specialized meanings determined by the particular problem.

***Example 3* (Geometry: Perimeter of a Square)** Express the following statement as a formula: The length of the perimeter of a square is obtained by multiplying the length of a side by 4.

SOLUTION Let P represent the length of the perimeter. Let s represent the length of a side.

FORMULA $P = 4s$

Example 4 Using the formula $P = 4s$, find P if $s = 3$ inches.

SOLUTION $P = 4(3) = 12$ inches

***Example 5* (Geometry: Area of a Square)** Express as a formula: The area A of a square is the square of the length of a side s.

SOLUTION $A = s^2$

Example 6 Evaluate $A = s^2$ for $s = 5$ centimeters.

SOLUTION $A = (5)^2 = 25$ square centimeters

***Example 7* (Uniform Motion)** Express as a formula: The distance d that an object travels is the product of the rate of speed r and the time t that is traveled.

SOLUTION $d = rt$

Example 8 Given $d = rt$, find d if $r = 30$ mph and $t = 2$ hours.

SOLUTION $d = 30 \cdot 2 = 60$ miles

***Example 9* (Resistance to an Electrical Circuit)** Write a formula to express the following relation: The resistance R, measured in ohms, of an electrical circuit is the quotient obtained by dividing the electromotive force E, measured in volts, by the intensity I of the current, measured in amperes.

SOLUTION $R = \dfrac{E}{I}$

Example 10 Using $R = \dfrac{E}{I}$, find R if $E = 60$ volts and $I = 15$ amperes.

SOLUTION $R = \dfrac{60}{15} = 4$ ohms

***Example 11* (Business Profit)** Express as a formula: The profit P of a business transaction is obtained by subtracting the cost C from the selling price S.

SOLUTION $P = S - C$

Example 12 Find the profit made on selling an end table for \$55 if the cost of the table was \$35.

SOLUTION Using $P = S - C$ with $S = 55$ and $C = 35$, $P = 55 - 35 = 20$ dollars.

François Viète: Lawyer, Mathematician

The French lawyer François Viète (1540–1603) was one of the first persons to use letters to represent numbers. In 1591 he used capital vowels, such as *A*, *E*, *I*, for the variables. Viète is often called the "father of algebra" because of the improvements he made in the symbolism of algebra.

It is said that Viète helped Henry IV of France in his war against Spain by deciphering a Spanish code that contained hundreds of characters. So certain was King Philip II of Spain that no one could break his code, he complained to the pope that the French were violating the doctrines of the Christian religion by using magic against his country.

Exercises 1.4

Evaluate the algebraic expressions in Exercises 1–50.

1. $x^2 + 5x + 8$ for $x = 6$
2. $3x^2 + 2x + 11$ for $x = 5$
3. $4x^3 + 5x^2 + 7x + 6$ for $x = 3$
4. $3x^3 + 2x^2 - 10x + 4$ for $x = 3$
5. $3(x - 4)^2$ for $x = 4$
6. $4(x - 2)^2$ for $x = 5$
7. $\dfrac{x^2 - 25}{x - 5}$ for $x = 7$
8. $\dfrac{x^3 + 4}{x + 4}$ for $x = 2$
9. $x(x^2 - x)$ for $x = 6$
10. $x^2(x - 3)$ for $x = 4$
11. $\dfrac{y(y - 1)}{2}$ for $y = 10$
12. $\dfrac{a(a + 1)}{3}$ for $a = 5$
13. $(n + 1)(n - 1)$ for $n = 11$
14. $(y + 3)(y - 3)$ for $y = 7$
15. $x^2 + 2xy + y^2$ for $x = 5$, $y = 3$
16. $x^2 + 3xy + 5y^2$ for $x = 5$, $y = 2$
17. $x^2 + y^2$ for $x = 5$, $y = 3$
18. $x^3 + y^3$ for $x = 2$, $y = 3$
19. $c^2 - a^2$ for $c = 13$, $a = 12$
20. $x^2 - y^2$ for $x = 25$, $y = 24$
21. $\dfrac{a^3 - b^3}{a - b}$ for $a = 10$, $b = 6$
22. $\dfrac{a^2 - b^2}{a + b}$ for $a = 3$, $b = 2$
23. $\dfrac{3x}{5(x - 7)}$ for $x = 10$
24. $ab(a + b^2)$ for $a = 3$, $b = 5$

25. $a^3 - \frac{a}{2}$ for $a = 6$

26. $\frac{6x}{2x + 4}$ for $x = 4$

27. $\frac{a^3 - 8}{a - 2}$ for $a = 3$

28. $y^2 - \frac{2y}{3}$ for $y = 9$

29. $(y - 2)(y^2 + 2y + 4)$ for $y = 3$

30. $\frac{y^3 + 8}{y + 2}$ for $y = 3$

31. $(x + y)^3$ for $x = 2$, $y = 3$

32. $(n - 3)(n^2 + 3n + 9)$ for $n = 5$

33. $\frac{xy + x^2}{xy + y^2}$ for $x = 8$, $y = 2$

34. $(x + y)^2$ for $x = 5$, $y = 3$

35. $\frac{x^3 + y^3}{x + y}$ for $x = 5$, $y = 2$

36. $\frac{xy}{x + y}$ for $x = 10$, $y = 15$

37. $5m^2 - ms + 6s^2$ for $m = 4$, $s = 3$

38. $\frac{a + b}{a - b}$ for $a = 6$, $b = 4$

39. $5ab^2 - 3a^2b$ for $a = 3$, $b = 5$

40. $3r^2 - rs - 2s^2$ for $r = 7$, $s = 3$

41. $(x + y)^2 + (x + y)(x - y) + (x - y)^2$ for $x = 4$, $y = 1$

42. $\frac{x}{y}(x^2 - y^2)$ for $x = 10$, $y = 5$

43. $\frac{x + y + z}{3}$ for $x = 10$, $y = 12$, $z = 20$

44. $x^2 + y^2 + z^2$ for $x = 3$, $y = 4$, $z = 12$

45. $\frac{a}{2}(b + c)$ for $a = 8$, $b = 6$, $c = 9$

46. $a(b - c)^2$ for $a = 5$, $b = 12$, $c = 3$

47. $r^2 - (s^2 + t^2)$ for $r = 3$, $s = 2$, $t = 1$

48. $(r - s)^2 + t^2$ for $r = 3$, $s = 2$, $t = 1$

49. $5(m - n)^2$ for $m = 10$, $n = 7$

50. $\frac{m^3 - n^3}{2m - n}$ for $m = 6$, $n = 5$

In Exercises 51–64 a relation is stated verbally, then written as a formula. Evaluate each formula using the indicated values.

51. The area A of a triangle is one half the product of its base b and its height h.

$$A = \frac{1}{2}bh$$

for $b = 18$ feet, $h = 20$ feet

52. The perimeter P of a rectangle is twice the sum of the base b and the height h.

$$P = 2(b + h)$$

for $b = 10$ centimeters, $h = 14$ centimeters

53. The area A of a trapezoid is one half the product of its height h and the sum of its bases a and b.

$$A = \frac{1}{2}h\,(a + b)$$

for $h = 7$ inches, $a = 13$ inches, $b = 5$ inches

54. The volume V of a pyramid with a square base is one third the product of the height h and the square of a side b of the base.

$$V = \frac{hb^2}{3}$$

for $h = 5$ meters, $b = 6$ meters

55. In psychology, the intelligence quotient Q is obtained by dividing 100 times the mental age M by the chronologic age C.

$$Q = \frac{100M}{C}$$

for $M = 18$, $C = 15$

56. The total resistance R in an electrical circuit with two resistances that have values S and T that are connected in parallel is obtained by dividing the product of the values of the two resistances by their sum.

$$R = \frac{ST}{S + T}$$

for $S = 20$ ohms, $T = 30$ ohms

57. The Celsius temperature C in degrees is equal to five ninths the difference obtained by subtracting 32 from the Fahrenheit temperature F in degrees.

$$C = \frac{5}{9}(F - 32)$$

for $F = 212$ degrees

58. The Fahrenheit temperature F in degrees is equal to the sum of nine fifths of the Celsius temperature C in degrees and 32.

$$F = \frac{9}{5}C + 32$$

for $C = 20$ degrees

59. The amount of money A that results from investing a sum P at a simple interest rate r for n interest periods is the sum of P and the product of P, r, and n.

$$A = P + Prn$$

for $P = \$500$, $r = 0.06$, $n = 8$

60. The percent efficiency E of an engine is equal to 100 times the quotient when the difference, heat input I minus heat output O, is divided by the heat input.

$$E = \frac{100(I - O)}{I}$$

for $I = 25$, $O = 20$

61. The distance d in meters that a dropped object falls in t seconds is equal to the product of 4.9 and the square of t.

$$d = 4.9t^2$$

for $t = 10$ seconds

62. The illumination E at a point on a surface in lumens is equal to the quotient obtained by dividing the intensity I of the source in candlepower by the square of the distance r between the point on the surface and the source of light.

$$E = \frac{I}{r^2}$$

for $I = 360$, $r = 6$

63. The sum S of all the natural numbers from 1 to n is one half the product of n and the sum of n and 1.

$$S = \frac{n(n + 1)}{2}$$

for $n = 200$

64. With the straight-line depreciation method, the annual depreciation D is obtained by dividing the difference between the original cost C and the probable scrap value S by the number n of years representing the probable useful life of the article.

$$D = \frac{C - S}{n}$$

for $C = \$3000$, $S = \$600$, $n = 12$ years

65. In anthropology, the cephalic index C is obtained by dividing the product of 100 and the width of a head W by the length of the head L.

$$C = \frac{100W}{L}$$

for $W = 18$ centimeters, $L = 24$ centimeters

66. The amount of medication C for a child over 1 year of age can be obtained by dividing the product of the age of the child y in years and the adult dosage A by the sum of y and 12.

$$C = \frac{yA}{y + 12}$$

for $y = 8$, $A = 10$

1.5 Real Numbers, Number Lines

Special Sets of Numbers

The answers to most number problems are found in a set of numbers called the set of real numbers. The set of real numbers and some of its important subsets will be introduced in this section. The set A is a **subset** of the set B if every element of A is an element of B. In symbols, $A \subset B$.

The counting numbers $\{1, 2, 3, 4, 5, \ldots\}$ were developed at a very early stage in our existence. By 3000 B.C., several numeral systems for these numbers had been invented. The early Egyptians had a simple grouping system something like our Roman numerals, and the early Babylonians had a positional system something like our modern Hindu-Arabic numerals.

Definition

The set of counting numbers, $\{1, 2, 3, 4, 5, \ldots\}$, is also called the set of **natural numbers,** which may be designated by the letter N.

$$N = \{1, 2, 3, 4, 5, \ldots\}$$

As civilization progressed, people's requirements rose above the necessity of simply recording numbers of objects. In particular, they needed a technique for measuring quantities that were opposite in nature.

For example, in order to measure temperature, someone invented a thermometer on which a scale was established. The starting point was called 0 (zero) and then a unit of measurement, such as 1 degree Fahrenheit, was selected. Then equal units of measurement were marked off on one side of the zero point to obtain this scale, as in Fig. 1.1.

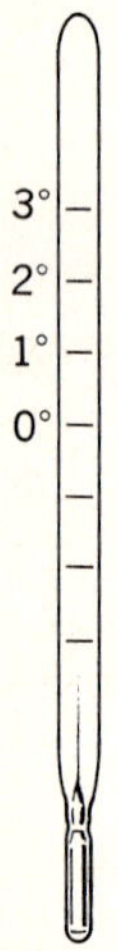

Figure 1.1

To record temperatures below 0 degrees—that is, when the mercury fell lower than the zero point—units were marked off in the opposite direction.

To distinguish between a temperature above zero and one below zero, the numbers above zero were tagged with the symbol + (plus), and the ones below zero were tagged with the symbol − (minus) (Fig. 1.2).

Thus negative numbers, such as -1, -2, -3, and so on, came into existence as measures of quantities opposite in nature to other quantities.

Some examples of opposite quantities that can be measured by using negative numbers are shown in Table 1.2.

Definition

The set of all natural numbers, their negatives, and the number zero is called the set of **integers,** which may be designated by the letter J.

$$J = \{\ldots, -3, -2, -1, 0, 1, 2, 3, \ldots\}$$

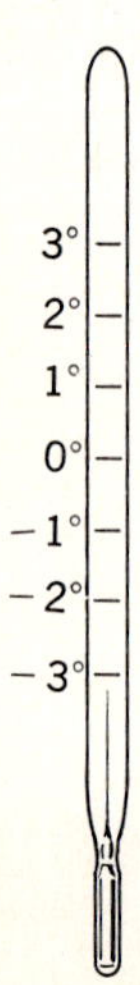

Figure 1.2

TABLE 1.2 **Quantities Measurable with Signed Numbers**

+30	−30
30 degrees above zero	30 degrees below zero
30 feet above sea level	30 feet below sea level
30 miles east	30 miles west
30 dollars bank deposit	30 dollars withdrawal

The natural numbers $\{1, 2, 3, \ldots\}$ are also called the **positive integers.**

The numbers in the set $\{-1, -2, -3, \ldots\}$ are called the **negative integers.**

The set of **even integers** is

$$\{\ldots, -4, -2, 0, 2, 4, \ldots\}.$$

The set of **odd integers** is

$$\{\ldots, -3, -1, 1, 3, \ldots\}.$$

The numbers in the set $\{0, 1, 2, 3, \ldots\}$ are called the **whole numbers,** or the **nonnegative integers.**

The set of natural numbers, N, is a subset of the set of integers, J. In symbols, $N \subset J$.

When one natural number is divided by another natural number, the resulting quotient is not always a natural number. In arithmetic, these nonintegral quotients are called **fractions,** and each fraction is thought of as the number of subunits of a basic unit.

Many visual representations of fractions are possible, such as those in Fig. 1.3.

In algebra this meaning of a fraction is retained. In addition to the fractions of arithmetic, called the *positive fractions*, their *negatives*, called the *negative fractions*, are also included in the number system of algebra.

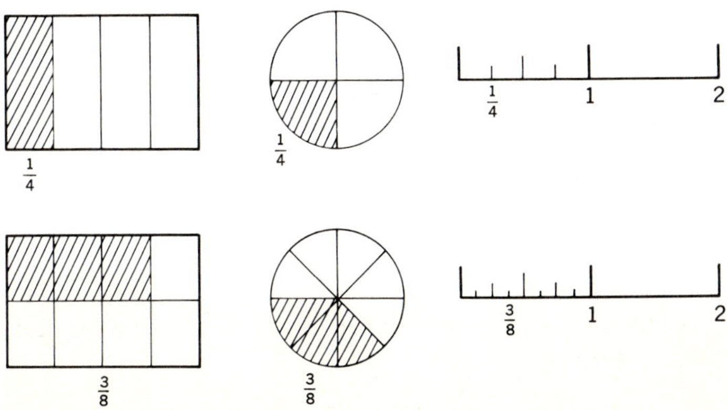

Figure 1.3

Definition

The set that includes all the integers and the positive and negative fractions is called the set of **rational numbers,** which may be designated by the letter Q.

The set Q may also be described as the set of numbers that can be expressed as the quotient of two integers, $\frac{p}{q}$, where p and q are integers and $q \neq 0$.

Examples of rational numbers are $\frac{2}{3}, \frac{5}{9}, \frac{-2}{3}, \frac{5}{-9}, -\frac{16}{17}$. In each of these numbers, the numerator and denominator are integers, and no denominator is zero.

The set of integers can be identified as a subset of the rationals—that is, those quotients $\frac{p}{q}$ with $q = 1$.

This means, for example, that $\frac{5}{1}$ and 5 are the same number.

Number Lines

A visual representation of numbers may be obtained by using the names of numbers as the names of points on a straight line.

A point is selected as the starting point. This point is called the **origin** and is given the name 0 (zero).

A unit of measurement and a direction, called the positive direction, are selected. With this unit of length, points are marked off in succession in the positive direction, which is to the right in Fig. 1.4.

The arrow at the right indicates that the line continues indefinitely in this direction.

The negative integers $\{-1, -2, -3, \ldots\}$ are assigned to the points in the opposite direction in a similar manner. This direction is to the left in Fig. 1.4.

The arrow at the left indicates that the line continues indefinitely in this direction.

A line whose points are named by using numbers is called a **number line.**

The number that names a point is called the **coordinate** of the point.

The point that is given a number name is called the **graph** of this number.

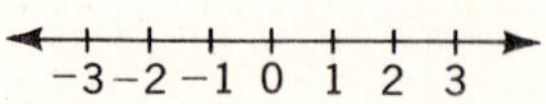

Figure 1.4 A horizontal number line.

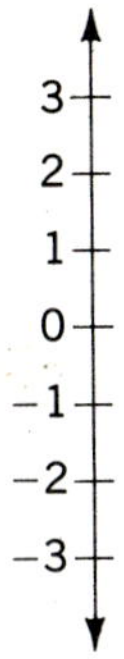

Figure 1.5 A vertical number line.

A number line does not have to be horizontal. A vertical arrangement is often used, as shown in Fig. 1.5. In this arrangement the positive direction is usually upward.

All the rational numbers can be identified as coordinates on a number line by considering subdivisions of the basic unit and by using the concept of opposites (Fig. 1.6).

Thus it is observed that each rational number can be made to correspond to exactly one point on a number line.

One might be tempted to believe that the rational numbers exhaust the number line. However, this is not the case. For example, $\sqrt{2}$, the positive square root of 2, is **not** a rational number. The number $\sqrt{2}$ is the number whose square is 2; that is, $(\sqrt{2})^2 = 2$. Geometrically, it can be identified as the length of a diagonal of a square whose sides are each 1 unit. See Fig. 1.6.

The real number $\sqrt{2}$ is called an *irrational* number. Some other examples of irrational numbers are $\sqrt{3}$, $\sqrt{5}$, $-\sqrt{2}$, $-\sqrt{3}$, and π.

Definition

An **irrational number** is a real number that is not rational—that is, it cannot be expressed as the quotient of two integers.

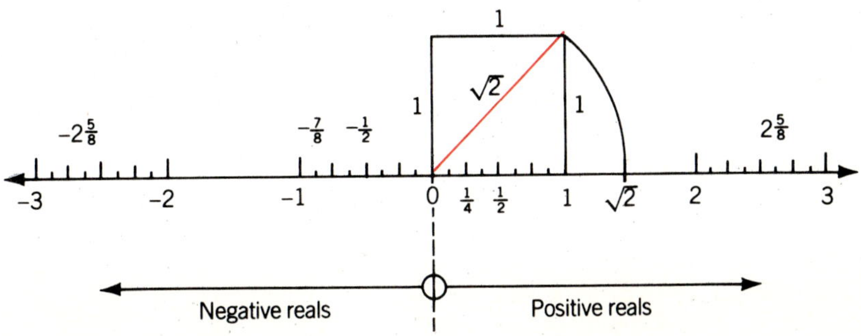

Figure 1.6 A real number line.

The totality of all the numbers that can be associated with points on a number line is called the set of **real numbers.** Thus a number line is called a **real number line.**

Each point on a real number line can be made to correspond to exactly one real number, and each real number can be made to correspond to exactly one point on a real number line. Thus there is a **one-to-one correspondence between the set of real numbers and the set of points on a real number line.** This important property of the set of real numbers is called the **axiom of completeness.** The set of real numbers is "complete" in the sense that all the real numbers are "used up" in naming the points on a number line, and every point has exactly one real number name.

The numbers associated with points to the right of zero are called the **positive real numbers,** and those to the left of zero are called the **negative real numbers.**

Numbers that are tagged with a + or − sign are often called **signed numbers.** The numbers x and $-x$ are called **opposites** of each other. For example, the positive number 5 and the negative number −5 are opposites of each other. They are also called **additive inverses.**

The "Absurd" and "Fictitious": Negative Numbers

The Chinese were familiar with negative numbers as early as 200 B.C. They were known to write positive numbers in red and negative numbers in black.

The Greek Diophantus (ca. 275), in his *Arithmetica,* calls the equation $4x + 20 = 4$ "absurd," since its solution would be −4. Diophantus did not seem to comprehend the abstract concept of a negative number.

The negative number is first mentioned as such in the works of the Hindu Brahmagupta (ca. 628). The Hindus indicated a negative number by placing a dot above a numeral or placing a circle above or around the numeral, such as $\dot{5}$, $\mathring{5}$, or ⑤.

Around 1225 the Italian Leonardo Fibonacci interpreted the negative solution of a financial problem to mean a loss.

In his *Ars magna* (1545), Cardano presented the elementary properties of negative numbers, although he was unable to give a clear interpretation of these numbers. He referred to the positive numbers as "true" numbers and to the negative numbers as "fictitious" numbers.

In 1544 Stifel stated that numbers such as 0 − 3 were "absurd."

The concept of the negative number became thoroughly understood in the Western world through the works of Viète, Fermat, Descartes, and other mathematicians.

Exercises 1.5

In Exercises 1–20, determine which of the following would best be measured by signed numbers and which by positive numbers alone.

1. Heights of people
2. Entries on a bank statement
3. Latitudes of cities
4. Historical dates from 4000 B.C. to now
5. Direction and speed of cars on a highway
6. Temperatures of cities
7. Number of calories in various foods
8. Areas of plots of land
9. Daily changes in barometric pressure
10. Scores in a card game in which points can be gained or lost
11. Daily minimum temperatures at Glacier National Park
12. Volumes of containers
13. Populations of cities
14. Daily changes in stock prices
15. Weekly changes in the weight of a person on a diet
16. Longitudes of cities
17. Ages of people
18. Electric charges on ions
19. Freezing points of chemical compounds, in degrees Celsius
20. Barometric pressures

If each phrase in Exercises 21–40 is represented by a positive number, state what is represented by the corresponding negative number.

21. 35 degrees north latitude
22. 250 A.D.
23. 20 degrees above 0 Fahrenheit
24. A score of 35 in a pinochle game in which points may be won or lost
25. A force of 20 pounds toward an object
26. An acceleration of 32 ft/sec^2
27. A gain in heat of 150 Btu
28. An increase in price of $5
29. A counterclockwise rotation of 40 degrees
30. An image distance of 30 centimeters in front of a mirror
31. 80 degrees east longitude
32. 50 dollars profit
33. 70 feet above sea level
34. A gain of 5 yards by a football team

35. A 15-mph south wind

36. A velocity of 2800 mph upward, of a rocket

37. 6 steps upward

38. 8 hours later

39. 35 pounds overweight

40. An increase in volume of 20 cubic centimeters

For Exercises 41–60, name the coordinate of the point on the number line indicated by the dot.

41. 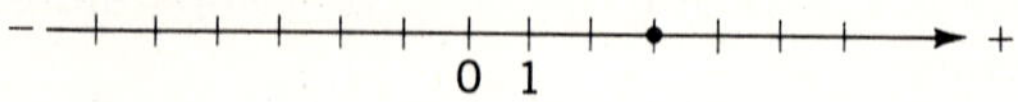

42. 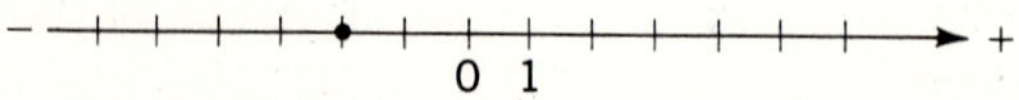

43. 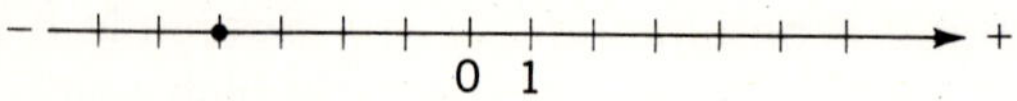

44. 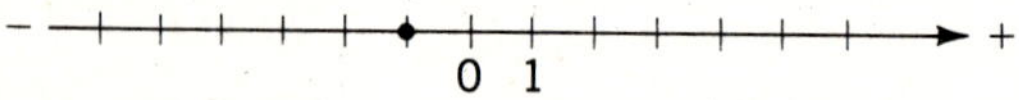

45. 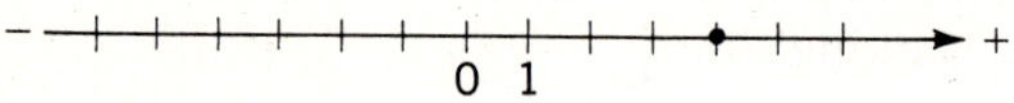

46. 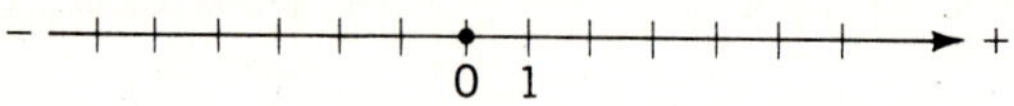

47. 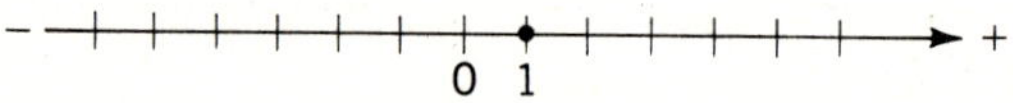

48. 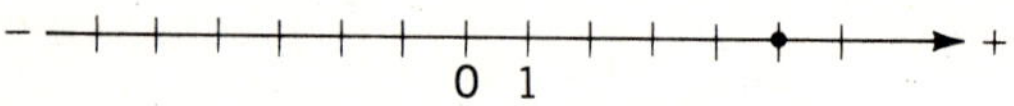

49. 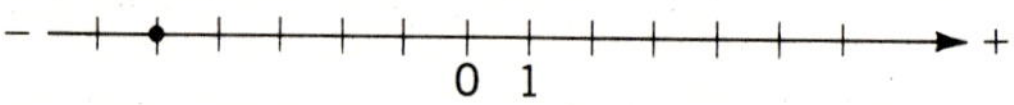

50. 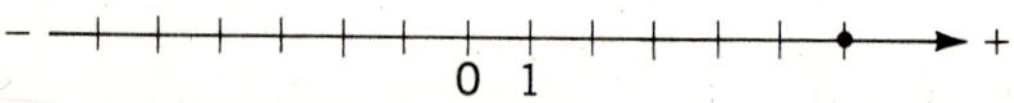

51. 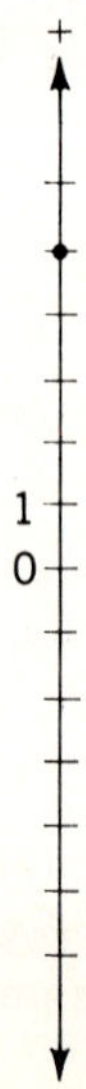

52. 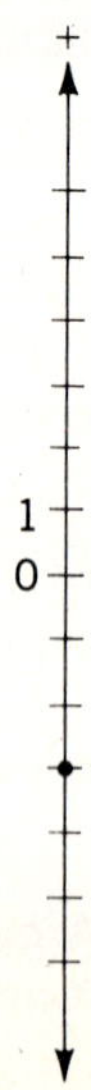

53. 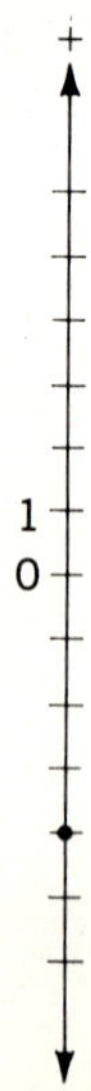

54. 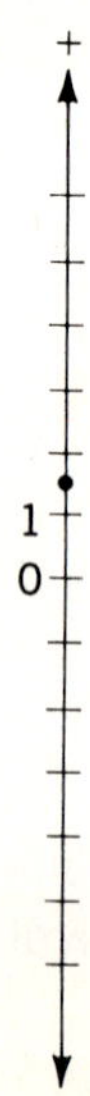

55. 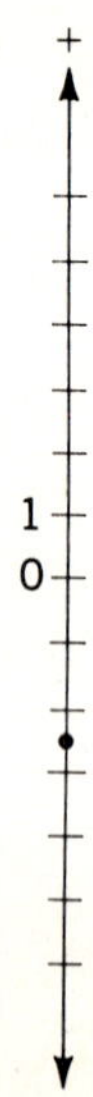

56. 57. 58. 59. 60.

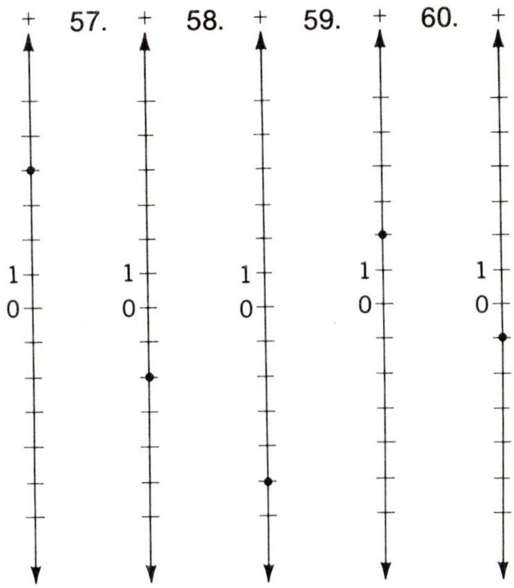

For Exercises 61–70, draw the graph of each of the following numbers on a horizontal number line whose positive direction is to the right.

61. 8
62. −8
63. 5
64. 0
65. −2
66. −4
67. 0
68. 4
69. −6
70. −10

For Exercises 71–80, draw the graph of each of the following on a vertical number line whose positive direction is upward.

71. −3
72. $2\frac{1}{2}$
73. $5\frac{5}{8}$
74. $-2\frac{1}{2}$
75. $4\frac{1}{2}$
76. $-1\frac{3}{4}$
77. $-2\frac{1}{3}$
78. $-\frac{3}{4}$
79. 0
80. $2\frac{3}{8}$

1.6 Absolute Value and Order Relations

Absolute Value

The number line provides a geometric model for the set of real numbers where points on the number line correspond to real numbers. Any point r and its opposite, or additive inverse, $-r$, are the same *distance* from the origin, but they are on opposite sides of the origin. The algebraic sign of a real number indicates on which side of the origin the corresponding point is located. The *distance* of the point from the origin, a positive number, regardless of the side on which the point is located, is called the *absolute value* of the number, denoted symbolically by $|r|$. The absolute value of a nonzero real number is always positive. As examples, $|6| = 6$ and $|-6| = 6$.

Definition of Absolute Value
If p is a **positive** real number, then $|p| = p$ and $|-p| = p$.
$|0| = 0$.

Example 1 Simplify $|-6| + |3|$.

SOLUTION $|-6| + |3| = 6 + 3 = 9$

Example 2 Simplify $8|-5|$.

SOLUTION $8|-5| = 8(5) = 40$

Example 3 Simplify $2|7 - 7|$.

SOLUTION $2|7 - 7| = 2|0| = 2(0) = 0$

Order Relations

The symbols $<$ and $>$ are used to indicate that one number is less than (smaller than) another number and that one number is greater than (larger than) another number.

For example, $3 < 5$ means "3 is less than 5," and $5 > 3$ means "5 is greater than 3."

The expression $r < s$ means "the number r is less than the number s."

The expression $r > s$ means "the number r is greater than the number s."

Note that the order symbol "opens up" toward the larger number.

The set of real numbers is an ordered set, and a number line provides a means to visualize this order. If points r and s correspond to two

TABLE 1.3 **Number Relations**

Symbolic Statement	Algebraic Meaning	Geometric Meaning (*for horizontal number line*)
1. $r < s$	r is less than s	r is to the left of s
2. $r = s$	r is equal to s	r is the same point as s
3. $r > s$	r is greater than s	r is to the right of s

real numbers, r and s, then there are exactly three ways in which the placement of r and s can occur on a horizontal number line:

1. r is to the left of s.
2. r is the same point as s.
3. r is to the right of s.

Symbolically, these three possibilities are expressed in Table 1.3.

Sometimes it is desirable to indicate that two numbers are related in one way or another. The combination of symbols $\leq$ is used to indicate that the number named on the left may be equal to or smaller than the one on the right. For example, $x + y \leq 25$ means that either the sum of x and y is equal to 25 or the sum is smaller than 25.

In general, $x \leq y$ means $x = y$ or $x < y$.

Similarly, $x \geq y$ means $x = y$ or $x > y$.

A slash drawn through an order symbol has the effect of the word "not." See Table 1.4.

Relation Symbols

The symbol $=$, which indicates equality, was introduced in 1557 by Robert Recorde in his work *The Whetstone of Witte* because, as he put it, "Noe 2 thynges can be moare equalle."

Descartes preferred $\propto$ or ∞, probably a modification of "ae," the first two letters of the Latin word *aequalis*, meaning "equal."

Thomas Harriot was the first to use the symbols $<$ and $>$. These symbols appeared in his work *Artis Analyticae Praxis,* published in 1631, 10 years after his death, but they were not immediately accepted, because other writers used ⊏ and ⊐, the symbols suggested by Oughtred, also in 1631.

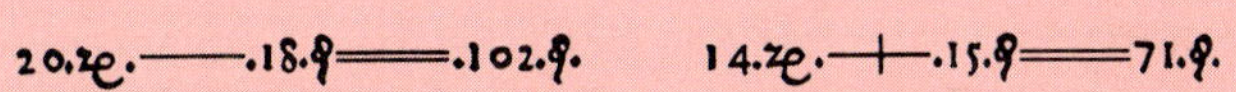

$20x \quad - \quad 18 \quad = \quad 102 \qquad 14x \quad + \quad 15 \quad = \quad 71$

TABLE 1.4 **Summary of Relation Symbols**

Symbol	Verbal Translation
$=$	Equals, is equal to
$\neq$	Does not equal
$<$	Is less than
$>$	Is greater than
$\not<$	Is not less than
$\not>$	Is not greater than
$\leq$	Is less than or is equal to
$\geq$	Is greater than or is equal to

Example 4 State the algebraic meaning and the geometric meaning of each of the following statements:
a. $2 > -5$ b. $-3 < -1$

SOLUTION
a. 2 is greater than -5; 2 is to the right of -5 on a horizontal number line.
b. -3 is less than -1; -3 is to the left of -1 on a horizontal number line.

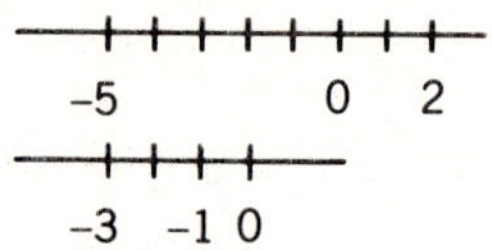

Figure 1.7

Example 5 Insert either the symbol $<$ or the symbol $>$ between each pair of numbers so that the resulting statement is true:
a. $-7, 3$ b. $-7, -8$

SOLUTION a. $-7 < 3$ b. $-7 > -8$

Graphs of Order Relations

In the preceding section it was shown that the number line can be used to graph points such as $x = 2$, $x = -\frac{1}{2}$, or $x = \sqrt{2}$.

Statements such as $x < 5$ or $x > 2$ can also be illustrated on a number line, but more than one point is involved. Consider the first statement: $x < 5$. Many numbers make this statement true: $3 < 5$, $0 < 5$, $-15 < 5$, $\frac{1}{2} < 5$; as a matter of fact, *all* numbers whose representation is to the left of 5 on a horizontal number line are less than 5. In the same manner it can be shown that *all* numbers to the right of 2 on a horizontal number line make $x > 2$ a true statement for x.

The following examples show how a set of real numbers described by an order relation can be graphed on a number line.

Example 6 Graph $x > 3$ on a number line.

SOLUTION The set of all numbers x such that $x > 3$ is graphed as shown in Fig. 1.8. The circle above the numeral 3 indicates that 3 is excluded from the solution set; the solution set is indicated by the half-line starting at 3 (but not including 3) and including all values greater than 3, as shown by the direction of the line.

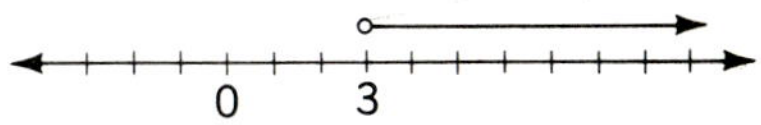

Figure 1.8

Example 7 Graph $x \leq -2$ on a number line.

SOLUTION The set of all numbers x such that $x \leq -2$ is graphed as shown in Fig. 1.9. This time a solid dot over the -2 coordinate indicates that -2 is included in the solution set as well as all points to the left of -2, since x is less than or equal to -2.

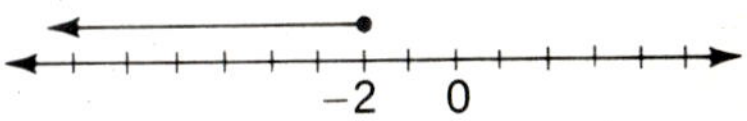

Figure 1.9

Often it is desirable to indicate that x can be any number *between* two given numbers. For example, "x is any number between 1 and 4" is expressed symbolically as

$$1 < x < 4$$

where 1 and 4 are called the *end points* of the interval containing x. If either end point or both end points are to be included, the $\leq$ symbol is used to indicate this fact. For example,

$$1 \leq x < 4$$

means "x is between 1 and 4 and x may equal 1."

Example 8 Graph $1 \leq x < 4$ on a number line.

SOLUTION The set of all numbers x such that $1 \leq x < 4$ is shown in Fig. 1.10. A solid dot over the 1 indicates that 1 is included, whereas the circle over the 4 indicates that 4 is excluded.

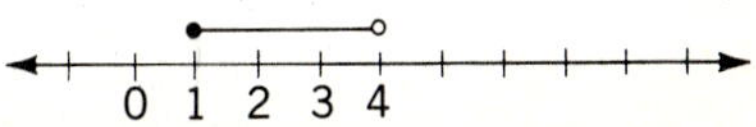

Figure 1.10

Exercises 1.6

Simplify Exercises 1–36. See Examples 1, 2, and 3.

1. $|3|$
2. $|-3|$
3. $\left|-2\frac{1}{2}\right|$
4. $-|-3|$
5. $\left|-\left|\frac{1}{3}\right|\right|$
6. $|-3| + |2|$
7. $|-3| - |2|$
8. $|3 - 3|$
9. $2(|3| + |-2|)$
10. $|-7| - |4|$
11. $|-4| \cdot |-2|$
12. $3|-2|$
13. $4 + |-2|$
14. $6 - |-2|$
15. $2(|-4| - |+4|)$
16. $|8 - 2|$
17. $-|-|-5||$
18. $-(|6| \cdot |-2|)$
19. $|14 - 2|$
20. $14 - 2|-3|$
21. $|5| - |-3|$
22. $|-4|^2$
23. $|0|$
24. $|-1| \cdot |-3|$
25. $\left|-1\frac{3}{4}\right|$
26. $\left|7\frac{1}{2}\right|$
27. $|-6| + |6|$
28. $|-7| + |-2|$
29. $|8| + |-3|$
30. $3|-5|$
31. $|-6| \cdot |-4|$
32. $|12| - |-5|$
33. $|-15| - |8|$
34. $2(|-9| - |-1|)$
35. $4(|-8| - |8|)$
36. $|10 - 6|$

In Exercises 37–62 insert the symbol $<$, the symbol $>$, or the symbol $=$ between each pair of numbers so that the resulting statement is true. See Examples 4 and 5.

37. 12, 5
38. 4, 9
39. -4, 2
40. 5, -6
41. 0, -3
42. 0, 8
43. -5, -7
44. -9, -7
45. -1, -2
46. 4, -1
47. -5, 0
48. 0, -2
49. $|-3|$, $|-6|$
50. $|3|$, $|6|$
51. $|-6|$, $|6|$
52. 6, $|-6|$
53. -6, $|6|$
54. $|6|$, -6
55. $|-5|$, 0
56. $|-2|$, $|1|$

57. $|-12|, |-4|$

58. $-4, |-12|$

59. $|-4|, -12$

60. $-12, -4$

61. $0, |-3|$

62. $0, -3$

Graph each of the sets of real numbers for Exercises 63–80 on a horizontal number line. See Examples 6, 7, and 8.

63. $x < 2$

64. $x > 2$

65. $x > -1$

66. $x \leq -3$

67. $x \leq -1$

68. $x > 0$

69. $x < 0$

70. $x > 4$

71. $x \geq 4$

72. $x < -5$

73. $x \geq 0$

74. $x \leq 0$

75. $-1 < x < 2$

76. $-1 \leq x < 2$

77. $2 < x \leq 5$

78. $-2 < x < 5$

79. $-5 \leq x \leq 3$

80. $0 < x \leq 2$

1.7 Sums of Signed Numbers

The meaning of the sum of two real numbers is motivated by the interpretation of a negative number as measuring a quantity opposite in nature to that measured by the corresponding positive number.

Since a \$5 gain followed by a \$3 gain results in a net gain of \$8, it is desirable to assign the meaning

$$5 + 3 = 8$$

Since a \$5 loss followed by a \$3 loss results in a net loss of \$8, it is desirable to assign the meaning

$$(-5) + (-3) = -8$$

Noting also that a \$3 gain followed by a \$5 gain results in a net gain of \$8 and that a \$3 loss followed by a \$5 loss results in a net loss of \$8, it follows that

$$5 + 3 = 3 + 5 = 8$$

and

$$(-5) + (-3) = (-3) + (-5) = -8$$

In other words, the order in which these numbers are added does not change the sum.

In general, the sum of two real numbers having the same sign is obtained by adding the absolute values of the numbers and prefixing their common sign.

Since a \$5 gain followed by a \$3 loss results in a net gain of \$2,

$$5 + (-3) = 2$$

Similarly, a \$3 loss followed by a \$5 gain results in a net gain of \$2,

$$-3 + 5 = 2$$

In other words,

$$5 + (-3) = (-3) + 5 = 2$$

On the other hand, since a \$5 loss followed by a \$3 gain results in a net loss of \$2,

$$-5 + 3 = -2$$

Also, since a \$3 gain followed by a \$5 loss results in a net loss of \$2,

$$3 + (-5) = -2$$

In other words,

$$-5 + 3 = 3 + (-5) = -(5 - 3) = -2$$

In general, the sum of two real numbers having opposite signs and unequal absolute values is obtained by subtracting their absolute values, the smaller from the larger, and prefixing the sign of the number having the larger absolute value.

Since the addition of zero can be interpreted as no change in measurement, for any real number x, $x + 0 = x$ and $0 + x = x$.

Addition Identity Axiom

For any real number x, there is exactly one real number 0 (zero) such that

$$x + 0 = x \quad \text{and} \quad 0 + x = x$$

When a number is added to its opposite (also called additive inverse), the result is zero. For example, $4 + (-4) = 0$ and $(-4) + 4 = 0$. In general, $x + (-x) = 0$ and $(-x) + x = 0$.

Addition Inverse Axiom

For each real number x, there is exactly one real number, $-x$, called the opposite (or additive inverse) of x such that

$$x + (-x) = 0 \quad \text{and} \quad (-x) + x = 0$$

Suppose that $x = -6$. Then its opposite $-x$ is $-(-6)$.

We know that 6 is the opposite of -6, since $6 + (-6) = 0$. We also know that -6 has exactly one opposite. Therefore, 6 is the opposite of -6 and $-(-6) = 6$. This property is true in general.

Opposite of an Opposite (Theorem)
For any real number x,

$$-(-x) = x$$

There are two other important properties concerning the addition of real numbers.

The Law of Order (Commutative Axiom)
The order in which real numbers are added does not change the sum. In symbols, $r + s = s + r$.

The Law of Grouping (Associative Axiom)
The way in which real numbers are grouped does not change the sum. In symbols, $(r + s) + t = r + (s + t)$.

Since $+5$ and 5 are simply different names for the same number, the $+$ sign is usually omitted for convenience in reading and writing. For example, the sum $(+9) + (-3)$ is usually written as $9 + (-3)$, and $(-8) + (+2)$ is usually written as $-8 + 2$.

Example 1 State the opposite (additive inverse) of each:
a. $+7$ b. -7 c. 0

SOLUTION
a. $-(+7) = -7$ because $-7 + 7 = 0$
b. $-(-7) = 7$ because $7 + (-7) = 0$
c. $-0 = 0$ because $0 + 0 = 0$

Example 2 Do the indicated additions:
a. $(+8) + (+2)$ b. $(-8) + (-2)$
c. $5 + 7$ d. $(-5) + (-7)$

SOLUTION When the signs are the same, add the absolute values and put the common sign before the sum.
a. $(+8) + (+2) = 8 + 2 = 10$
b. $(-8) + (-2) = -(8 + 2) = -10$
c. $5 + 7 = 12$
d. $(-5) + (-7) = -(5 + 7) = -12$

Example 3 Do the indicated additions:
a. $(+8) + (-2)$ b. $(-8) + (+2)$
c. $5 + (-7)$ d. $-5 + 7$

SOLUTION When the signs are different, subtract the absolute values, smaller from larger, and put the sign of the number having the larger absolute value before the difference.
a. $(+8) + (-2) = +(8 - 2) = 6$
b. $(-8) + (+2) = -(8 - 2) = -6$
c. $5 +(-7) = -(7 - 5) = -2$
d. $-5 + 7 = +(7 - 5) = 2$

Example 4 Add:

	a.	b.	c.	d.
	7	−4	−3	−8
	−9	−8	+9	8
SOLUTION	−2	−12	6	0

Note that addition may be written vertically as well as horizontally.

Example 5 Find the indicated sum: $5 + [(-8) + 3]$

SOLUTION $5 + [(-8) + 3] = 5 + (-5) = 0$

Example 6 Find the indicated sum: $[(-6) + 6] + (-7)$

SOLUTION $[(-6) + 6] + (-7) = 0 + (-7) = -7$

Example 7 Find the indicated sum: $20 + (-30) + (-40)$

SOLUTION $20 + (-30) + (-40)$
$= [(+20) + (-30)] + (-40)$
$= [-(30 - 20)] + (-40)$
$= (-10) + (-40)$
$= -(10 + 40) = -50$

Since the associative axiom for the addition of real numbers states that the way in which the numbers are grouped does not affect the sum, the sum may be obtained by grouping one way, and then this result may be checked by regrouping the other way.

CHECK $20 + (-30) + (-40)$
$= 20 + [(-30) + (-40)]$
$= 20 + [-(30 + 40)]$
$= 20 + (-70)$
$= -(70 - 20)$
$= -50$

Exercises 1.7

1–10. Simplify.

1. $-(+4)$
2. $-(+3)$
3. $-(-4)$
4. $-(-3)$
5. $-(-12)$
6. $-(-15)$
7. $-(+20)$
8. $-(+30)$
9. -0
10. $-(-0)$

11–82. Do each indicated addition.

11. $(+9) + (+3)$
12. $(+4) + (+10)$
13. $(-9) + (-3)$
14. $(-4) + (-10)$
15. $(+9) + (-3)$
16. $(+4) + (-10)$
17. $(-9) + (+3)$
18. $(-4) + (+10)$
19. $-5 + 0$
20. $0 + (-8)$
21. $0 + (-7)$
22. $-9 + 0$
23. $-4 + 4$
24. $-7 + 7$

25. $6 + (-6)$
26. $8 + (-8)$
27. $-(-9) + 9$
28. $-(-2) + (-2)$
29. $-(-5) + (-5)$
30. $-(-3) + 3$
31. $-12 + (-18)$
32. $-23 + (-17)$
33. $-12 + 18$
34. $-23 + 17$
35. $14 + (-9)$
36. $23 + (-17)$
37. $-25 + 15$
38. $30 + (-45)$
39. $-75 + (-25)$
40. $-35 + (-65)$

41. $\begin{array}{r} +12 \\ -7 \\ \hline \end{array}$ 42. $\begin{array}{r} +20 \\ -35 \\ \hline \end{array}$ 43. $\begin{array}{r} 15 \\ -40 \\ \hline \end{array}$ 44. $\begin{array}{r} 13 \\ -9 \\ \hline \end{array}$

45. $\begin{array}{r} -15 \\ -35 \\ \hline \end{array}$ 46. $\begin{array}{r} -99 \\ -6 \\ \hline \end{array}$ 47. $\begin{array}{r} -8 \\ 0 \\ \hline \end{array}$ 48. $\begin{array}{r} 0 \\ -5 \\ \hline \end{array}$

49. $\begin{array}{r} -28 \\ +28 \\ \hline \end{array}$ 50. $\begin{array}{r} +75 \\ -75 \\ \hline \end{array}$ 51. $\begin{array}{r} 14 \\ -14 \\ \hline \end{array}$ 52. $\begin{array}{r} -9 \\ +9 \\ \hline \end{array}$

53. $(-6) + (-9) + (-5)$
54. $-12 + 4 + (-2)$
55. $50 + (-70) + 90$
56. $25 + (-35) + (-45)$
57. $-9 + (-6 + 6)$
58. $[7 + (-7)] + 8$
59. $8 + (-10 + 2)$
60. $-5 + (-4 + 9)$
61. $[20 + (-14)] + (-10)$
62. $20 + [(-14) + (-10)]$
63. $(-3) + [(-5) + (-2)]$
64. $[(-4) + (-5)] + (-6)$
65. $[10 + (-20)] + (-30)$
66. $15 + [(-25) + (-40)]$
67. $[(-8) + 8] + (-7)$
68. $9 + [(-12) + 3]$
69. $-(-15 + 7)$
70. $-(-20 + 6)$
71. $-(-7 + 8 + 14)$
72. $-(-30 + 50 + 0)$
73. $5 + (-10) + (-15)$
74. $(-7) + 8 + (-9)$
75. $12 + 4 + (-20)$
76. $(-60) + (-70) + 90$
77. $(-11) + (-9) + (-7)$
78. $18 + (-15) + 17$
79. $-1 + 1 + (-1)$
80. $5 + (-3) + (-2)$
81. $-2 + 5 + (-3)$
82. $-78 + 49 + 78$

83. A bank statement contains the following entries (a positive number means a deposit; a negative number, a withdrawal):

$+220, \ +70, \ -95, \ -22, \ +15, \ -84$

If the account originally contained \$300, how much was in the account after the above deposits and withdrawals were made?

84. In the card game of pinochle, it is possible to gain points and to lose points. Find the final score of a player whose scores on successive hands are listed as follows:

$+10, \ -25, \ +32, \ +8, \ -34$

85. The weekly changes in weight for 6 weeks of a person on a diet are recorded below:

    $+2, \ -4, \ -2, \ +3, \ -5, \ +1$

    If the person originally weighed 71 kilograms, find his weight at the end of the 6-week period.

86. Determine the net yardage for the following sequence of football plays, where a positive number (+) means a gain in yardage and a negative number (−) means a loss in yardage:

    $+5, \ -3, \ -6, \ +15, \ +2, \ -10, \ +4$

87. Find the resulting speed of an object if its original speed was 40 mph and it was subject to the following successive accelerations (+) and decelerations (−):

    $+15, \ -30, \ -10, \ +15, \ +20$

88. The following is a treasurer's report for a certain organization (a positive number indicates dollars received, and a negative number indicates dollars paid out):

    Dues, +150; Stationery, −8; Stamps, −15; Food sale, +60; Gifts, −80; Party expenses, −75

    Find the amount of money in the treasury.

89. A piston is moving upward and downward from a central position *C*. A positive number indicates the number of centimeters above *C*, and a negative number indicates the number of centimeters below *C*. Find the position of the piston with respect to the central position *C* after the following motions have taken place (the piston started at *C*):

    $-5, \ -3, \ +2, \ +6, \ -4, \ -1$

In chemistry the valence of an atom (or a set of atoms) is an integer that determines how the atom (or set) will unite chemically compared with the hydrogen atom, assigned a valence of +1. *Find the valence (the sum of the component valences) for the ions in Exercises 90–94 whose component valences are given.*

90. Sulfate ion, SO_4, where the valence of S = +6, of $O_4 = -8$

91. Hydroxyl ion, OH, where the valence of O = −2, of H = +1

92. Ammonium ion, NH_4, where the valence of N = −3, of H_4 = +4

93. Hydronium ion, H_3O, where the valence of H_3 = +3, of O = −2

94. Nitrate ion, NO_3, where the valence of N = +5, of O_3 = −6

95. A chemical compound is such that the sum of the valences of its components is zero. Find the valence of the component whose valence is not given.
a. Baking soda, $NaHCO_3$; Na = +1, H = +1, O_3 = −6
b. Ammonium sulfate (fertilizer), NH_4HSO_4; H_4 = +4, H = +1, S = +6, O_4 = −8
c. Jade (jadeite), $NaAlSi_2O_6$; Na = +1, Si_2 = +8, O_6 = −12

96. At the end of a week, the closing price of a certain stock was quoted in the newspaper as $57\frac{1}{4}$ (meaning $57\frac{1}{4}$ dollars, or $57.25). The changes in price for the next week were quoted as follows, in dollars and fractions of dollars:

Monday $-1\frac{1}{4}$,

Tuesday $+\frac{3}{4}$,

Wednesday $+\frac{3}{8}$,

Thursday $-1\frac{1}{8}$,

Friday $-\frac{3}{4}$

Find the closing price of the stock at the end of this week.

1.8 Differences of Signed Numbers

In arithmetic the difference between two natural numbers (the numbers of arithmetic) is defined in terms of addition. For example, $12 - 3 = 9$, because $9 + 3 = 12$. In words, we ask: "What number, when *added* to 3, yields 12?" The answer, of course, is 9. The difference between two *real* numbers is defined in the same way.

For example, when looking for the answer to the problem $3 - 12$, we ask what number must be added to 12 to yield the answer 3. Since $12 + (-9) = 3$, $3 - 12 = -9$.

As another example,

$$-3 - 12 = -15 \quad \text{since} \quad 12 + (-15) = -3$$

Similarly,

$$3 - (-12) = 15 \quad \text{because} \quad -12 + 15 = 3$$

Addition and subtraction are called **inverse operations.** There is an easy way to do subtraction problems by noting this relationship between addition and subtraction.

Note that

$$\begin{aligned} 12 - 3 &= 12 + (-3) = 9 \\ 3 - 12 &= 3 + (-12) = -9 \\ -3 - 12 &= -3 + (-12) = -15 \\ 3 - (-12) &= 3 + (+12) = 15 \\ -3 - (-12) &= -3 + (+12) = 9 \end{aligned}$$

As a generalization of these special examples, the following rule can be stated.

Subtraction of Signed Numbers

To subtract a signed number, add the opposite of this number. In symbols,

$$r - s = r + (-s) \quad \text{and} \quad r - (-s) = r + s.$$

Example 1 Do the indicated subtractions:

a. $3 - 7$ b. $-3 - 7$ c. $3 - (-7)$ d. $-3 - (-7)$

SOLUTION
a. $3 - 7 = 3 + (-7) = -4$
b. $-3 - 7 = -3 + (-7) = -10$
c. $3 - (-7) = 3 + 7 = 10$
d. $-3 - (-7) = -3 + 7 = 4$

Example 2 Subtract and check:

a. $15 - 8$ b. $15 - (-8)$ c. $-15 - 8$ d. $-15 - (-8)$

SOLUTION a. $15 - 8 = 7$ (The answer is the same as it is in arithmetic.)

b. $15 - (-8) = 15 + 8 = 23$

c. $-15 - 8 = -15 + (-8) = -23$

d. $-15 - (-8) = -15 + 8 = -7$

CHECK b. $23 + (-8) = 15$

c. $-23 + 8 = -15$

d. $-7 + (-8) = -15$

Example 3 Subtract:

a. $0 - 6$ b. $0 - (-6)$ c. $-6 - 6$ d. $-6 - (-6)$

SOLUTION a. $0 - 6 = 0 + (-6) = -6$

b. $0 - (-6) = 0 + 6 = 6$

c $-6 - 6 = -6 + (-6) = -12$

d. $-6 - (-6) = -6 + 6 = 0$

Example 4 Write as a single integer: $-(2 - 3)$.

SOLUTION
$$\begin{aligned} -(2 - 3) &= -(2 + [-3]) \\ &= -(-1) \\ &= 1 \end{aligned}$$

Exercises 1.8

In Exercises 1–40, subtract and check.

1. $18 - 12$
2. $-9 - 4$
3. $10 - (-4)$
4. $(-15) - (-19)$
5. $(-17) - (-23)$
6. $0 - (-7)$
7. $2 - 9$
8. $-3 - 8$
9. $0 - 6$
10. $-2 - (-7)$
11. $10 - 9$
12. $6 - 10$
13. $50 - 15$
14. $40 - (-25)$
15. $(-3) - (-3)$
16. $5 - (-7)$
17. $7 - (-7)$
18. $(-7) - (-7)$
19. $0 - (-3)$
20. $0 - 3$
21. $7 - 3$
22. $3 - 7$
23. $7 - (-3)$
24. $-3 - 7$
25. $-6 - 4$
26. $-6 - (-4)$
27. $6 - 6$
28. $-4 - (-4)$
29. $-7 - 7$
30. $8 - (-8)$

31. $-20 - 8$
32. $-15 - (-10)$
33. $22 - 12$
34. $17 - 19$
35. $0 - 2$
36. $0 - (-5)$
37. $-8 - (-15)$
38. $8 - 15$
39. $-14 - 9$
40. $-12 - 7$

41–60. Subtract the lower number from the upper number.

41. $\begin{array}{r} 9 \\ \underline{-2} \end{array}$	42. $\begin{array}{r} 10 \\ \underline{-\ 6} \end{array}$	43. $\begin{array}{r} -9 \\ \underline{-2} \end{array}$	44. $\begin{array}{r} -10 \\ \underline{-\ 6} \end{array}$
45. $\begin{array}{r} -15 \\ \underline{25} \end{array}$	46. $\begin{array}{r} -70 \\ \underline{90} \end{array}$	47. $\begin{array}{r} -15 \\ \underline{-25} \end{array}$	48. $\begin{array}{r} -70 \\ \underline{-90} \end{array}$
49. $\begin{array}{r} 17 \\ \underline{9} \end{array}$	50. $\begin{array}{r} 30 \\ \underline{18} \end{array}$	51. $\begin{array}{r} 9 \\ \underline{17} \end{array}$	52. $\begin{array}{r} 18 \\ \underline{30} \end{array}$
53. $\begin{array}{r} 23 \\ \underline{23} \end{array}$	54. $\begin{array}{r} 54 \\ \underline{54} \end{array}$	55. $\begin{array}{r} -23 \\ \underline{-23} \end{array}$	56. $\begin{array}{r} -54 \\ \underline{-54} \end{array}$
57. $\begin{array}{r} -37 \\ \underline{37} \end{array}$	58. $\begin{array}{r} -19 \\ \underline{19} \end{array}$	59. $\begin{array}{r} 13 \\ \underline{-13} \end{array}$	60. $\begin{array}{r} 28 \\ \underline{-28} \end{array}$

Do the indicated operations in Exercises 61–80.

61. $(-5 + 5) - 9$
62. $(-1) - [(-2) - (-3)]$
63. $2 + (8 - 20)$
64. $0 - (6 - 12)$
65. $27 + [59 + (-27)]$
66. $(-38) + (38 - 26)$
67. $(-65) + [46 - (-65)]$
68. $(-6) + (7 - 10)$
69. $50 - (20 - 20)$
70. $[-(-8) - 2] - 3$
71. $20 - (10 - 5)$
72. $(20 - 10) - 5$
73. $-12 - (14 - 17)$
74. $-8 - (-4 + 4)$
75. $35 - [(-20) - (-20)]$
76. $[-65 - (-65)] - 40$
77. $-3 - (4 + [-2])$
78. $-3 - (4 - [-2])$
79. $(-3 - 4) + (-2)$
80. $(-3 - 4) - (-2)$

On a certain day a newspaper recorded the high and low temperatures (in degrees Celsius) for six cities, as shown in Exercises 81–86. Find the difference between the high temperature and the low temperature for each city.

	CITY	HIGH	LOW
81.	San Francisco	15	9
82.	New York	−4	−17
83.	Chicago	10	−6
84.	Fairbanks	−26	−40

85. Denver	−20	−25
86. Honolulu	24	22

Determine the difference in latitudes between each pair of cities in Exercises 87–90 (a positive number means north of the equator, and a negative number means south of the equator).

87. Sacramento, California +39 degrees
    Lima, Peru −12 degrees
88. Los Angeles, California +34 degrees
    Hanoi, North Vietnam +21 degrees
89. Johannesburg, South Africa −26 degrees
    Perth, Australia −32 degrees
90. Moscow, U.S.S.R. +56 degrees
    Seattle, Washington +48 degrees

Find the difference in elevation between each pair of geographic locations whose elevations are given in Exercises 91–95. A positive number indicates an elevation above sea level, and a negative number indicates an elevation below sea level.

91. Mount Kilimanjaro (highest point in Africa) +19,321 feet
    Qattara Depression, Egypt (lowest point in Africa) − 436 feet
92. Mount Whitney +14,495 feet
    Death Valley − 282 feet
    (highest and lowest points in California)
93. Qattara Depression − 436 feet
    Death Valley − 282 feet
94. Brawley, California − 119 feet
    El Centro, California − 45 feet
95. Dead Sea (surface level) − 1,292 feet
    Mount Everest (highest point in the world) +29,028 feet

For Exercises 96–100 determine the number of years that elapsed between each pair of dates. A positive number means time A.D.*, and a negative number means time* B.C. *(Note: There was no year* 0*, since the number* 0 *had not been invented when the calendar was reformed. In other words, the year* A.D. 1 *followed the year* 1 B.C. *For Exercises 98–100, calculate the difference in the usual way and then subtract one from the result to obtain the answer.)*

96. −2900, the building of the Great Gizeh Pyramid, to −1650, date of the Rhind papyrus (mathematical work)

97. -550, during lifetime of Pythagoras, to -300, during lifetime of Euclid
98. -212, fall of Syracuse to Rome, to $+400$, fall of Rome (western part)
99. -425, Golden Age of Athens, to $+500$, Golden Age of India
100. -2400, first historical records, to $+1980$

1.9 Products of Signed Numbers

Although there are many practical applications that involve products of signed numbers, it is not easy to provide a simple geometric illustration of such a product.

In arithmetic the product 3(5), often read as "three fives," means the sum $5 + 5 + 5$, or 15. The rules for multiplying signed numbers are made to keep this meaning and to keep the laws of order and grouping.

The Law of Order for Multiplication (Commutative Axiom)

The order in which numbers are multiplied does not change the product. In symbols, $rs = sr$.

The Law of Grouping for Multiplication (Associative Axiom)

The way in which the factors of a product are grouped does not change the product. In symbols, $(rs)t = r(st)$.

Considering $3(-5)$ as the sum of three negative fives,

$$3(-5) = (-5) + (-5) + (-5) = -15$$

Now $-3(5) = 5(-3)$ by the law of order, and therefore,

$$\begin{aligned} -3(5) &= 5(-3) \\ &= (-3) + (-3) + (-3) + (-3) + (-3) = -15 \end{aligned}$$

Note in these special cases that the product of a negative number and a positive number is negative.

The product $-3(5)$ can also be thought of as the subtraction of three fives, that is, as

$$-5 - 5 - 5 = (-5) + (-5) + (-5) = -15$$

On the basis of this idea, we can think of $(-3)(-5)$ as the subtraction of three negative fives, that is, as

$$-(-5) - (-5) - (-5) = 5 + 5 + 5 = 15$$

Accordingly, $(-3)(-5)$ is 15.

These results can be summarized and generalized in the following rule.

Multiplication of Signed Numbers

1. The product of two nonzero real numbers having the same sign is positive.
2. The product of two nonzero real numbers having different signs is negative.

In symbols, if r and s are any real numbers,

$$r(-s) = -rs, \quad -r(s) = -rs, \quad \text{and} \quad (-r)(-s) = rs$$

Girolamo Cardano: Doctor, Gambler, Mathematician

Girolamo Cardano (1501–1576) was born in Pavia, Italy, the illegitimate son of a professor of jurisprudence and medicine. His father educated him to be a doctor of medicine, and he did achieve fame as a physician. He was a professor of mathematics and medicine at Milan and later at Pavia and Bologna.

Cardano's life was stormy. He was an inveterate gambler who wrote a book on gambling, *Liber de ludo aleae* ("The Book on Games of Chance"). (An English version of this work, translated by Sydney H. Gould, was published by Holt, Rinehart and Winston, Inc., in 1961.) Deeply interested in astrology, Cardano also published a horoscope of the life of Christ, for which he was sent to prison. When he was released, he resigned his professorship, moved to Rome, and became astrologer to the papal court, which post he held until he died.

The numbers 1 and 0 are such special numbers that their properties need to be stated.

Products Involving 1 and 0

The product of any real number and 1 is that number. In symbols,

$$x \cdot 1 = x \quad \text{and} \quad 1 \cdot x = x.$$

The product of any real number and 0 is 0. In symbols,

$$x \cdot 0 = 0 \quad \text{and} \quad 0 \cdot x = 0.$$

The number 1 is called the **multiplication identity** because of its special property

$$x1 = 1x = x$$

Example 1 Find the indicated products:

a. $7(4)$ b. $7(-4)$ c. $-7(4)$ d. $(-7)(-4)$

SOLUTION
a. $7(4) = 28$
b. $7(-4) = -28$
c. $-7(4) = -28$
d. $(-7)(-4) = 28$

Example 2 Multiply:

a. $6(-1)$ b. $(-6)(1)$ c. $(0)(-6)$ d. $(-6)(0)$

SOLUTION
a. $6(-1) = -6$
b. $(-6)(1) = -6$
c. $(0)(-6) = 0$
d. $(-6)(0) = 0$

Example 3 Find the product of $(-4)(59)(-25)$ in two different ways.

SOLUTION

$$(-4)(59)(-25) = (-236)(-25) = 5900$$

$$\begin{aligned}(-4)(59)(-25) &= (59)(-4)(-25)\\ &= 59(100) = 5900\end{aligned}$$

Example 4 Do the indicated operations:

a. -3^2 b. $(-3)^2$

SOLUTION

a. -3^2 means $-(3^2)$, or the negative of 3^2, which is $-(3 \cdot 3)$; therefore, $-3^2 = -9$.

b. $(-3)^2$ means $(-3)(-3) = +9$; therefore, $(-3)^2 = 9$.

Example 5 Do the indicated operation: $(-9)^3$.

SOLUTION

$$\begin{aligned}(-9)^3 &= (-9)(-9)(-9)\\ &= [(-9)(-9)](-9)\\ &= (+81)(-9)\\ &= -729\end{aligned}$$

The Birth of Symbols

During the Middle Ages abbreviations began to be used to indicate the operations. The symbols gradually developed after the appearance of printed works. (The printing press was invented in 1438.)

The plus sign, +, is a contraction of the Latin word, *et*, meaning "and," When *et* is written rapidly, it looks like *et*.

The minus sign, −, was used as an equivalent for "$\overline{m}$" and "m," abbreviations of the word minus.

The marks + and − were used in medieval warehouses on sacks, crates, or barrels to indicate whether the contents were more or less than what they were supposed to be. The signs + and − made their first appearance in print in Widman's *Commercial Arithmetic,* published in 1489.

The dot for multiplication was introduced in 1583 by Christopher Clavius of Germany (1537–1612) and established by the German mathematician Gottfried Wilhelm Leibniz (1646–1716).

The division notation of algebra can be traced to the Hindus through the works of Bhaskara, who in 1150 wrote $\begin{matrix}1\\3\end{matrix}$ for $\frac{1}{3}$. The bar appears in the works of Viète, who also indicated multiplication by juxtaposition.

The symbol $\sqrt{\ }$ for root extraction was introduced in 1525 by Christoff Rudolff because it resembles a small "r," the initial letter of *radix,* the Latin word for root.

Viète wrote *A*, *Aq*, and *Acu* for our modern x, x^2, and x^3. Pierre Herigone, in his *Cursus Mathematicus* (1634–1637), used *a*2, *a*3, and *a*4 for a^2, a^3, and a^4. Our modern notation is first found in Descartes' *La géométrie* (1637), in which he used x^3, x^4, x^5, and so on. However, Descartes used both *xx* and x^2 to indicate the square of x. Until the latter part of the eighteenth century, it was common to find *xx* for x^2 and *xxx* for x^3.

Exercises 1.9

Do the indicated operations in Exercises 1–60.

1. $(5)(7)$
2. $(5)(-7)$
3. $(-5)(-7)$
4. $(-8)(2)$
5. $(-8)(-2)$
6. $(4)(-25)$
7. $(-8)(125)$
8. $(-25)(-4)$
9. $(-125)(8)$
10. $(3)(-5)(-7)$
11. $(8)(125)$
12. $(11)(-8)$
13. $(-8)(11)$
14. $(-17)(-13)$
15. $(-35)^2$
16. $(-20)^3$
17. $(10)(-25)$
18. $(-25)(-100)$
19. $(0)(-19)$
20. $(-3)(6-6)$
21. $(-3)(6)+(-3)(-6)$
22. $(5)(2)(7)$
23. $(-2)(-5)(9)$
24. $(-5)(-8)(-2)$
25. $(4)(-25)(7)$
26. $(-4)(9)(-25)$
27. $4(-5)^2$
28. $4(-5)^3$
29. $(10-4)(4-10)$
30. $(5-8)(6-6)$
31. $(-2)(-4)(-6)$
32. $(-5)(46)(-2)$
33. $(9)(-2)(3)$
34. $(-4)(-4)(-4)$
35. $(-5)^3$
36. $(-6)^2$
37. $(-2)^3(-2)^2$
38. $4(-3)^2$
39. $2(-7)^3$
40. $(-2)(5-8)$
41. $3(4-9)$
42. $(7-8)(6-10)$
43. $(-4+8)(-7+7)$
44. $5[-3-(-5)]$
45. $(3-9)(9-3)$
46. $[-2-(-2)]\cdot(3-9)$
47. $(-3)(7-15)$
48. $(-3)(7)-(-3)(15)$
49. $8-4(3-9)$
50. $(8-4)(3-9)$
51. $8(4-9)$
52. $8(4)-8(9)$
53. $(-7)(10-15)$
54. $(-7)(10)-(-7)(15)$
55. $(-2)(-2)(-2)(-2)$
56. $(-2)(-5)(-2)(-5)(-2)(-5)$
57. $[(-2)(-5)]^3$
58. $(-7+3)^2-(-7-3)^2$
59. $-(6-5)(6-4)(6-3)$
60. $(10-1)(10-2)(10-3)(10-4)$

In physics the moment M of a force F is given by $M = Fd$, where d is the directed distance of the force from a fixed position. Find M in Exercises 61–65.

61. $F = +200$ pounds, $d = +15$ feet
62. $F = +200$ pounds, $d = -10$ feet

63. $F = -70$ pounds,
$d = +8$ feet

64. $F = -120$ pounds,
$d = -6$ feet

65. $F = -500$ grams,
$d = -15$ centimeters

In chemistry a subscript at the lower right of a symbol for an atom indicates the number of atoms. For example, sulfuric acid is expressed in symbols as H_2SO_4*, meaning* 2 *atoms of hydrogen* (H), 1 *atom of sulfur* (S), *and* 4 *atoms of oxygen* (O). *The valence of an ion, a positive or negative integer indicating how the ion combines chemically, is the sum of the valences of the atoms composing the ion. For example, the valence of the sulfate ion,* SO_4*, equals (valence of* S) + 4(*valence of* O) $= (+6) + 4(-2) = 6 - 8 = -2$. *Find the valence of the ions in Exercises 66–70, given the valence of the component atoms.*

66. Nitrate ion, NO_3; valence of N $= +5$, valence of O $= -2$

67. Phosphate ion, PO_4; valence of P $= +5$, valence of O $= -2$

68. Bichromate ion, Cr_2O_7; valence of Cr $= +5$, valence of O $= -2$

69. Ammonium ion, NH_4; valence of N $= -3$, valence of H $= +1$

70. Permanganate ion, MnO_4; valence of Mn $= +7$, valence of O $= -2$

1.10 Quotients of Signed Numbers

Division may be defined for real numbers just as it is defined in arithmetic as the inverse operation to multiplication. For example, the quotient of 12 divided by 3 is defined to be 4, $\frac{12}{3} = 4$, because 4 is the only number such that $12 = 3 \cdot 4$.

Definition of Division

If r and s are any real numbers such that $s \neq 0$, then $\frac{r}{s} = t$ if and only if there is exactly one real number t so that $r = st$.

Example 1 Find the indicated quotients:

a. $\frac{12}{3}$ b. $\frac{-12}{-3}$ c. $\frac{12}{-3}$ d. $\frac{-12}{3}$

SOLUTION

a. $\frac{12}{3} = 4$ because $12 = (3)(4)$

b. $\frac{-12}{-3} = 4$ because $-12 = (-3)(4)$

c. $\frac{+12}{-3} = -4$ because $12 = (-3)(-4)$

d. $\frac{-12}{3} = -4$ because $-12 = (3)(-4)$

Example 2 Find the indicated quotients, if they exist:

a. $\frac{0}{5}$ b. $\frac{0}{-5}$ c. $\frac{5}{0}$ d. $\frac{0}{0}$

SOLUTION

a. $\frac{0}{5} = 0$ because $0 = (5)(0)$

b. $\frac{0}{-5} = 0$ because $0 = (-5)(0)$

c. $\frac{5}{0}$ is undefined. There is no number t such that $5 = 0 \cdot t$, since $0 \cdot t = 0$ for all real numbers t.

d. $\frac{0}{0}$ is undefined. By definition, if $\frac{0}{0} = t$, then there must be exactly one real number t such that $0 = 0 \cdot t$. In this case, there are many possibilities, since $0 = 0 \cdot t$ for all real numbers t.

Now note that $2\left(\frac{1}{2}\right) = \frac{2}{2} = 1$ and that $5\left(\frac{1}{5}\right) = \frac{5}{5} = 1$.

$\frac{1}{2}$ is called the **reciprocal** of 2 because $2\left(\frac{1}{2}\right) = 1$

$\frac{1}{5}$ is called the **reciprocal** of 5 because $5\left(\frac{1}{5}\right) = 1$

Similarly,

$\frac{-1}{2}$ is the **reciprocal** of -2, since $-2\left(\frac{-1}{2}\right) = 1$

In general, every real number except zero has a **reciprocal,** also called a **multiplication inverse.**

Multiplication Inverse (Axiom)

For every nonzero real number x, there is exactly one real number $\frac{1}{x}$ called the **reciprocal** (or **multiplication inverse**) of x so that

$$x\left(\frac{1}{x}\right) = \frac{1}{x}(x) = 1$$

Since every nonzero real number has a reciprocal, there is an alternative definition of division as a multiplication by the reciprocal of the divisor.

Alternative Definition of Division

$$\text{If } s \neq 0, \text{ then } \frac{r}{s} = r\left(\frac{1}{s}\right)$$

The special cases shown in the preceding examples can be generalized as follows.

Quotients of Signed Numbers

The quotient $\frac{0}{s} = 0$ for any real number s different from zero.

The quotients $\frac{s}{0}$ and $\frac{0}{0}$ are undefined.

The quotient of two nonzero real numbers having the **same** sign is **positive.**

The quotient of two nonzero real numbers having **different** signs is **negative.**

Example 3 Do the indicated operations, if possible:

a. $\frac{7 - 22}{-5}$ b. $\frac{6 - 6}{3}$ c. $\frac{4 - (-4)}{4 - 4}$ d. $\frac{8 - 2}{2 - 8}$

SOLUTION

a. $\frac{7 - 22}{-5} = \frac{-15}{-5} = 3$

b. $\frac{6 - 6}{3} = \frac{0}{3} = 0$

c. $\frac{4 - (-4)}{4 - 4} = \frac{4 + 4}{4 + (-4)} = \frac{8}{0}$, undefined

d. $\frac{8 - 2}{2 - 8} = \frac{6}{-6} = -1$

Exercises 1.10

In Exercises 1–80, do the indicated operations, if possible. If the expression is undefined, write "undefined."

1. $\frac{63}{9}$
2. $\frac{63}{-9}$
3. $\frac{-63}{9}$
4. $\frac{-63}{-9}$
5. $\frac{-1000}{125}$
6. $\frac{-1000}{-8}$
7. $\frac{100}{-5}$
8. $\frac{0}{-60}$
9. $\frac{-50}{0}$
10. $\frac{56}{7}$
11. $\frac{-56}{7}$
12. $\frac{56}{-4}$
13. $\frac{21}{-7}$
14. $\frac{-21}{7}$
15. $\frac{-21}{-7}$
16. $\frac{21}{7}$
17. $\frac{0}{-6}$
18. $\frac{-6}{0}$
19. $\frac{-56}{8}$
20. $\frac{63}{-7}$
21. $\frac{-55}{-11}$
22. $\frac{42}{-7}$
23. $\frac{-28}{4}$
24. $\frac{36}{4}$
25. $\frac{-132}{12}$
26. $\frac{169}{-13}$
27. $\frac{-100}{-4}$
28. $\frac{-1000}{8}$
29. $\frac{-125}{-25}$
30. $-\frac{15}{0}$
31. $\frac{0}{25}$
32. $\frac{0}{0}$
33. $\frac{7-2}{2-7}$
34. $\frac{4-12}{-4}$
35. $\frac{-120}{(-4)+6}$
36. $\frac{8-8}{8-(-8)}$

37. $\dfrac{120}{5-17}$

38. $\dfrac{10-30}{40-45}$

39. $\dfrac{25-15}{25-35}$

40. $\dfrac{0-9}{0+(-9)}$

41. $\dfrac{-132}{-22}$

42. $\dfrac{8-8}{8}$

43. $\dfrac{8-(-8)}{-8}$

44. $\dfrac{-3}{-3-(-3)}$

45. $\dfrac{18-25}{25-18}$

46. $\dfrac{96-39}{39-96}$

47. $\dfrac{(-5)^2}{-5}$

48. $\dfrac{(-5)^3}{-5}$

49. $\dfrac{40-55}{55-58}$

50. $\dfrac{60}{\left(\dfrac{-6}{2}\right)}$

51. $\dfrac{\left(\dfrac{60}{-6}\right)}{2}$

52. $\dfrac{-60}{\left(\dfrac{-10}{-2}\right)}$

53. $\dfrac{\left(\dfrac{-60}{-10}\right)}{-2}$

54. $\dfrac{(-4)-(-4)}{(-4)+(-4)}$

55. $\dfrac{(-1)+(-1)}{(-1)-(-1)}$

56. $\dfrac{(-7)^2-(-3)^2}{(-7)-(-3)}$

57. $\dfrac{12}{\left(\dfrac{-6}{2}\right)}$

58. $\dfrac{\left(\dfrac{12}{-6}\right)}{2}$

59. $\dfrac{-54}{\left(\dfrac{-6}{-3}\right)}$

60. $\dfrac{\left(\dfrac{-54}{-6}\right)}{-3}$

61. $\dfrac{(-10)^2}{-5}$

62. $\dfrac{56-89}{89-56}$

63. $\dfrac{95-48}{48-95}$

64. $\dfrac{(-10)^3}{125}$

65. $\dfrac{7-(-7)}{-7}$

66. $\dfrac{2(-6)}{6-(-6)}$

67. $\dfrac{(-7)^2-(5)^2}{-7-5}$

68. $\dfrac{(-3)^3-(-2)^3}{-3-(-2)}$

69. $\dfrac{(-10)^3 + (-5)^3}{(-10) + (-5)}$

70. $\dfrac{(5-1)(5-2)(5-3)}{(-1)(-2)(-3)}$

71. $\dfrac{(8-2)(2-4)(2-6)}{(-2)(-4)(-6)}$

72. $\dfrac{(-5)^2 + 2(-5)(-3) + (-3)^2}{(-5) + (-3)}$

73. $-\dfrac{-72}{-8}$

74. $-\dfrac{60}{-15}$

75. $-\dfrac{-154}{11}$

76. $\dfrac{(2-8)^3}{(8-2)^3}$

77. $(26-19)\dfrac{(2-6)}{(26-19)}$

78. $\dfrac{(3-9)(4-7)}{(4-7)(3-9)}$

79. $\dfrac{(5-1)(6-1)}{(1-6)(1-5)}$

80. $\dfrac{(-6)^2 - 2(-6)(-4) + (-4)^2}{(-6) - (-4)}$

1.11 Combined Operations: Evaluation

Since the associative axiom for addition states that the way in which the numbers are grouped does not affect the sum, it is not necessary to use parentheses to indicate a sum of three or more terms. In other words,

$$x + y + z = (x + y) + z = x + (y + z)$$

for all real numbers x, y, and z.

On the other hand, subtraction is *not* associative. For example,

$$(10 - 5) - 2 \neq 10 - (5 - 2)$$

However, since by definition $x - y = x + (-y)$, it is convenient to use this idea and adopt certain conventions that reduce the number of symbols that must be written.

Conventions

For all real numbers r, s, and t

$$r + s - t = r + s + (-t)$$

$$r - s + t = r + (-s) + t$$

$$r - s - t = r + (-s) + (-t)$$

Example 1 Use the preceding convention to calculate each of the following expressions in two ways, thus showing that the result is independent of the way in which the terms are grouped:

a. $7 + 6 - 5$ b. $12 - 8 + 3$ c. $10 - 5 - 2$

continued

SOLUTION

a. $7 + 6 - 5 = 7 + 6 + (-5)$
$= (7 + 6) + (-5)$
$= 13 + (-5) = 8$
$7 + 6 - 5 = 7 + 6 + (-5)$
$= 7 + [6 + (-5)]$
$= 7 + 1 = 8$

b. $12 - 8 + 3 = 12 + (-8) + 3$
$= [12 + (-8)] + 3$
$= 4 + 3 = 7$
$12 - 8 + 3 = 12 + (-8) + 3$
$= 12 + [(-8) + 3]$
$= 12 + (-5) = 7$

c. $10 - 5 - 2 = 10 + (-5) + (-2)$
$= [10 + (-5)] + (-2)$
$= 5 + (-2) = 3$
$10 - 5 - 2 = 10 + (-5) + (-2)$
$= 10 + [(-5) + (-2)]$
$= 10 + (-7) = 3$

Certain expressions and formulas require the substitution of one or more negative values. This is shown in the following examples.

Example 2 Evaluate $F = \frac{9C}{5} + 32$ for $C = -20$ degrees.

SOLUTION

$$F = \frac{9C}{5} + 32$$

$$F = \frac{9(\quad)}{5} + 32$$

$$F = \frac{9(-20)}{5} + 32$$

$$F = 9(-4) + 32 = -36 + 32 = -4 \text{ degrees}$$

Example 3 Evaluate $\frac{100 + 2x - y}{5}$ for $x = -35$, $y = -15$.

SOLUTION

$$\frac{100 + 2x - y}{5} = \frac{100 + 2(-35) - (-15)}{5}$$

$$= \frac{100 + (-70) + 15}{5}$$

$$= \frac{30 + 15}{5}$$

$$= \frac{45}{5} = 9$$

Exercises 1.11

Do the indicated operations in Exercises 1–26.

1. $(8 - 2) + (2 - 8)$
2. $5 - 2(3 - 4)$
3. $8 - (2 + 2) - 8$
4. $(5 - 2)(3 - 4)$
5. $3(-2)^2 + 4(-2)$
6. $2(-3)^2 - 4(-3)$
7. $8 - 4 + 2$
8. $15 - 8 - 3$
9. $9 + 6 - 5$
10. $2 - 5 - 7$
11. $1 - 2 + 3$
12. $4 + 8 - 2 - 6$
13. $12 - 3 - 15$
14. $1 - 6 + 3$
15. $20 + 40 - 30$
16. $-15 - 25 + 5$
17. $4 - 7 + 3$
18. $6 - 5 - 4 - 3$
19. $15 - 17 + 19 - 21$
20. $8 - 2 + 2 - 8$
21. $50 - 40 - 30 - 20$
22. $75 - 64 + 25 - 36$
23. $27 - 3 - 2 + 24$
24. $15 - 22 - 6 - 2$
25. $5 - 2 + 3 - 6$
26. $-8 - 10 + 5 + 7$

Evaluate in Exercises 27–50.

27. $x + y + z$ for $x = 7$, $y = -5$, $z = -9$
28. $x - y - z$ for $x = 12$, $y = 3$, $z = -8$
29. $x - y + z$ for $x = -6$, $y = -7$, $z = 8$
30. $x + y - z$ for $x = -15$, $y = 19$, $z = 21$
31. $a^2 + b^2 - c^2$ for $a = 2$, $b = -2$, $c = -1$
32. $a^2 - ab + b^2$ for $a = 7$, $b = -6$
33. $\dfrac{rs}{r - s - 1}$ for $r = 10$, $s = 7$
34. $\dfrac{r^2 + s^2 + 1}{r - s + 1}$ for $r = 4$, $s = -2$
35. $\dfrac{x + y + z}{3}$ for $x = 20$, $y = -17$, $z = -15$
36. $2x + 2y - 1$ for $x = 5$, $y = -4$
37. $\dfrac{c}{a} + \dfrac{d}{b} + \dfrac{a}{b}$ for $a = -6$, $b = -3$, $c = +18$, $d = -15$
38. $c^2 - a^2 - b^2$ for $c = 11$, $a = 2$, $b = 6$
39. $\dfrac{ac + bd}{a + b}$ for $a = 8$, $c = 3$, $b = 10$, $d = -6$
40. $a^2 - ab + b^2$ for $a = -4$, $b = -5$
41. $\dfrac{p + q - w}{2}$ for $p = 20$, $q = -5$, $w = -25$
42. $\dfrac{xyz}{x - y - z}$ for $x = 50$, $y = 60$, $z = -20$
43. $xy + xz + yz$ for $x = 5$, $y = -4$, $z = -2$
44. $a^2 + b^2 + c^2$ for $a = 2$, $b = -3$, $c = -4$

45. $(n + 1)^2 - (n + 1)(n - 1) + (n - 1)^2$ for $n = -5$

46. $b^2 - 4ac$ for $a = -3$, $b = -5$, $c = -4$

47. $b^2 - 4ac$ for $a = 1$, $b = 3$, $c = -4$

48. $b^2 - 4ac$ for $a = 2$, $b = -6$, $c = -3$

49. $b^2 - 4ac$ for $a = 2$, $b = -4$, $c = 2$

50. $b^2 - 4ac$ for $a = -5$, $b = -1$, $c = -1$

51. (Temperature scales: Fahrenheit to Celsius)

$$C = \frac{5(F - 32)}{9}$$

Find C for $F = -4$ degrees

52. (Temperature scales: Celsius to Fahrenheit)

$$F = \frac{9C}{5} + 32$$

Find F for $C = -15$ degrees

53. (Center of mass)

$$x = \frac{MD + md}{M + m}$$

Find x for $M = 150$ grams, $D = 40$ centimeters, $m = 50$ grams, $d = -20$ centimeters.

54. Find x in Exercise 53 for $M = 25$ pounds, $D = -3$ feet, $m = 15$ pounds, $d = -11$ feet.

In Exercises 55–56 (mirrors),

$$f = \frac{pq}{p + q}$$

where

f = *focal length of mirror (positive for concave mirrors and negative for convex mirrors)*
p = *object distance from mirror (positive when object is in front of the mirror)*
q = *image distance from mirror (positive when image is in front of the mirror and negative when image is behind the mirror)*

55. Find f for $p = 45$ centimeters, $q = -180$ centimeters.

56. Find f for $p = 6$ feet, $q = -3$ feet.

57. (Thermodynamics)

$$E = Q - W$$

where

E = change in internal energy
Q = change in heat (positive when added to the system and negative when removed)
W = work done (positive when done by the system and negative when done on the system)

Find E for each of the following cases:

a. $Q = 0$, $W = -6$ ft-lb
b. $Q = 0$, $W = 50$ joules
c. $Q = 400$ joules, $W = -175$ joules
d. $Q = -20$ Btu, $W = 0$
e. $Q = 810$ Btu, $W = 470$ Btu

58. (Corrective lenses)

$$P = \frac{100(p + q)}{pq}$$

where

P = power in diopters (positive for farsighted persons and negative for nearsighted)
p = object distance from lens (positive) in centimeters
q = image distance from lens (positive when on the opposite side of the lens as the object and negative when on the same side) in centimeters

Find P for each of the following cases:

a. $p = 20$, $q = -50$
b. $p = 300$, $q = -75$

59. (Flow of liquids)

$$P = \frac{d}{2}(V^2 - v^2)$$

where

P = change in pressure (positive for increase and negative for decrease)
V = original speed of the liquid
v = new speed of the liquid
d = density of the liquid

60. (Acoustics)

$$F = \frac{fV}{V + v}$$

where

F = observed frequency (pitch) of a moving source of sound waves
V = speed of sound in air
v = speed of the moving source (positive if moving away from the

Find P for each of the following cases:

a. $d = 62.5$, $V = 3$ feet per second, $v = 5$ feet per second

b. $d = 50$, $V = 4$ feet per second, $v = 2$ feet per second

observer and negative if moving toward the observer)

Find F for $f = 460$ cycles per second, $V = 1100$ feet per second, $v = -88$ feet per second.

Review Exercises

In Exercises 1–10 do the indicated operations, if possible.

1. $6[-3 - (-5)]$

2. $(-2)(7 - 10)$

3. $-2(7) - 10$

4. $(3 - 5)(5 - 3)$

5. $3 - 5(5 - 3)$

6. $3 - (5)(5) - 3$

7. $\dfrac{(-2) - (-2)}{(-2) + (-2)}$

8. $\dfrac{(-2) + (-2)}{(-2) - (-2)}$

9. $\dfrac{8 - 3}{3 - 8}$

10. $(-2)^2 + (-2)^3$

Express the verbal expressions in Exercises 11–15 in symbols.

11. The sum of x and y.

12. The product of x and y.

13. The difference when 4 is subtracted from the product of 5 and x.

14. The quotient when twice the product of a and b is divided by the sum of a and b.

15. The sum of the cube of x and the square of x.

16–20. List each subset of the set of natural numbers.

16. The factors of 70

17. The even factors of 70

18. The odd factors of 70

19. The prime factors of 70

20. The composite factors of 70

In Exercises 21–32, evaluate each.

21. $(x + 2)(x - 3)$ for a. $x = 5$, b. $x = 0$, c. $x = -5$

22. $\dfrac{4x}{6 - x}$ for a. $x = 8$, b. $x = 0$, c. $x = -2$

23. $\dfrac{x^2 - 16}{x - 4}$ for a. $x = 2$, b. $x = 0$, c. $x = -5$

24. $3x - (7 - x)$ for a. $x = 4$, b. $x = 0$, c. $x = -2$

25. $x - [x - (x - 2)]$ for a. $x = 7$, b. $x = 0$, c. $x = -1$

26. $10 - 2[x - (5 - 3x)]$ for a. $x = 1$, b. $x = 3$, c. $x = -2$

27. $\dfrac{(x-1)(x-2)(x-3)}{(-1)(-2)(-3)}$ for a. $x = 9$, b. $x = 1$, c. $x = -3$

28. $\dfrac{x^2 + 3xy - 10y^2}{x - 2y}$ for a. $x = 3, y = 2$

b. $x = -4, y = -5$

29. $x - y + z$ for a. $x = 7, y = 9, z = 11$

b. $x = 5, y = -2, z = -6$

30. $x - y - z$ for a. $x = 25, y = 15, z = 35$

b. $x = -5, y = -10, z = +15$

31. xyz for a. $x = 7, y = -4, z = -25$

b. $x = -4, y = -3, z = -2$

32. $\dfrac{xy}{xz}$ for a. $x = -5, y = -8, z = +2$

b. $x = -6, y = -18, z = -9$

In Exercises 33–42, answer true or false.

33. $4 < 10$

34. $-4 < -10$

35. $8 > -3$

36. $0 > -6$

37. $-12 > -15$

38. $-7 > -6$

39. $|-5| < 0$

40. $|-5| > 0$

41. $|-8| = 8$

42. $|7 - 3| = |3 - 7|$

For Exercises 43–52, graph each on a horizontal number line.

43. $x < 4$

44. $x \leq -2$

45. $x > 0$

46. $x > -3$

47. $x \geq 5$

48. $x \leq 0$

49. $3 < x < 6$

50. $-2 \leq x < 0$

51. $1 < x \leq 5$

52. $-2 \leq x \leq 3$

53. (Slope of a line)

$$m = \frac{Y - y}{X - x}$$

Find m for $Y = 4$, $y = 10$, $X = -1$, $x = -7$.

54. (Flow of liquids)

$$P = p + \frac{d}{2}(v^2 - V^2)$$

Find P for $p = 50$, $d = 1$, $v = 20$, $V = 40$.

55. (Medication for a child)

$$C = \frac{WA}{150}$$

Find C for $W = 60$ pounds, $A = 200$ milligrams.

56. (Radio and television)

$$r = \frac{R(E - G)}{E}$$

Find r for $R = 8$ megohms, $E = -12$ volts, $G = -9$ volts.

2 Linear Equations

Objectives

OBJECTIVES	TYPICAL PROBLEMS	LOCATIONS
To rewrite the terms of a sum or the factors of a product in conventional form.	1–2. Rewrite in conventional form. 1. $(b - 7) + (a + 3)$ 2. $(-7y)(-2xy)$	*Section 2.1* *Example 1* *Exercises 1–20*
To use the distributive axiom to remove parentheses and collect like terms.	3–4. Remove parentheses and collect like terms. 3. $8(x - 2) - 3(x - 4)$ 4. $2x(x + 3) - 5(x + 3)$	*Section 2.1* *Examples 2–18* *Exercises 21–96*
To solve simple linear equations in one variable.	5–10. Solve and check. 5. $x + 7 = 3$ 6. $5 - t = 8$ 7. $3y - 8 = 7$ 8. $\frac{x + 5}{5} = -6$ 9. $x - 6 = -(6 - x)$ 10. $x - 6 = x + 6$	*Section 2.2* *Examples 1–13* *Exercises 1–60*
To solve a general linear equation in one variable.	11–12. Solve and check. 11. $7(y - 2) - 3(y - 4) = 18$ 12. $x - 2(x + 2) = 16 + 2(x - 4)$	*Section 2.3* *Examples 1–4* *Exercises 1–52*
To solve a linear inequality in one variable.	13–14. Solve and check. 13. $5 - 2x \leq x - 4$ 14. $6x - (x + 7) < 2 - (x - 3)$	*Section 2.4* *Examples 1–4* *Exercises 1–40*
To solve a literal equation.	15. Solve $2x - 5y = 10$ for y 16. Solve $S = a + (n - 1)d$ for n	*Section 2.5* *Examples 1–6* *Exercises 1–60*
To solve a stated number problem.	17. Three times the difference when 5 is subtracted from a number is 9 more than the number. Find the number.	*Section 2.6* *Examples 1, 2* *Exercises 1–12*
	18. The sum of two consecutive odd integers is 56. Find the integers.	*Section 2.6* *Examples 3, 4* *Exercises 13–26*

19. Sara's aunt is 5 times as old as Sara. In 3 years Sara's aunt will be 4 times as old as Sara. Find the present ages of Sara and her aunt.

Section 2.6
Example 5
Exercises 27–32

To solve a stated geometric problem.

20. The length of a rectangle is 6 feet less than 3 times its width. The perimeter of the rectangle is 68 feet. Find the width and length of the rectangle.

Section 2.7
Examples 1, 2
Exercises 1–24

To solve a stated mixture problem.

21. Walnuts worth 70¢ per pound are mixed with almonds worth $1.10 per pound to produce 200 pounds of nuts worth 80¢ per pound. How many pounds of walnuts and how many pounds of almonds are used?

Section 2.8
Examples 1–5
Exercises 1–20

To solve a stated uniform motion problem.

22. Two trains start from the same station and travel in opposite directions. One train travels 20 mph faster than the other train. At the end of 7 hours, they are 630 miles apart. Find the rate of the slower train.

Section 2.9
Examples 1, 2
Exercises 1–4, 13, 14

23. A boat traveled downstream at 18 mph and then returned at 12 mph. The total travel time was 5 hours. Find the distance the boat traveled downstream.

Section 2.9
Example 3
Exercises 5–8, 17, 18

24. Al's running speed is 1 meter per second slower than Bill's running speed. When Bill gives Al a 10-meter head start, Al finishes a certain race in 140 seconds and Bill finishes the race in 120 seconds. Find Bill's running speed in meters per second.

Section 2.9
Examples 4, 5
Exercises 9–12, 15, 16

Answers to Typical Problems

1. $a + b - 4$
2. $14xy^2$
3. $5x - 4$
4. $2x^2 + x - 15$
5. -4
6. -3
7. 5
8. -35
9. R
10. $\varnothing$
11. 5
12. -4
13. $x \geq 3$
14. $x < 2$
15. $y = \dfrac{2x - 10}{5}$
16. $n = \dfrac{S - a + d}{d}$
17. $3(x - 5) = x + 9$; 12
18. $x + (x + 2) = 56$; 27, 29
19. $5x + 3 = 4(x + 3)$; 9, 45
20. $2W + 2(3W - 6) = 68$
    10 feet, 24 feet
21. $70x + 110(200 - x) = 80(200)$
    150 pounds of walnuts
    50 pounds of almonds
22. $7x + 7(x + 20) = 630$; 35 mph
23. $18x = 12(5 - x)$
    $x = 2$ hrs; $d = 18 \cdot 2 = 36$ miles
24. $140(x - 1) + 10 = 120x$
    $20x = 130$
    $x = 6\frac{1}{2}$ meters per second

One of the most important skills needed for success in mathematics is the solution of equations. It is the aim of this chapter to introduce you to the solution of the simplest type of equation: the linear equation in one variable. In preparation for this, some preliminary axioms and theorems must be stated, some of which were introduced in the previous chapter.

There are three major categories of general statements in mathematics: definitions, axioms, and theorems. A **definition** is a precise explanation of the meaning of a term. An **axiom** is a statement that is assumed to be true. A **theorem** is a statement that is proved.

2.1 Some Axioms

Commutative and Associative Axioms

The **commutative axioms** (the laws of order) for addition and multiplication state that the order in which two numbers are added does not change the sum and that the order in which two numbers are multiplied does not change the product.

The **associative axioms** (the laws of grouping) for addition and multiplication state that both the terms of a sum and the factors of a product may be regrouped without changing the result.

Two important consequences result from combining the commutative axiom with the associative axiom for the same operation:

1. The **terms of a sum can be rearranged** in any order without changing the value of the sum.
2. The **factors of a product can be rearranged** in any order without changing the value of the product.

There are certain conventions for writing the terms of a sum or the factors of a product. These conventions, stated as follows, are especially useful for comparing the results obtained by another person (such as checking the answers in the back of the book).

Conventions

The terms of a sum are written so that

1. A numerical term is to the right of all literal terms.
2. The letters involved are written in alphabetical order whenever possible.

The factors of a product are written so that

1. A numerical factor is to the left of all literal factors.
2. The letters involved are written in alphabetical order whenever possible.

Example 1 Rearrange the terms of each sum and the factors of each product according to the sum and product conventions. Simplify each result if possible.

GIVEN EXPRESSION	SOLUTION REARRANGEMENT	SIMPLIFICATION
a. $y + 5 + x$	$x + y + 5$	None possible
b. $-6 + n + 6$	$n + 6 + (-6)$	n
c. $a - 4 - b$	$a - b - 4$	None possible
d. $6x(-2)$	$(-2)(6)x$	$-12x$
e. $(-4xy)(-7x)$	$(-4)(-7)xxy$	$28x^2y$
f. $y(3x)$	$3xy$	None possible

The Distributive Axiom

The distributive axiom is important for many reasons. Three of these reasons are

1. It links the operations of addition and multiplication.
2. It permits the collecting of "like terms" to simplify expressions, such as $2x + 3x$, which can be written as $5x$.
3. It permits the "removal of parentheses" needed in simplifying expressions and in solving equations.

The Distributive Axiom
For all real numbers r, s, and t,

$$r(s + t) = rs + rt$$

In other words, multiplication "distributes" over addition. The factor r is distributed to *each* term of the sum.

Example 2 Use the distributive axiom to remove the parentheses in $3(x + 2)$.

SOLUTION $3(x + 2) = 3 \cdot x + 3 \cdot 2$
$= 3x + 6$

Example 3 Express $5(y + z)$ as a sum by applying the distributive axiom.

SOLUTION $5(y + z) = 5y + 5z$

The following immediate consequences of the distributive axiom are useful:

1. $a(b - c) = ab - ac$
2. $ab + ac = a(b + c)$
3. $ac + bc = (a + b)c$
4. $ac - bc = (a - b)c$
5. $ad + bd + cd = (a + b + c)d$
6. $-(b + c) = (-1)(b + c) = -b - c$
7. $-(b - c) = (-1)(b - c) = -b + c$

Example 4 Express $2(x - 5)$ as a sum, using the distributive axiom or one of its consequences.

SOLUTION
$2(x - 5) = 2(x) - 2(5)$ (By consequence 1)
$= 2x - 10$

Example 5 Express $2x + 3x$ as a product by using consequence 3.

SOLUTION $2x + 3x = (2 + 3)x = 5x$

Example 6 Simplify $5z - 7z$.

SOLUTION Expressing $5z - 7z$ as a product by using consequence 4,

$$5z - 7z = (5 - 7)z = -2z$$

Example 7 Simplify $3y + 2y - 4y$.

SOLUTION

$$\begin{aligned} 3y + 2y - 4y &= (3 + 2 - 4)y \\ &= 1y \\ &= y \text{ (By consequence 5)} \end{aligned}$$

Example 8 Remove the parentheses: $-(2x + 8)$.

SOLUTION Using consequence 6,

$$-(2x + 8) = -2x - 8$$

Example 9 Remove the parentheses: $-(5 - 3x)$.

SOLUTION Using consequence 7,

$$\begin{aligned} -(5 - 3x) &= -5 + 3x \\ &= 3x + (-5) \\ &= 3x - 5 \end{aligned}$$

Example 10 Remove the parentheses: $-4(2x + 5)$.

SOLUTION $-4(2x + 5) = -8x - 20$

One of the purposes of the distributive axiom, as stated at the beginning of this section, is to "collect *like* terms."

Definitions

Like terms are terms with identical literal factors.

A **literal factor** is a factor denoted by a letter.

As examples, $5x$ and $2x$ are like terms; $16x^2$ and $-4x^2$ are like terms; $39xy^2z$ and $-\frac{1}{2}xy^2z$ are like terms.

The terms $3x$ and $3y$ are *not* like terms, because the literal factors, x and y, are not identical. Similarly, $5x^2$ and $5x$ are *not* like terms, and xy and x are *not* like terms. Terms that are not like are called *unlike* terms.

The distributive axiom can be used to collect like terms, as shown in the preceding examples and also in the following examples. For convenience, the distributive axiom and all its consequences will be referred to simply as "the distributive axiom" from now on.

Simplify Examples 11–16 by collecting like terms.

Example 11 $2x - 3x$

SOLUTION $2x - 3x = (2 - 3)x = -x$

Example 12 $4x^2 + 6x^2 - 2x^2$

SOLUTION

$$\begin{aligned} 4x^2 + 6x^2 - 2x^2 &= (4 + 6 - 2)x^2 \\ &= 8x^2 \end{aligned}$$

Example 13 $2x + 3 + 4x$

SOLUTION

$$\begin{aligned} 2x + 3 + 4x &= (2x + 4x) + 3 \\ &= (2 + 4)x + 3 \\ &= 6x + 3 \end{aligned}$$

Note that $6x$ and 3 are unlike terms.

Example 14 $x^2 + 2x^2$

SOLUTION

$$\begin{aligned} x^2 + 2x^2 &= 1x^2 + 2x^2 \\ &= (1 + 2)x^2 \\ &= 3x^2 \end{aligned}$$

Example 15 $xy - 2y + xy + x$

SOLUTION

$$\begin{aligned} xy - 2y + xy + x &= (xy + xy) + x - 2y \\ &= 2xy + x - 2y \end{aligned}$$

Note that $2xy$, x, and $2y$ are unlike terms.

Example 16 $x^2 + 5x - 6x + 2$

SOLUTION

$$\begin{aligned} x^2 + 5x - 6x + 2 &= x^2 + (5x - 6x) + 2 \\ &= x^2 + (5 - 6)x + 2 \\ &= x^2 + (-1)x + 2 \\ &= x^2 - x + 2 \end{aligned}$$

Example 17 Simplify $3(x + 2y) + 5(x - y)$.

SOLUTION

$$\begin{aligned} 3(x + 2y) + 5(x - y) &= 3x + 6y + 5x - 5y \\ &= (3x + 5x) + (6y - 5y) \\ &= (3 + 5)x + (6 - 5)y \\ &= 8x + y \end{aligned}$$

Example 18 Simplify $2(x + y) - (x - y)$.

SOLUTION

$$\begin{aligned} 2(x + y) - (x - y) &= 2x + 2y - x + y \\ &= (2x - x) + (2y + y) \\ &= x + 3y \end{aligned}$$

With practice, you should be able to omit many of the steps shown in the examples.

The Rule of False Position

The ancient Egyptians used a method called the *Rule of False Position* for solving certain types of linear equations. The following is a translated and simplified version of Problem 26 in the Rhind papyrus.

> A quantity and its fourth added together become 15. What is the quantity?
>
> Try 4. Then the sum of 4 and $\frac{1}{4}$ of 4 is 5.
>
> As many times as you must multiply 5 to get 15, so must you also multiply your trial number.
> Thus 4 times 3 is 12. Behold, see it is 12.
>
> 12 plus $\frac{1}{4}$ of 12 is 12 plus 3, 15.

In our modern notation this problem reads

$$x + \frac{x}{4} = 15$$

A modern solution could be

$$5\left(\frac{x}{4}\right) = 15; \ \frac{x}{4} = 3; \ x = 12$$

Exercises 2.1

Rearrange the terms of each sum and the factors of each product in Exercises 1–20 according to the sum and product conventions stated in this section. Simplify each result if possible.

1. $y + 4 + x$
2. $4 + x - 7$
3. $(-8 - x) + 8$
4. $(y - 8) + (x + 4)$
5. $(-4xy)(-9xy)$
6. $(2 - x) + (x - 7)$
7. $8 + b(2a) - 12$
8. $3b(-4) + (-5a)(-6)$
9. $n - 7 + m$
10. $6 - n + m - 7$
11. $(3x)(-5x)$
12. $(-9x)(-3y)$
13. $(y - 6) + (x - 3)$
14. $(-6x)(5xy)$
15. $(x - 5) + (8 - x)$
16. $4 - y + x - 3$
17. $(4c)(-3b)(-2a)$
18. $d - b(-5a) - 4c$
19. $(-na)(-nb)(-nc)$
20. $(ka)(-kb)(-5b)(-7)$

For Exercises 21–40, remove parentheses by applying the distributive axiom or one of its consequences.

21. $2(x + 5)$
22. $-3(y + 2)$
23. $5(n - 7)$
24. $-6(2x - 3)$
25. $-(x - 5)$
26. $-(3x + 4)$
27. $-(6 - y)$
28. $3(x + y - 2)$
29. $-\frac{1}{2}(2x - 6y)$
30. $(a + b)x$
31. $(2a - b)3y$
32. $a - (b - c)$
33. $x - (y + 5)$
34. $-4(-2x + 3y - 4)$
35. $a - 2[b - (c - 3)]$
36. $y - [4z - 2(x + 5)]$
37. $3(a - 2b) + 4(x + 2y)$
38. $0 - (c - d)$
39. $4x^2(x - 1) - (x - 1)$
40. $a(x + 5) - 2(x + 5)$

Combine like terms and simplify by using the distributive axiom or one of its consequences in Exercises 41–66.

41. $3x + 2x$
42. $x + 4x + 5x$
43. $2x - x$
44. $3y - 3y$
45. $-4xy + 2xy$
46. $3z + 5z - 7z$
47. $3x + 5 - x$
48. $10k - 5k + 3k$
49. $6 - 4t - 3t$
50. $5x^2 + 5x - 8x - 2$
51. $3z^2 - z + 6z + 4$
52. $3p - q - 2p$
53. $6x^2 + 12xy - 5xy - 10y^2$
54. $x^2 - xy + xy - y^2$

55. $3z + 3z^2 - z$
56. $2k + 7 - 5k$
57. $x^2 + 2x - 3x + 5$
58. $2x + 5 + 3x - 7$
59. $4 - x + 2$
60. $5n + 4m + n$
61. $2s - t + 3s + 2t$
62. $2p + 3q + 3p$
63. $3x^2 - 6xy + 5xy - 10y^2$
64. $5xy - 6xz - 4xy - 2zx$
65. $9x^2 + 12xy - 12xy - 16y^2$
66. $ab + 2ac - 4ab - ac$

In Exercises 67–96, remove parentheses and collect like terms.

67. $3(x + 2) - 5$
68. $-(y - 3) + 2y$
69. $3x^2 - (5x - 2x) + 3$
70. $y^2 - (y + 3y) + 2(y^2 + 1)$
71. $6(x + y) + 2(x + y)$
72. $-7(c + d) - (c + d)$
73. $5(a - b) - 4(a - b)$
74. $3x(2x - 1) - 2x(3x + 2)$
75. $4x(2y - 3) - 3y(2x - 5)$
76. $10 - (x + 5)$
77. $6k - (4 + 5k)$
78. $2a(b - 8) - b(a - 6)$
79. $2(x + x) + 5$
80. $-(4 - 2k) + 3k$
81. $4y^2 - (3y + 5y) + 9$
82. $x^2 - (x - 2) + 3(x^2 + 5)$
83. $7(x + y) - 4(x + y)$
84. $-8(a - b) - (a - b)$
85. $2(x + y) - 2(x - y)$
86. $6x(5y - 2) - 3y(5x - 1)$
87. $7y(5y - 2) - 5y(7y + 2)$
88. $8x - (3x - 4)$
89. $6x - (6 + 7x)$
90. $a(b + 1) - 3b(5 - a)$
91. $5k - 2[4(2 - k) - (6 - k)]$
92. $x - 3[x - 3(x - 3)]$
93. $2(m - 4n) - 6(m - 4n) + (m - 4n)$
94. $2n - [3(n - 2) - 5(n + 3)]$
95. $2x - [2 - (x - 2) - 2(x - 2)]$
96. $3(c + 4d) - 5(c + 4d) - (c + 4d)$

2.2 Equations and Solutions

Basic Concepts

An equation is a statement having the form $A = B$. However, an equation does not have to be a true statement. Consider the following examples of equations:

$2 + 4 = 6$	(True)
$7 - 3 = 8$	(False)
$x + 5 = 5 + x$	(Open, true for all values of x)
$x + 5 = x + 1$	(Open, false for all values of x)
$x + 5 = 9$	(Open, true for some value of x and false for some value of x)

An **open equation** is an equation containing one or more variables. An open equation is not classified as true or false but becomes true or false when each variable is replaced by a numerical value.

The equation $2 + 4 = 6$ is called a *true equation*, since it does not contain any variables and since it is known that the sum of 2 and 4 is 6.

The equation $7 - 3 = 8$ is called a *false equation*, since it contains no variables and since it is known that the difference between 7 and 3 is 4 and not 8.

The open equation $x + 5 = 5 + x$ becomes true for all replacements of the variable. Such an equation is also called an **identity.**

The open equation $x + 5 = x + 1$ becomes false for all replacements of the variable. Equations such as these are sometimes called *contradictions*.

The open equation $x + 5 = 9$ is neither true nor false but becomes true or false when the variable is replaced by a numerical value.

For example, if $x = 2$, then $x + 5 = 9$ becomes $2 + 5 = 9$, a false equation.

If $x = 4$, then $x + 5 = 9$ becomes $4 + 5 = 9$, a true equation.

An open equation that is sometimes true and sometimes false is also called a **conditional equation** because the equation becomes true on the condition that the variable is replaced by a certain numerical value. The conditional equation $x + 5 = 9$ is true on the condition that x is replaced by 4.

The number 4 is called a **solution** or **root** of the equation $x + 5 = 9$ because $x + 5 = 9$ becomes true when x is replaced by 4. The number 4 is also said to **satisfy** the equation $x + 5 = 9$.

An important concern of algebra is the process of finding all the solutions of an open equation. This process is called **solving the equation.**

Definitions

A **solution** or **root** of an open equation in one variable is a number that makes the equation true when the variable is replaced by this number.

The **solution set** of an open equation is the set of all solutions of the equation.

To **solve an equation** means to find the solution set of the equation.

One primitive method for solving an equation is the trial-and-error method. This method is illustrated in the following example.

Example 1 Solve $2x^2 + 5x = 3$.

TRIAL-AND-ERROR SOLUTION First, try positive integers.

If $x = 1$, then $2x^2 + 5x = 2 \cdot 1^2 + 5 \cdot 1 = 2 + 5 = 7$; $7 = 3$ is false.

If $x = 2$, then $2x^2 + 5x = 2 \cdot 2^2 + 5 \cdot 2 = 8 + 10 = 18$; $18 = 3$ is false.

If $x = 3$, then $2x^2 + 5x = 2 \cdot 3^2 + 5 \cdot 3 = 18 + 15 = 33$; $33 = 3$ is false.

Since the resulting numbers are getting larger and larger, it seems useless to continue trying positive integers. Thus negative integers are tried next.

If $x = -1$, then $2x^2 + 5x = 2(-1)^2 + 5(-1) = 2 - 5 = -3$; $-3 = 3$ is false.

If $x = -2$, then $2x^2 + 5x = 2(-2)^2 + 5(-2) = 8 - 10 = -2$; $-2 = 3$ is false.

If $x = -3$, then $2x^2 + 5x = 2(-3)^2 + 5(-3) = 18 - 15 = 3$; $3 = 3$ is true.

At last a solution has been found! $x = -3$

Have all the solutions to Example 1 been found? There is no way to know for certain by this method because all the integers have not been tried nor have any of the nonintegral rational numbers been tried.

As a matter of fact, it may be shown that $x = \frac{1}{2}$ is also a solution.

If $x = \frac{1}{2}$, then $2x^2 + 5x = 2\left(\frac{1}{2}\right)^2 + 5\left(\frac{1}{2}\right) = \frac{1}{2} + \frac{5}{2} = \frac{6}{2} = 3$; $3 = 3$ is true.

The solution set of $2x^2 + 5x = 3$ is $\left\{\frac{1}{2}, -3\right\}$. However, at this point there is no guarantee that the solution set has been found, that is, that $x = \frac{1}{2}$ and $x = -3$ are the *only* replacements that make the equation $2x^2 + 5x = 3$ true.

Finally, it seems reasonable that there must be a better method than the trial-and-error method for solving equations. Indeed, the next sections are devoted to the development of procedures that will be less time-consuming than the trial-and-error method and that will give assurance that the solution set has indeed been found.

Equivalence Theorems

Definition

Equivalent equations are equations that have the same solution set.

For example, all the following equations are equivalent because each has $\{5\}$ for its solution set—that is, 5 is the only solution of each equation:

$$x + 2 = 7$$
$$2(x - 3) = 4$$
$$\frac{x}{3} = \frac{5}{3}$$
$$x = 5$$

The process of solving an equation involves the replacement of one equation by a simpler equivalent equation.

For example, a procedure is developed so that $2x - 6 = 4$ is replaced by the simpler equivalent equation $x = 5$. Not only does this technique find the solution, it also guarantees that all the solutions have been found, since it is known that 5 is the only replacement for x that makes $x = 5$ a true statement.

The following equivalence theorems are fundamental to the process of solving equations and are stated without proof.

The Equivalence Theorems

1. ***Addition theorem***
   If the same number is added to each side of an equation, the resulting sums are again equal, and the two equations are equivalent.
2. ***Subtraction theorem***
   If the same number is subtracted from each side of an equation, the resulting differences are equal, and the two equations are equivalent.

3. ***Multiplication theorem***
   If each side of an equation is multiplied by the same nonzero real number, the resulting products are equal, and the two equations are equivalent.
4. ***Division theorem***
   If each side of an equation is divided by the same nonzero real number, the resulting quotients are equal, and the two equations are equivalent.

The following examples illustrate the use of the equivalence theorems.

Example 2 Solve $x - 4 = 3$.

SOLUTION

$$\begin{aligned} x - 4 &= 3 \text{ if and only if} \\ x - 4 + 4 &= 3 + 4 \quad \text{(Addition theorem—add 4 to both sides)} \\ x + 0 &= 7 \\ x &= 7 \end{aligned}$$

Thus 7 is the solution, and $\{7\}$ is the solution set.

CHECK For $x = 7$, $x - 4 = 3$ becomes $7 - 4 = 3$, a true statement.

Example 3 Solve $x + 2 = 4$.

SOLUTION

$$\begin{aligned} x + 2 &= 4 \text{ if and only if} \\ x + 2 - 2 &= 4 - 2 \quad \text{(Subtraction theorem—subtract 2 from each side)} \\ x + 0 &= 2 \\ x &= 2 \end{aligned}$$

Thus 2 is the solution. The solution set is $\{2\}$.

CHECK For $x = 2$, $x + 2 = 4$ becomes $2 + 2 = 4$, true.

Example 4 Solve $\frac{x}{2} = 3$.

SOLUTION

$$\begin{aligned} \frac{x}{2} &= 3 \text{ if and only if} \\ 2\left(\frac{x}{2}\right) &= 2(3) \quad \text{(Multiplication theorem—multiply both sides by 2)} \\ x &= 6 \end{aligned}$$

Thus 6 is the solution.

CHECK For $x = 6$, $\frac{x}{2} = 3$ becomes $\frac{6}{2} = 3$, true.

Example 5 Solve $4x = 20$.

SOLUTION

$$4x = 20 \text{ if and only if}$$
$$\frac{4x}{4} = \frac{20}{4} \quad \text{(Division theorem—divide both sides by 4)}$$
$$x = 5$$

Thus 5 is the solution.

CHECK For $x = 5$, $4x = 20$ becomes $4(5) = 20$, true.

Equations such as $2x - 6 = 4$ and $5x + 10 = 0$ are called **linear equations.**

Definition

A **linear equation in one variable, *x*,** is an equation that can be expressed in the form $ax + b = 0$, where $a \neq 0$, and a and b and x are real numbers.

By using the equivalence theorems, any linear equation can be solved by transforming the given equation into a simpler equation.

Example 6 Solve $2x - 6 = 4$.

SOLUTION

$$2x - 6 = 4$$
$$2x - 6 + 6 = 4 + 6 \quad \text{(Addition theorem—add 6 to both sides)}$$
$$2x = 10$$
$$\frac{2x}{2} = \frac{10}{2} \quad \text{(Division theorem—divide each side by 2)}$$
$$x = 5$$

Thus 5 is the solution.

CHECK For $x = 5$, $2x - 6 = 4$ becomes $2(5) - 6 = 4$

$$10 - 6 = 4$$
$$4 = 4, \text{ true.}$$

Since $2x - 6 = 4$ is true for $x = 5$, 5 is the solution of $2x - 6 = 4$.

Example 7 Determine if the indicated value of the variable is a solution of the given equation: $3(x - 2) = 5x + 2$; $x = 4$.

SOLUTION For $x = 4$,

$$3(x - 2) = 3(4 - 2) = 3(2) = 6$$
$$5x + 2 = 5(4) + 2 = 20 + 2 = 22$$

Since $6 \neq 22$, 4 is *not* a solution.

Solving Linear Equations

The examples of the previous section have shown how to find the solution of a linear equation systematically. The technique involves the use of the equivalence theorems to transform the given equation into one having the form $x = a$, where a is the solution desired.

The **operations** that are to be used in transforming the equation are determined by analyzing the original equation. The symbols indicate the order in which the operations were done on the number named by the variable. The solution requires that the **inverse** operations be done in the **reverse** order, as shown in the table in Example 8.

Example 8 Solve $\frac{3(x-5)}{2} + 4 = 10$.

ANALYSIS

(This work is usually done mentally.)

Order of Operations on x	Inverse Operations	Reversed Order
1. 5 was subtracted from x.	1. Add 5.	1. Subtract 4.
2. Result was multiplied by 3.	2. Divide by 3.	2. Multiply by 2.
3. Result was divided by 2.	3. Multiply by 2.	3. Divide by 3.
4. 4 was added.	4. Subtract 4.	4. Add 5.

SOLUTION

(This work is written.)

$$\frac{3(x-5)}{2} + 4 = 10$$

$$\frac{3(x-5)}{2} = 10 - 4 = 6 \quad \text{(Subtract 4 from each side)}$$

$$3(x-5) = 2(6) = 12 \quad \text{(Multiply each side by 2)}$$

$$x - 5 = \frac{12}{3} = 4 \quad \text{(Divide each side by 3)}$$

$$x = 4 + 5 = 9 \quad \text{(Add 5 to each side)}$$

The solution is 9.

CHECK

If $x = 9$, then

$$\frac{3(x-5)}{2} + 4 = \frac{3(9-5)}{2} + 4$$

$$= \frac{3(4)}{2} + 4$$

$$= \frac{12}{2} + 4$$

$$= 6 + 4 = 10$$

Example 9 Solve $3 - y = 6$.

SOLUTION

$$
\begin{aligned}
3 - y &= 6 \\
-3 + 3 - y &= -3 + 6 \quad \text{(Add } -3 \text{ to each side)} \\
-y &= 3 \\
(-1)(-y) &= (-1)(3) \quad \text{(Multiply each side by } -1) \\
y &= -3
\end{aligned}
$$

The solution is -3.

CHECK If $y = -3$, then $3 - y = 3 - (-3) = 3 + 3 = 6$. Thus $3 - y = 6$ is true for $y = -3$.

Note in Example 9 that both sides were multiplied by -1 so that the final equivalent equation would have the form $y = a$ and *not* $-y = a$.

Example 10 Solve $4(k + 3) = 10$.

SOLUTION

$$
\begin{aligned}
4(k + 3) &= 10 \\
4k + 12 &= 10 \quad \text{(Distributive axiom)} \\
4k + 12 - 12 &= 10 - 12 \quad \text{(Subtraction theorem—subtract 12 from each side)} \\
4k &= -2 \\
\frac{4k}{4} &= \frac{-2}{4} \quad \text{(Division theorem—divide each side by 4)} \\
k &= -\frac{1}{2}
\end{aligned}
$$

The solution is $-\frac{1}{2}$.

CHECK

$$
\begin{aligned}
4(k + 3) &= 10 \\
4\left(-\frac{1}{2} + 3\right) &= 10 \\
4\left(\frac{5}{2}\right) &= 10 \\
10 &= 10, \text{ true}
\end{aligned}
$$

Example 11 Solve $8 = 8 - 5p$.

SOLUTION

$$
\begin{aligned}
8 &= 8 - 5p \\
8 + 5p &= 8 - 5p + 5p \quad \text{(Addition theorem—add } 5p \text{ to each side)} \\
8 + 5p &= 8 \\
5p &= 0 \quad \text{(Subtraction theorem—subtract 8 from each side)} \\
\frac{5p}{5} &= \frac{0}{5} \quad \text{(Division theorem—divide each side by 5)} \\
p &= 0
\end{aligned}
$$

continued

The solution is 0. The solution set is $\{0\}$. Note that this is *not* the empty set.

CHECK

$$8 = 8 - 5p$$
$$8 = 8 - 5(0)$$
$$8 = 8 - 0$$
$$8 = 8, \text{ true}$$

Example 12 Solve $x + 2 = x + 5$.

SOLUTION

$$x + 2 = x + 5$$
$$x + 2 - 2 = x + 5 - 2$$
$$x = x + 3$$
$$x - x = x - x + 3$$
$$0 = 3$$

But 0 can never equal 3. Therefore, there is no value of x that makes the equation true, and the solution set is the empty set, $\varnothing$.

The Hindu Method of Inversion

Inversion was a favorite method among the Hindus for solving certain types of problems, such as those found in the works of the elder Aryabhata (ca. A.D. 500). The following problem, although much simpler than those found in the Hindu works, illustrates this method.

> Beaming maiden with beaming eyes, tell me, as you understand the right method of inversion, what is the number which when multiplied by 3, then decreased by 29, then divided by 4, and then increased by 7, gives the number 26?

Operation	Inverse Operation	Reversed Order	Calculations
			26
Multiply by 3	Divide by 3	Subtract 7	-7
			19
Subtract 29	Add 29	Multiply by 4	$4(19) = 76$
Divide by 4	Multiply by 4	Add 29	$76 + 29 = 105$
Add 7	Subtract 7	Divide by 3	$\frac{105}{3} = 35$

The solution is 35.

Note that our modern solution of $\frac{3x - 29}{4} + 7 = 26$ parallels this method.

Example 13 Solve $5(x - 4) = 5x - 20$.

SOLUTION

$$5(x - 4) = 5x - 20$$
$$5x - 20 = 5x - 20$$

Since the left side and the right side of the equation are identical, the statement is true regardless of the value of x; in other words, all real numbers satisfy the equation and the solution set is R, the set of real numbers.

Exercises 2.2

Solve and check the equations in Exercises 1–60.

1. $x + 3 = 5$
2. $x - 3 = 5$
3. $3 - x = 5$
4. $5 - x = 3$
5. $y + 3 = -4$
6. $z - 1 = -6$
7. $x + 5 = 12$
8. $x - 5 = 12$
9. $5 - x = 12$
10. $12 - x = 5$
11. $2 - k = 2$
12. $4 - m = 4$
13. $4 - p = 30$
14. $30 - p = 4$
15. $2m = 6$
16. $3p = 15$
17. $5x = -10$
18. $7x = 105$
19. $2p + 3 = 7$
20. $2y - 1 = 7$
21. $2 - 3x = 8$
22. $1 = t + 9$
23. $3m + 2 = 11$
24. $2x - 3 = 7$
25. $2 = 2s + 6$
26. $3 - 2n = 5$
27. $q + 2 = 0$
28. $2 - x = 0$
29. $t + 7 = 0$
30. $x - 5 = x$
31. $4 - x = 0$
32. $-2 = s + 3$
33. $4 - 3y = 0 - 3y$
34. $6 - 2p = 2p - 6$
35. $5(t + 2) = 15$
36. $2(n - 1) = 8$
37. $3(x + 4) = -12$
38. $2(x + 3) = 16$
39. $3(k - 4) = 12$
40. $-2(y + 1) + 2 = 6$
41. $2 + 3(x + 1) = 5$
42. $3 - 2(x + 4) = 11$
43. $\frac{x}{4} = 5$
44. $\frac{z}{3} = 5$

45. $\frac{x}{5} = 1$

46. $\frac{x + 4}{2} = -1$

47. $2(4 - x) = 8 - 2x$

48. $\frac{x}{3} - 2 = 7$

49. $\frac{x + 3}{2} - 6 = 1$

50. $5 - x = -(x - 5)$

51. $\frac{x + 2}{3} = -2$

52. $\frac{x - 1}{2} = 6$

53. $2x - 5 = -(5 - 2x)$

54. $x + 4 = x - 4$

55. $\frac{2(x + 5)}{3} = 4$

56. $2 - 2x = 2x - 2$

57. $-3(x - 1) = 12$

58. $\frac{-2(x + 1)}{5} + 1 = 3$

59. $2x + 2 = 2x - 2$

60. $2(x + 2) = 2x + 4$

2.3 More Linear Equations

The following examples illustrate the solution of more complicated linear equations in one variable.

Note in Examples 1 and 2 that the additions and subtractions are done vertically. In the previous section, they were done horizontally. Either method may be used. Usually, fewer errors are made when additions and subtractions are done vertically.

Example 1 Solve $2x + 9 = 5x - 12$.

SOLUTION

$$
\begin{array}{rcrl}
2x + 9 & = & 5x - 12 & \\
-9 & & -9 & \text{(Subtract 9 from both sides)} \\
\hline
2x & = & 5x - 21 & \\
-5x & & -5x & \text{(Subtract } 5x \text{ from both sides)} \\
\hline
-3x & = & -21 & \\
\frac{-3x}{-3} & = & \frac{-21}{-3} & \text{(Divide both sides by } -3\text{)} \\
x & = & 7 &
\end{array}
$$

CHECK Left side. $2x + 9 = 2 \cdot 7 + 9 = 14 + 9 = 23$
Right side. $5x - 12 = 5 \cdot 7 - 12 = 35 - 12 = 23$
Thus $23 = 23$, and $\{7\}$ is the solution set.

Example 2 Solve $4y - (y + 14) = 5(2 - y)$.

SOLUTION

$$\begin{aligned} 4y - (y + 14) &= 5(2 - y) \\ 4y - y - 14 &= 10 - 5y && \text{(Use the distributive axiom to remove parentheses and collect like terms)} \\ 3y - 14 &= 10 - 5y \\ +14 \quad &\ \ +14 && \text{(Add 14 to both sides)} \\ 3y &= 24 - 5y \\ +5y \quad &\qquad\ \ +5y && \text{(Add } 5y \text{ to both sides)} \\ 8y &= 24 \\ \frac{8y}{8} &= \frac{24}{8} && \text{(Divide both sides by 8)} \\ y &= 3 \end{aligned}$$

CHECK

$$\begin{aligned} 4y - (y + 14) &= 5(2 - y) \\ 4(3) - (3 + 14) &= 5(2 - 3) \\ 12 - 17 &= 5(-1) \\ -5 &= -5 \end{aligned}$$

Since the resulting statement is true, $\{3\}$ is the solution set.

Example 3 Solve $3x + 25 + 4x = 1 + x$.

SOLUTION

It is usually desirable to simplify each side as much as possible before attempting the solution.

The left side of the equation is $3x + 25 + 4x$. Combine like terms to obtain $7x + 25$. Thus

$$\begin{aligned} 3x + 25 + 4x &= 1 + x \\ 7x + 25 &= 1 + x && \text{(Combine like terms of left side)} \\ 7x + 25 - 25 &= 1 + x - 25 && \text{(Subtract 25 from each side)} \\ 7x &= -24 + x && \text{(Simplify right side)} \\ 7x - x &= -24 + x - x && \text{(Subtract } x \text{ from each side)} \\ 6x &= -24 \\ \frac{6x}{6} &= \frac{-24}{6} && \text{(Divide each side by 6)} \\ x &= -4 \end{aligned}$$

CHECK

$$\begin{aligned} 3x + 25 + 4x &= 1 + x \\ 3(-4) + 25 + 4(-4) &= 1 + (-4) \\ -12 + 25 - 16 &= 1 - 4 \\ -28 + 25 &= -3 \\ -3 &= -3, \text{ true} \end{aligned}$$

continued

The solution set is $\{-4\}$.

Note that the check is always made in the original equation.

Example 4 Solve $y + 7 - 4(y - 8) = 2(y + 2)$.

SOLUTION

$$y + 7 - 4(y - 8) = 2(y + 2)$$

$$y + 7 - 4y + 32 = 2y + 4 \quad \text{(Distributive axiom to remove parentheses)}$$

$$-3y + 39 = 2y + 4 \quad \text{(Combine like terms on left side)}$$

$$-3y + 39 - 39 = 2y + 4 - 39 \quad \text{(Subtract 39 from each side)}$$

$$-3y = 2y - 35 \quad \text{(Simplify)}$$

$$-3y - 2y = 2y - 2y - 35 \quad \text{(Subtract 2y from each side)}$$

$$-5y = -35 \quad \text{(Simplify)}$$

$$\frac{-5y}{-5} = \frac{-35}{-5} \quad \text{(Divide each side by } -5\text{)}$$

Thus, $y = 7$, and $\{7\}$ is the solution set.
The check is left for you to do.

General Procedure for Solving Linear Equations in One Variable

1. Use the distributive axiom to remove parentheses and to collect like terms on each side of the equation.
2. Use the addition and subtraction theorems to add or subtract the same term from each side of the equation.
   Goal: To get the form $ax = b$, if possible.
3. a. For the form $ax = b$ with $a \neq 0$, use the division theorem and divide each side by a.
   b. For the form $0 = 0$, the solution set is R, the set of real numbers.
   c. For the form $0 = k$ or $k = 0$, where $k \neq 0$, the solution set is the empty set, $\varnothing$.
4. Check the solution set.

Exercises 2.3

Solve and check the equations in Exercises 1–52.

1. $5x + 1 = 4x - 4$
2. $x - 1 = 2x + 3$
3. $3x + 4 = 10 - 3x$
4. $2x - 2 = 3x + 5$
5. $7a + 7 = 4a - 2$
6. $3t + 1 - 2t = 0$
7. $4y - 3(y - 5) = 25$
8. $10p - 275 = 25(p + 1)$
9. $3y + 3(y - 1) = 21$
10. $2(m + 3) = 3(2 - m)$
11. $x - 20 = 5 + x$
12. $x + 2(x + 2) = 3(x + 3)$
13. $x - 2(x + 1) + 1 = 0$
14. $5(m - 4) - 6(m + 1) = 4$
15. $10 + 2(3x + 4) = 3x + 21$
16. $4n + 9 - 5n = 0$
17. $2(y + 7) - 5 = -1 + 4(y - 2)$
18. $2(x - 1) + 4x - 3 = 6x - 5$
19. $3p - 6 = 3(p - 2)$
20. $2(x - 3) - 2(3 - x) = 0$
21. $2x - 5 - (4 - 3x) = 11$
22. $3(x + 20) - 2(x + 30) = 2x + 5$
23. $8x + (5 - x) + 30 = 0$
24. $4(t - 20) - 5(t - 22) = 2t$
25. $13y - 6 = 19y + 42$
26. $10t - 2(3t - 1) = 23 - 3t$
27. $4x - 1 = 2x - 5$
28. $2x - 4 = 2(x - 2)$
29. $2a = 7 - 3(a - 1)$
30. $3x + 4 = 5x - 2$
31. $x + 2(3x - 5) = 4$
32. $12 - 5(t - 2) = 2 - t$
33. $2x - 16 = 3(13 - 3x)$
34. $2n - (6 - n) = 4n - 13$
35. $5 + 3(2y + 1) = 6y + 8$
36. $5(3 - y) = 121 - 6(25 - y)$
37. $5(4p - 3) + 3(2p + 2) = 7p + 29$
38. $5(12 - x) = 7 - 2(6 - 4x)$
39. $30 - 3(2x - 3) = 5(3 - x)$
40. $12 = 8z - 3(7 - z)$
41. $4(2y - 5) - 8(y + 5) = 7$
42. $7 - 3x - (5 + 2x) = 12$
43. $p - 7 = 7 - p$
44. $p - 7 = p + 7$
45. $x - [1 - (x - 1)] = 3x$
46. $x - 2[x - 2(x - 2)] = 4$
47. $4t + 2(t - 6) = 0$
48. $4 = 2x + 10 + 4x$
49. $4x + (3 - 2x) = (4x + 3) - 2x$
50. $6(3y + 4) - 3(10y + 16) = 8(2y - 17)$
51. $6(y - 3) - 5(2y - 6) = 0$
52. $-2(t + 1) = 4(1 - 2t)$

2.4 Linear Inequalities

Statements such as $x < 4$, $x \geq 7$, $2x - 1 \leq 9$, $3x > 2x + 5$, and $10 - (x - 3) < 24$ are called **inequalities.** There are many practical applications that involve inequalities.

Inequalities are said to have the **same sense** if they have the same order symbol. For example, $x < 4$ and $5 < 7$ are inequalities of the same sense.

On the other hand, $x < 7$ and $8 > 3$ are inequalities of the **opposite sense,** since their order symbols are different.

A **solution of an inequality** is a number that makes the inequality true when its variable is replaced by this number.

The **solution set** of an inequality is the set of all solutions of the inequality.

Two inequalities are equivalent if and only if they have the same solution set.

Solving an inequality is similar to solving an equation, but there are two important exceptions. One has to do with exchanging sides, and the other has to do with multiplying (or dividing) by a negative number.

Theorem on Exchanging Sides

If one inequality is obtained from another one by exchanging sides and by changing the order symbol, then these two inequalities are equivalent.

For example,

$3 < 8$ and $8 > 3$ are equivalent.

$x \geq 9$ and $9 \leq x$ are equivalent.

$12 - x > 3x$ and $3x < 12 - x$ are equivalent.

Addition Theorem

If the same number is added to each side of an inequality, then the resulting inequality having the same sense is equivalent to the original one.

For example,

$3 < 8$ and $3 + 4 < 8 + 4$ are equivalent.

$x - 4 > 5$ and $x - 4 + 4 > 5 + 4$ (that is, $x > 9$) are equivalent.

Multiplication by a Positive Number Theorem

If each side of an inequality is multiplied by the same positive number, then the resulting inequality having the same sense is equivalent to the original one.

For example,
$5 > 2$ and $5(4) > 2(4)$ are equivalent.
$\frac{x}{4} \leq 6$ and $x \leq 24$ are equivalent.

Multiplication by a Negative Number Theorem
If each side of an inequality is multiplied by the same negative number and if the order symbol is changed, then the resulting inequality, having the opposite sense, is equivalent to the original one.

For example,
$-2 < 8$ and $-2(-3) > 8(-3)$ (that is, $6 > -24$) are equivalent.
$-2x \geq 8$ and $x \leq -4$ are equivalent.

Example 1 Solve the inequality $2x + 3 < x + 5$ and graph the solution set. Check the solution.

SOLUTION

$$2x + 3 < x + 5$$
$$2x < x + 2 \quad \text{(Addition theorem—adding } -3\text{)}$$
$$x < 2 \quad \text{(Addition theorem—adding } -x\text{)}$$

See Fig. 2.1.

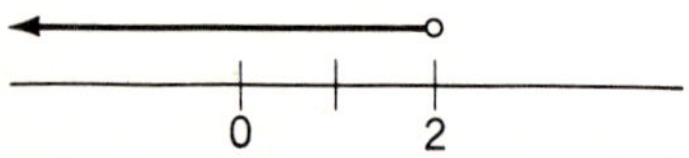

Figure 2.1

CHECK

1. Select a value in the interval obtained and substitute in the original inequality. For $x < 2$, select $x = 1$, since $1 < 2$.

   Then $2x + 3 < x + 5$ becomes
   $$2(1) + 3 < 1 + 5$$
   $$5 < 6, \text{ true}$$

2. Check the endpoint value in the equation corresponding to the original inequality. For $x < 2$, the endpoint value is 2.

   For $2x + 3 = x + 5$
   $$2(2) + 3 = 2 + 5$$
   $$7 = 7, \text{ true}$$

Note that 2 is **not** in the solution set, since $7 < 7$ is false.

Example 2 Solve for x and graph the solution set. Check.

SOLUTION

$$2 - x \leq 14 + 3x$$
$$2 - 4x \leq 14 \qquad \text{(Adding } -3x \text{ to each side)}$$
$$-4x \leq 12 \qquad \text{(Adding } -2 \text{ to each side)}$$
$$x \geq -3 \qquad \text{(Multiplying each side by } -\tfrac{1}{4} \text{ and changing the order symbol)}$$

See Fig. 2.2.

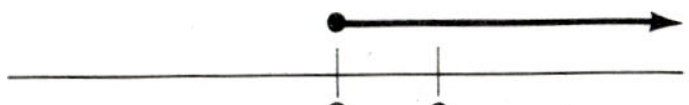

Figure 2.2

CHECK

1. Select $x = 1$.

   Then $2 - x \leq 14 + 3x$ becomes $2 - 1 \leq 14 + 3(1)$
   $1 \leq 17$, true

2. $x = -3$

   $2 - x = 14 + 3x$ becomes $2 - (-3) = 14 + 3(-3)$
   $2 + 3 = 14 - 9$
   $5 = 5$, true

Note that -3 **is** in the solution set, since $5 \leq 5$ is true.

Example 3 Solve $2x - 3 > 2x$.

SOLUTION

$$2x - 3 > 2x$$
$$-3 > 0 \qquad \text{(Adding } -2x \text{ to each side)}$$

This is a false statement.
The solution set is $\varnothing$, the empty set.

Example 4 Solve $x - 5 \leq 5 + x$.

SOLUTION

$$x - 5 \leq 5 + x$$
$$-5 \leq 5 \qquad \text{(Adding } -x \text{ to each side)}$$

This is a true statement, since $-5 \leq 5$.
The solution set is R, the set of real numbers.

Exercises 2.4

Solve Exercises 1–40 and graph the solution sets. Check.

1. $2x - 4 > 0$
2. $2x + 4 > 0$
3. $3x - 2 > 1 + 2x$
4. $3x + 5 < x + 7$
5. $2x + 1 < x - 5$
6. $6x + 10 < 2 + 4x$
7. $4 - 2x \leq 0$
8. $8 - 2x \geq 0$
9. $15 - 3x \geq 0$
10. $15 - 3x \leq 0$
11. $1 < 2x + 3$
12. $1 \leq 2x - 3$
13. $2x - 3(x + 1) > 0$
14. $2x - 3(x + 1) \leq 0$
15. $5 - x \leq 3 - (x - 2)$
16. $4x + 2 \leq 4x$
17. $x + 4 - 3x \leq 2x + 4$
18. $3x - (x + 4) \leq 4 - 2x$
19. $-2(x + 3) \geq 4(2x + 1)$
20. $2(x + 3) < -4(2x + 1)$
21. $33 - 9x \leq 2x$
22. $x - 12 \geq 0$
23. $12 - x \geq 0$
24. $\frac{x + 3}{2} < 0$
25. $\frac{x + 3}{5} \geq 0$
26. $2x - \frac{1}{3} \leq 0$
27. $\frac{x - 6x}{2} < -20$
28. $\frac{x - 6}{2} < -20$
29. $\frac{6 - x}{2} < -20$
30. $2(x + 3) \geq 8(2 - x)$
31. $\frac{x}{3} < 0$
32. $\frac{x}{3} > 0$
33. $x > 4 + x$
34. $x < 4 + x$
35. $x - 2 < 2 - x$
36. $x - 2 < 2 + x$
37. $-3(2 - 3x) < 15 - (x + 1)$
38. $15 - (x + 2) \leq 0$
39. $-2(x + 4) \geq -2(x - 4)$
40. $3 \leq \frac{x + 2}{5}$

2.5 Literal Equations

A formula is a general equation for the solution of a specific problem. For example, a formula for finding the area of a rectangle is $A = LW$, where A represents area and L and W the length and width of the rectangle. The circumference of a circle can be expressed as $C = 2\pi r$, where C represents the circumference and r the radius of the circle. You are familiar with many formulas, some of which were presented in Chapter 1.

Often a formula is not in the desired form or is not convenient for the solution of a problem. Since a formula is an equation, it can be treated as such and changed to a desired equivalent equation by the methods shown in the preceding sections.

Example 1 The formula for the perimeter of a rectangle is

$$P = 2L + 2W$$

where P stands for the perimeter, L stands for the length, and W stands for the width. Solve this formula for W in terms of the other letters.

SOLUTION The technique for solution is to treat W as a variable and all other numerals and letters as constants.

$$\begin{aligned} P &= 2L + 2W \\ 2W + 2L &= P \\ 2W + 2L - 2L &= P - 2L && \text{(Subtract } 2L \text{ from each side)} \\ 2W &= P - 2L && \text{(Simplify)} \\ \frac{2W}{2} &= \frac{P - 2L}{2} && \text{(Divide each side by 2)} \\ W &= \frac{P - 2L}{2} \end{aligned}$$

Now the equation is solved for W.

An equation involving more than one letter, some of which are constant, is called a **literal equation.**

Example 2 Solve the formula for the circumference of a circle, $C = 2\pi r$, for r.

SOLUTION

$$\begin{aligned} C &= 2\pi r \\ 2\pi r &= C \\ \frac{2\pi r}{2\pi} &= \frac{C}{2\pi} && \text{(Divide each side by } 2\pi\text{)} \\ r &= \frac{C}{2\pi} \end{aligned}$$

Thus the radius of any circle is equal to the circumference of the circle divided by 2π.

Example 3 $C = \frac{5}{9}(F - 32)$ is the formula for converting Fahrenheit temperature to degrees Celsius. Solve the formula for F—that is, rewrite it so that it can be readily used to convert Celsius temperature to Fahrenheit.

SOLUTION

$$\begin{aligned}
C &= \frac{5}{9}(F - 32) \\
9C &= 9\left(\frac{5}{9}\right)(F - 32) && \text{(Multiply each side by 9)} \\
9C &= 5(F - 32) && \text{(Simplify)} \\
9C &= 5F - 160 && \text{(Distributive axiom)} \\
9C + 160 &= 5F - 160 + 160 && \text{(Add 160 to each side)} \\
5F &= 9C + 160 \\
F &= \frac{9C + 160}{5} && \text{(Divide each side by 5)} \\
F &= \frac{9}{5}C + 32 && \text{(Further simplification of right side)}
\end{aligned}$$

Thus $F = \frac{9}{5}C + 32$.

Example 4 If $mx - y + b = 0$, solve for each of the following:
a. x b. y c. m d. b

SOLUTION In each instance, consider all letters constant except the letter for which you are solving:

a. Solve for x:

$$\begin{aligned}
mx - y + b &= 0 \\
mx - y + y + b &= 0 + y && \text{(Add } y \text{ to each side)} \\
mx + b - b &= y - b && \text{(Subtract } b \text{ from each side)} \\
mx &= y - b && \text{(Simplify)} \\
x &= \frac{y - b}{m} && \text{(Divide each side by } m\ [m \neq 0])
\end{aligned}$$

b. Solve for y:

$$\begin{aligned}
mx - y + b &= 0 \\
mx - y + y + b &= 0 + y && \text{(Add } y \text{ to each side)} \\
mx + b &= y && \text{(Simplify)} \\
y &= mx + b
\end{aligned}$$

c. Solve for m:

$$\begin{aligned}
mx - y + b &= 0 \\
mx - y + b - b &= 0 - b && \text{(Subtract } b \text{ from each side)} \\
mx - y + y &= -b + y && \text{(Add } y \text{ to each side)}
\end{aligned}$$

continued

$$mx = y - b \qquad \text{(Commutative axiom, addition)}$$

$$m = \frac{y - b}{x} \qquad \text{(Divide each side by } x \; [x \neq 0])$$

d. Solve for b:

$$mx - y + b = 0$$

$$mx + b = y \qquad \text{(Add } y \text{ to each side)}$$

$$b = y - mx \qquad \text{(Subtract } mx \text{ from each side)}$$

Example 5 Solve $2x + y = 10$ for y.

SOLUTION

$$2x + y = 10$$

$$-2x + 2x + y = 10 - 2x$$

$$y = 10 - 2x$$

or

$$y = -2x + 10$$

Example 6 Solve $4x - y = 12$ for y.

SOLUTION

$$4x - y = 12$$

$$-4x + 4x - y = -4x + 12$$

$$-y = -4x + 12$$

$$(-1)(-y) = (-1)(-4x + 12)$$

$$y = 4x - 12$$

Exercises 2.5

Solve the formulas in Exercises 1–40 for the indicated variables.

1. $A = LW$ (Area of rectangle); L
2. $V = LWH$ (Volume of a parallelepiped—box); W
3. $V = \pi r^2 h$ (Volume of a circular cylinder); h
4. $P = 2L + 2W$ (Perimeter of a rectangle); L
5. $D = rt$ (Uniform motion formula); t
6. $y = mx + b$ (Slope-intercept equation of a line); x
7. $A = \frac{bh}{2}$ (Area of a triangle); h
8. $A = \frac{1}{2}h(a + b)$ (Area of a trapezoid); a
9. $A = P + Prt$ (Simple interest); r
10. $W = Fd$ (Work); d
11. $C = 2\pi r$ (Circumference of a circle); r
12. $V = \frac{1}{3}bh$ (Volume of a cone); b

13. $A = a(a + 2s)$ (Surface area of a square pyramid); s

14. $P = I^2r$ (Electric power); r

15. $F = \frac{9}{5}C + 32$ (Temperature conversion); C

16. $A = \frac{1}{2}h(a + b)$ (Area of a trapezoid); h

17. $A + B + C = 180$ (Sum of angles of a triangle); B

18. $P = a + b + c$ (Perimeter of a triangle); b

19. $D = A(n - 1)$ (Physics: law of small prisms); n

20. $D = A(n - 1)$ (Physics: law of small prisms); A

21. $Q = \frac{100M}{C}$ (Psychology); C

22. $E = \frac{I - O}{I}$ (Engineering); O

23. $R = \frac{C - S}{n}$ (Economics: depreciation); n

24. $R = \frac{C - S}{n}$ (Economics: depreciation); S

25. $\frac{S}{s} = \frac{d}{D}$ (Pulleys and gears); d

26. $\frac{S}{s} = \frac{d}{D}$ (Pulleys and gears); D

27. $\frac{PV}{T} = \frac{pv}{t}$ (Boyle's gas law); t

28. Solve Exercise 27 for P.

29. Solve Exercise 27 for p.

30. Solve Exercise 27 for V.

31. $C = \frac{100W}{L}$ (Anthropology: cephalic index); W

32. Solve Exercise 31 for L.

33. $r = c(x + a)$ (Biology and chemistry: growth rate); a

34. Solve Exercise 33 for x.

35. $C = \frac{yA}{y + 12}$ (Medical dosage); A

36. $C = \frac{WA}{150}$ (Medical dosage); W

37. $p = \frac{s}{s + f}$ (Probability; life insurance); f

38. $d = \frac{r}{1 + nr}$ (Discount rate); n

39. $r = \frac{R(E - G)}{E}$ (Radio, television); G

40. Solve Exercise 39 for R.

Solve the equations in Exercises 41–60 for the specified variables.

41. $x + y = 8$; y

42. $y - 2x = 5$; y

43. $3x - y = 6$; y

44. $5x - y = 20$; y

45. $x - 2y = 9$; x

46. $x + 7y = 4$; x

47. $2y - x + 4 = 0$; x

48. $8 - x - 4y = 0$; x

49. $2x - 7y = 8$; x
50. $5x + 4y = 20$; y
51. $y - 5x = 10$; y
52. $4x + y = 12$; y
53. $x + 4y = 12$; x
54. $x - 6y = 18$; x
55. $x + y + 1 = 0$; x
56. $x - 2y + 6 = 0$; x
57. $3x + 2y = 6$; y
58. $2x - 5y = 10$; x
59. $ax + by + c = 0$; y
60. $ax + by + c = 0$; x

2.6 Applications: Numerical Problems, Age Problems, Consecutive Integers

Problems stated in words and requiring a number for a solution occur in almost every walk of life. We shall examine some of these in this section and the following sections.

Many problems fall into categories for which there are standard methods that yield an equation or equations that provide a solution. If the verbal problem can be translated into a linear equation in one variable, then it can be solved by the methods of the preceding sections.

Problems about numbers alone may have few practical applications, but they are useful in developing methods and techniques for solving applied problems. One needs to know how to translate "twice a number less 5 equals" no matter what the number represents. For example, a problem from chemistry might state, "Twice the number of grams of potassium sulfate less 5 grams yields . . ."

Techniques that are useful in translating a verbal problem into an equation are summarized as follows.

VERBAL PROBLEM TECHNIQUES

1. Read the problem slowly at least two times.
2. Illustrate the problem with a simple picture or diagram, if possible.
3. Designate one of the unknown numbers by a variable, such as x.
4. Represent the other unknown numbers, if any, in terms of the variable selected.
5. Summarize the numerical information in a chart, if possible.
6. List any formula or formulas that might apply.
7. Find an equation relating the numbers of the problem.
8. Solve the equation.
9. Check the solution.

The expressions in Table 2.1 occur frequently in verbal problems.

TABLE 2.1 **Verbalization of Algebraic Terms**

$a + b$	*a* plus *b* The sum of *a* and *b* *a* added to *b* *a* more than *b* *a* greater than *b* *a* increased by *b*	$a - b$	*a* minus *b* The difference between *a* and *b* (when *b* is subtracted from *a*) *b* subtracted from *a* *b* less than *a* *a* decreased by *b*
ab	*a* times *b* The product of *a* and *b* *a* multiplied by *b*	$\frac{a}{b}$	*a* divided by *b* The quotient when *a* is divided by *b*
$2x$	Twice *x* The double of *x*	$\frac{x}{2}$	One half of *x*
$\frac{3x}{5}$	Three fifths of *x*	$\frac{x}{3}$	One third of *x*
$=$	Equals, equal, is, is equal to, are equal to, is the same as, results in, becomes, was, will be		

Example 1 Twice the difference obtained when 7 is subtracted from a certain number results in 20. Find the number.

SOLUTION

Let x = the unknown number. Then

$$2(x - 7) = 20$$
$$2x - 14 = 20$$
$$2x = 34$$
$$x = 17$$

CHECK

$$2(17 - 7) = 20$$
$$2(10) = 20$$
$$20 = 20$$

Example 2 A man invests twice as much money in stock A as in stock B and $740 more in stock C than in stocks A and B combined. If his total investment is $5000, how much did he invest in each stock?

SOLUTION

Let x = the number of dollars invested in stock B. Then

$2x$ = the number of dollars invested in stock A

and

$x + 2x + 740$ = the number of dollars invested in stock C

The total investment, \$5000, $= x + 2x + (x + 2x + 740)$.

continued

$$x + 2x + (x + 2x + 740) = 5000$$
$$6x + 740 = 5000$$
$$6x = 4260$$
$$x = 710$$

He invested \$710 in stock B.
He invested 2(710) = \$1420 in stock A.
He invested \$1420 + \$710 + \$740 = \$2870 in stock C.

CHECK 710 + 1420 + 2870 = 5000

Another type of application involves consecutive numbers—specifically, integers. For example, if x stands for an integer, then $x + 1$ is the next or consecutive integer, followed by $x + 2$, $x + 3$, ... So if $x = 3$, then $x + 1 = 4$, $x + 2 = 5$, and $x + 3 = 6$, and 3, 4, 5, 6 are consecutive integers. Consecutive *even* integers are numbers such as 8, 10, 12, 14, whereas consecutive *odd* integers are numbers such as 9, 11, 13, 15, 17. Table 2.2 is a guide, assuming that x is an integer.

Example 3 Find three consecutive integers whose sum is 48.

SOLUTION

Let x = first integer
$x + 1$ = second integer
$x + 2$ = third integer

$$x + (x + 1) + (x + 2) = 48$$
$$3x + 3 = 48$$
$$3x = 45$$
$$x = 15$$

Therefore, $x = 15$, $x + 1 = 16$, and $x + 2 = 17$; the three consecutive integers are 15, 16, 17.

CHECK 15 + 16 + 17 = 48

TABLE 2.2 **Consecutive Integers**

$x, x + 1$	Two consecutive integers
$x, x + 1, x + 2$	Three consecutive integers
$x, x + 1, x + 2, x + 3$	Four consecutive integers
$x, x + 2$	Two consecutive even integers
$x, x + 2$	Two consecutive odd integers
$x, x + 2, x + 4$	Three consecutive even integers
$x, x + 2, x + 4$	Three consecutive odd integers
$x, x + 2, x + 4, x + 6$	Four consecutive even (or odd) integers

Example 4 If 3 times the smallest of three consecutive *even* integers is 36 less than 5 times the largest, find the three integers.

SOLUTION

Let x = smallest integer

Then

$x + 2$ = next consecutive *even* integer

and

$x + 4$ = third consecutive *even* integer (the largest)

$$\begin{aligned} 3x &= 5(x+4) - 36 \\ 3x &= 5x + 20 - 36 \\ -2x &= -16 \\ x &= 8 \\ x + 2 &= 8 + 2 = 10 \\ x + 4 &= 8 + 4 = 12 \end{aligned}$$

Thus the three consecutive even integers are 8, 10, 12.

Problems involving ages of people or objects follow a pattern similar to that of consecutive integers.

Example 5 Four years ago a father was twice as old as his son. The difference in their ages is 18 years. How old is each today?

SOLUTION

	Now	4 Years Ago
Son	x	$x - 4$
Father	$x + 18$	$(x + 18) - 4$

Father's age 4 years ago = twice (son's age 4 years ago)

$$\begin{aligned} (x + 18) - 4 &= 2(x - 4) \\ x + 14 &= 2x - 8 \\ x &= 22 \\ x + 18 &= 40 \end{aligned}$$

The father is 40 years old, and the son is 22.

The following chart may be helpful.

Age Problem Expressions		Examples	
$x + k$	k years older than x	5 years older than	$x + 5$
$x - k$	k years younger than x	5 years younger than	$x - 5$
$x + a$	Age a years from now, present age x	Age 10 years from now	$x + 10$
$x - a$	Age a years ago, present age x	Age 8 years ago	$x - 8$

Exercises 2.6

Solve and check Exercises 1–32.

1. Twice the sum of a number and 5 is 13 more than the number. Find the number.

2. Three more than twice a certain number is 7 less than 3 times the number. Find the number.

3. When 4 times a number is decreased by 24, the difference is 18 less than 5 times the number. Find the number.

4. The sum of 19 plus 3 times a number is 4 times the sum of the number increased by 6. Find the number.

5. Five times the difference when 4 is subtracted from a number is 16 more than the number. Find the number.

6. The sum of two numbers is 28. Five times the smaller number is 4 less than 4 times the larger number. Find the numbers.

7. In a football game, the winners won by 4 points. Twice the number of points scored by the winning team was 14 less than 3 times the number of points scored by the losing team. What was the final score of the game?

8. A jar of 62 coins contains pennies, nickels, and dimes. The number of dimes is 3 times the number of nickels. The number of pennies is 6 more than the number of dimes. How many coins of each type are in the jar? How much money is in the jar?

9. A resort owner has 100 rooms for rent. Some of the rooms are double rooms, and the rest are single rooms. At the end of one day, the number of double rooms rented was 3 more than twice the number of single rooms rented. Seven single rooms were unrented. How many rooms of each type are there for rent at the resort?

10. A small agency in a seaside resort has bicycles and dune buggies for rent. The total number of bicycles and dune buggies is 165. For maximum profit, the number of bicycles is 9 less than twice the number of dune buggies. How many bicycles and how many dune buggies does the agency have?

11. A rod 80 meters long is cut into two pieces. One piece is 10 meters shorter than twice the length of the other piece. Find the length of each piece.

12. A son and a daughter inherit a total of \$20,000. The will states that the son is to receive twice the difference of the daughter's share decreased by \$2000. How much does each receive?

13. The sum of two consecutive integers is 75. Find these integers.

14. Two numbers are consecutive integers. Three times the smaller number is 10 more than twice the larger. Find the numbers.

15. The sum of two consecutive even integers is 90. Find the integers.

16. The sum of two consecutive odd integers is 48. Find the integers.

17. Two numbers are consecutive odd integers. Four times the larger number is 9 less than 5 times the smaller number. Find the numbers.

18. Two numbers are consecutive even integers. Twice the larger number is 4 less than 3 times the smaller number. Find the numbers.

19. The sum of three consecutive integers is 36. Find the integers.

20. Let x be the middle integer of three consecutive integers.
    a. Write algebraic expressions involving x for the smallest and the largest of the three integers.
    b. If the sum of the three consecutive integers is 159, find the integers.

21. Three numbers are consecutive integers. The sum of the smallest and the largest is 10 more than the middle integer. Find the three integers.

22. The sum of three consecutive integers is equal to the difference obtained by subtracting the largest from 5 times the smallest. Find the integers.

23. The sum of three consecutive even integers is 114. Find the integers.

24. The sum of three consecutive odd integers is 129. Find the integers.

25. Find three consecutive odd integers so that the sum of twice the largest and 3 times the smallest is 73.

26. Find three consecutive even integers so that the sum of 3 times the largest and twice the smallest is 132.

27. Twelve years from now, the roof on a house will be 3 times as old as it is now. How old is the roof now?

28. At a certain college, the library is now 3 times as old as the cafeteria. Seven years from now, the library will be twice as old as the cafeteria. Find the present age of each.

29. A woman is 3 years younger than her husband. Four years ago, when they were married, the sum of their ages was 45. Find the present age of each.

30. A boy is 2 years older than his sister. Seven years ago, the boy was twice as old as his sister. Find the present ages of the boy and his sister.

31. A mother is 5 times as old as her daughter. The father is 2 years older than the mother. The sum of the ages of the mother, father, and daughter is 68. Find the age of each.

32. A boy's sister is 3 years younger than he is. His father is 6 times as old as he is. The sum of the ages of the boy, his sister, and his father is 53. Find the age of each.

2.7 Applications: Geometric Problems

The solution of geometric problems uses formulas about geometric figures. In this type of problem, it is useful to draw the figure discussed and mark it with the numerical information. The equation is obtained by substituting the numerical information into the appropriate formula.

Some geometric figures and their formulas are shown in Fig. 2.3.

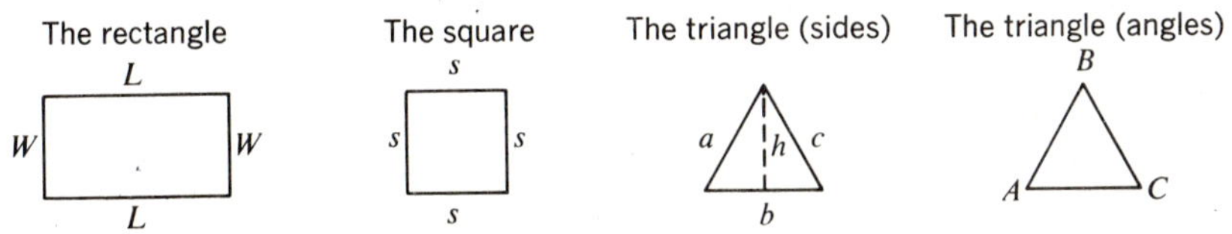

The rectangle: $P = 2L + 2W$, $A = LW$

The square: $P = 4s$, $A = s^2$

The triangle (sides): $P = a + b + c$, $A = \frac{bh}{2}$

The triangle (angles): $\angle A + \angle B + \angle C = 180°$

Figure 2.3

Example 1 A flower bed has the shape of a rectangle. The length of the rectangle is 4 times its width. The perimeter is unchanged if the width is doubled and the length is decreased by 9 feet. What are the original dimensions of the flower bed?

SOLUTION

Let W = the original width

Then

$4W$ = the original length

See Fig. 2.4.

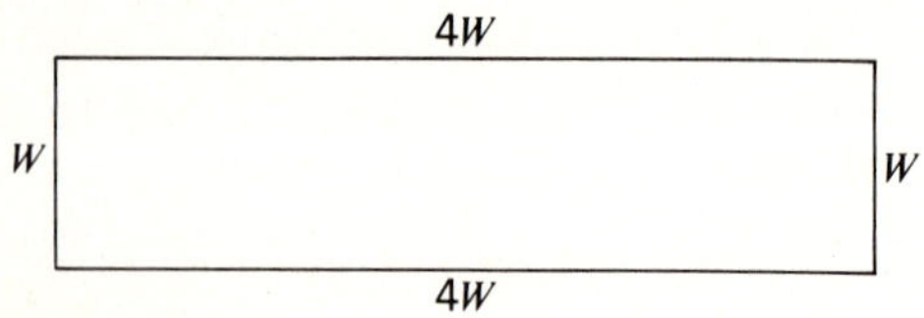

Figure 2.4

EQUATION $2 \cdot 2W + 2(4W - 9) = 2W + 2 \cdot 4W$; $W = 9$ feet, $L = 36$ feet.

Example 2 One angle of a triangle is 10 degrees more than the second angle, and the third angle is 10 degrees less than the second angle. How many degrees are in each angle of the triangle?

SOLUTION

Let x = the number of degrees in the second angle
Then $x + 10$ = measure of first angle
$x - 10$ = measure of third angle

See Fig. 2.5.

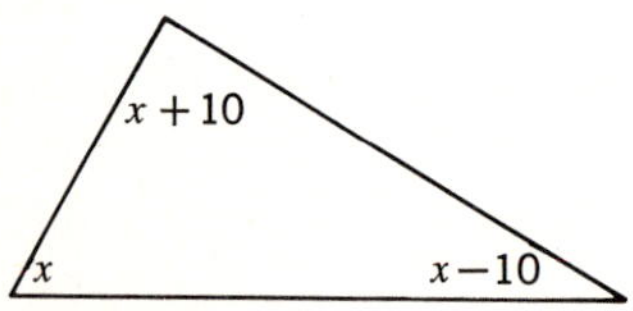

Figure 2.5

EQUATION $x + (x + 10) + (x - 10) = 180$;
$x = 60$ degrees, $x + 10 = 70$ degrees, and $x - 10 = 50$ degrees.

Exercises 2.7

Solve and check Exercises 1–24.

1. The length of a rectangle is 3 times the width. The perimeter of the rectangle is 80 meters. Find the width and length of the rectangle.

2. The length of a rectangle is 3 inches longer than the width. The perimeter of the rectangle is 30 inches. Find the width and length of the rectangle.

3. The length of a rectangle is 5 feet less than twice the width. The perimeter of the rectangle is 62 feet. Find the width and length of the rectangle.

4. The length of a rectangle is 3 times the width. If 20 centimeters are subtracted from the length and 8 centimeters are added to the width, the perimeter of the resulting rectangle is 296 centimeters. Find the width and length of the original rectangle.

5. The perimeter of a square is 32 meters. Find the area of the square.

6. Find the area of a square whose perimeter is 40 inches.

7. The length of a rectangle is twice its width. If 8 yards are subtracted from the length and 8 yards are added to the width, the resulting figure is a square.
   a. Find the width and length of the original rectangle.
   b. Find the perimeter of the original rectangle.
   c. Find the area of the original rectangle.
   d. Find the perimeter of the square.
   e. Find the area of the square.

8. The length of a rectangle is 4 times its width. If the width is doubled and the length is decreased by 50 meters, the resulting figure is a square.
   a. Find the width and length of the original rectangle.
   b. Find the perimeter of the original rectangle.
   c. Find the area of the original rectangle.
   d. Find the perimeter of the square.
   e. Find the area of the square.

9. One side of a triangle is 5 inches longer than the second side. The length of the third side is 5 inches less than twice the length of the second side. The perimeter of the triangle is 60 inches. Find the length of each side.

10. A piece of copper tubing 48 centimeters long is bent into the shape of a triangle. One side of the triangle is 5 centimeters longer than the second side. The length of the third side is 7 centimeters less than 3 times the length of the second side. Find the length of each side of the triangle.

11. One angle of a triangle is twice as large as the smallest angle of the triangle. The largest angle of the triangle is 3 times as large as the smallest angle. Find the number of degrees in each angle of the triangle.

12. Two angles of a triangle are equal. The third angle is twice as large as one of the two equal angles. Find the number of degrees in each angle.

13. The measurements of the angles of a triangle can be obtained by multiplying each of three consecutive integers by 10. Find the number of degrees in each angle, recalling that the sum of the angles is 180 degrees.

14. The measurements of the angles of a triangle can be obtained by multiplying each of three consecutive even integers by 10. Find the number of degrees in each angle.

15. The width of each of two rectangles is 40 inches. The length of the second rectangle is 20 inches more than the length of the first rectangle. The sum of the areas of the rectangles is 5600 square inches. Find the length of each rectangle.

16. Two rectangular carpets have widths of 2.5 meters and 3 meters, respectively. If the length of the first carpet is 1 meter longer than that of the second carpet and if the sum of their areas is 24.5 square meters, find the length of each carpet.

17. A 12-inch by 18-inch rectangular picture is to have a frame of uniform width. The perimeter of the framed picture determines the length of wood needed for the frame. Find the width of the frame if the perimeter of the framed picture is 84 inches. See Fig. 2.6.

18. The width and length of each page of a book are 18 centimeters and 24 centimeters, respectively. There is a 2-centimeter margin along each of the 18-centimeter sides. The margins along the other two sides of the page are equal in width. The printing area is to be 200 square centimeters. How wide should each of the other two margins be? See Fig. 2.7.

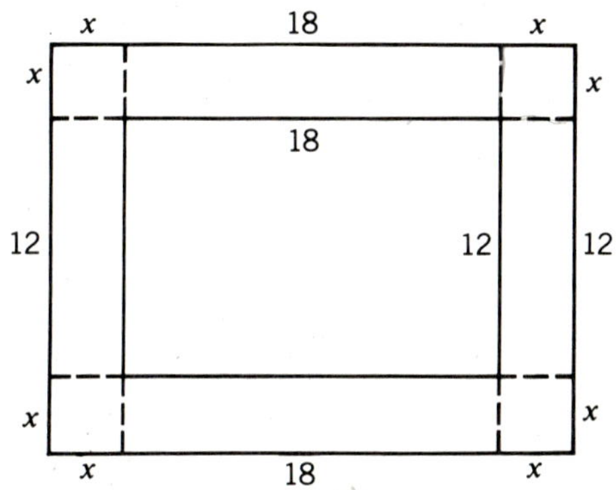

Figure 2.6

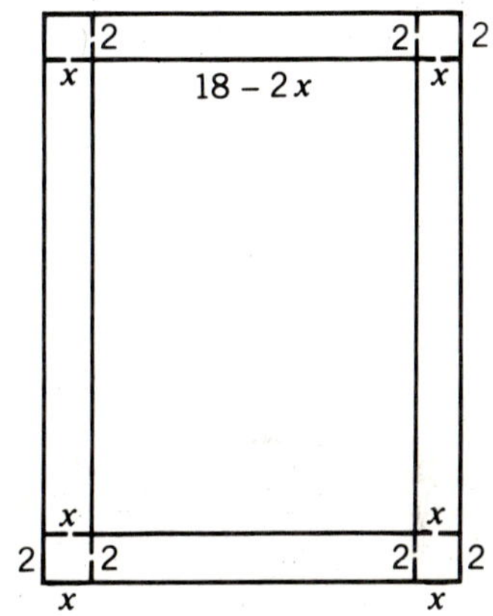

Figure 2.7

19. A farmer has 78 lineal feet of fencing to be used to enclose a rectangular area of land. He wants the length to be 6 feet less than twice the width. Find the width and length of the area.

20. It takes 18 meters of luminous tape to go around the edges of a rectangular sign. The length of the sign is 1 meter more than the width. Find the width and length of the sign.

21. The cross-section of an irrigation ditch has the shape of a rectangle whose depth is 1 foot less than its width. If the ditch is dug 2 feet deeper, the area of the cross-section can be increased by 6 square feet. Find the width and the depth of the ditch before digging. See Fig. 2.8.

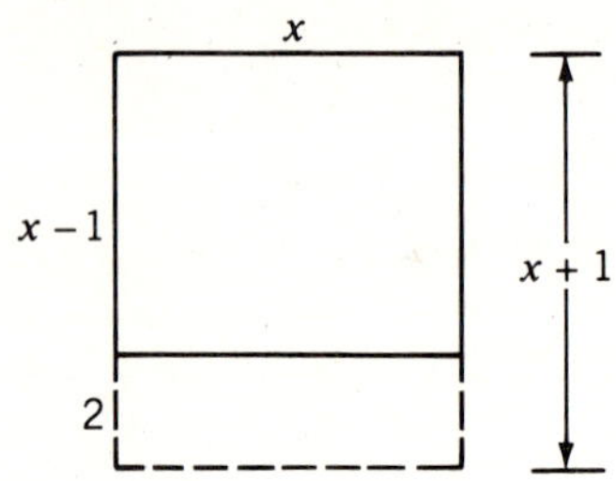

Figure 2.8

22. An owner building his home specified a rectangular window whose height was 2 feet shorter than its base. However, in order to meet the city lighting requirements, it was necessary to increase the base by 4 feet to permit an increase in area of 12 square feet. Find the base and height of the original window and of the new window. See Fig. 2.9.

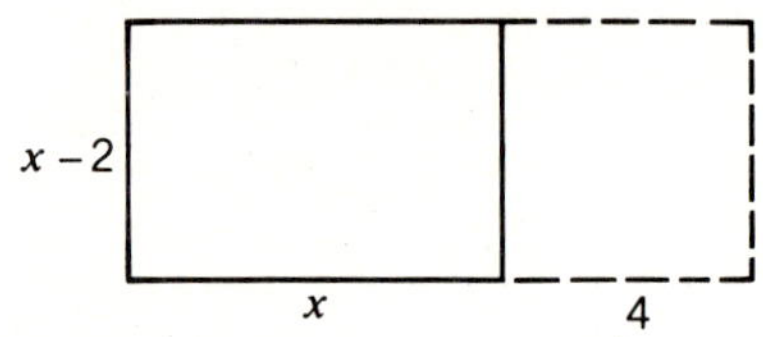

Figure 2.9

23. A wood template has the shape of a rectangle surmounted by an equilateral triangle (three equal sides), as in Fig. 2.10. The height of the rectangle is 3 inches less than the base of the rectangle (also a side of the triangle). It takes a total length of 34 inches of metal stripping to go around the template. Find the length of a side of the triangle.

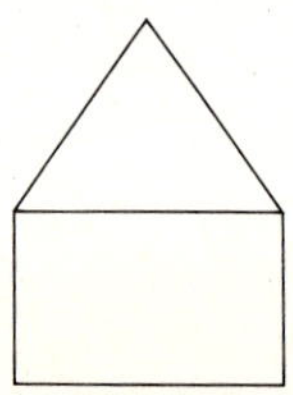

Figure 2.10

24. The length of the base of a triangle is 10 inches. The length of the base of a second triangle is 12 inches. The height of the second triangle is 2 inches more than that of the first. The area of the second triangle is 18 square inches more than the area of the first. Find the height of each triangle.

2.8 Applications: Mixture Problems

Many problems can be classified as mixture problems. The essential feature of this type of problem is that two or more items each having a specified unit value are combined to form a mixture. The sum of the values of the components must equal the value of the final mixture.

In this type of problem, the chart or table is a very useful device for summarizing the information. From it the equation may be obtained very easily. The final entry in the last column is equal to the sum of the other entries in this column.

Example 1 A piggy bank contains 23 coins, consisting of dimes and quarters. The total value is \$3.35. How many coins of each kind are there?

SOLUTION

Let d = the number of dimes.
Then $23 - d$ = the number of quarters.

Item	Unit Value ·	Number of Items =	Value
Dimes	10¢	d	$10d$
Quarters	25¢	$23 - d$	$25(23 - d)$
Mixture		23	335¢

EQUATION $10d + 25(23 - d) = 335$;
$d = 16$ dimes and $23 - d = 7$ quarters

Example 2 How many pounds of candy worth 75¢ a pound should be mixed with 20 pounds of candy worth \$1.30 a pound to obtain a mixture worth \$1.00 a pound?

SOLUTION

Let x = the number of pounds of 75-cent candy.

Item	Unit Value ·	Number of Items =	Value
Candy	75¢	x pounds	$75x$
Candy	130¢	20 pounds	$20(130)$
Mixture	100¢	$x + 20$	$100(x + 20)$

EQUATION

$75x + 20(130) = 100(x + 20)$; $x = 24$ pounds

Example 3 A businessman makes two types of investments: type A, which yields a 15% annual return, and type B, which yields a 12% return. If his total investment is \$10,000, and his income from the investment at the end of 1 year is \$1320, how much money did he invest in type A?

SOLUTION

Let d = amount invested in type A.

Item	Unit Value (Percent Return) ·	Number of Items (Amount Invested) in Dollars =	Value (Income)
Type A	0.15	d	$0.15d$
Type B	0.12	$10{,}000 - d$	$0.12(10{,}000 - d)$
Mixture (Totals)		10,000	1320

EQUATION

$0.15d + 0.12(10{,}000 - d) = 1320$ and $d = \$4000$

Example 4 Tickets to a certain event cost 75¢ for children and \$2 for adults. A total of \$2430 was collected. If there were 3 times as many adult tickets sold as children's tickets, how many adult tickets were sold?

SOLUTION

Let x = the number of children's tickets

Then

$3x$ = the number of adult tickets

Item	Unit Value ·	Number of Items =	Value
Children's tickets	75¢	x	$75x$
Adult tickets	200¢	$3x$	$200(3x)$
Mixture			243,000

EQUATION

$75x + 200(3x) = 243{,}000;\ x = 360,\ 3x = 1080$

Example 5 How many cubic centimeters of distilled water must be added to 30 cubic centimeters of a 20% solution of ammonium sulfate to obtain a 15% solution of ammonium sulfate?

SOLUTION

Let x = the number of cubic centimeters of water.

Item	Unit Value (Percentage of Sulfate) ·	Number of Items (Number of cc) =	Value
Water	0	x	0
20% solution	0.20	30	0.20(30)
15% solution	0.15	$x + 30$	$0.15(x + 30)$

EQUATION

$0.15(x + 30) = 0.20(30)$; $x = 10$ cubic centimeters

Exercises 2.8

Solve and check Exercises 1–20.

1. How many liters of high-octane gasoline worth 40¢ a liter should be blended with 300 liters of low-octane gasoline worth 35¢ a liter to produce a mixture worth 37¢ a liter?

2. A grocer mixes 45 pounds of ordinary coffee beans worth $2.80 a pound with rare coffee beans worth $3.60 a pound. How many pounds of rare beans should he use to obtain a mixture worth $3.00 a pound?

3. A farmer combines seed worth 50¢ a pound with seed worth 90¢ a pound to obtain a 100-pound mixture worth 60¢ a pound. How many pounds of each type of seed does he use?

4. Concrete is a mixture of cement and sand. How many pounds of cement worth 72¢ a pound and how many pounds of sand worth 12¢ a pound have been mixed to obtain 120 pounds of concrete worth 36¢ a pound?

5. A cashier in a ticket booth starts the day with $100 worth of change. The change is in quarters, dimes, and dollar bills. The number of dimes is 50 times the number of dollar bills. The number of quarters is 60 more than 10 times the number of dollar bills. How many dimes, quarters, and dollar bills did the cashier start with?

6. A jar at the checkout counter of a grocery store has a sign asking for contributions to a charitable cause. At the end of the week, the jar contains coins worth $7.40. The number of nickels was 2 more than twice the number of dimes. The number of quarters was 4 more than the number of dimes. These were the only types of coins. How many quarters were in the jar?

7. How many doughnuts worth 25¢ each and how many worth 20¢ each are combined to produce an assortment of 3 dozen doughnuts costing $7.95?

8. At a theater one evening, there were 420 paid admissions. The reserved seats sold for $5 each. The general admission seats sold for $3 each. The total amount collected for the evening was $1480. How many tickets of each kind were sold?

9. An ore containing 60% tin is combined with 3 times as much ore containing 40% tin to produce 90 kilograms of tin. How many kilograms of each type of ore were used?

10. The student association of a certain college decides to invest its building fund, since it cannot start construction of a student union for 3 more years. Some of the money is invested at 12% per year. Twice this amount is invested at 10% per year. The total income at the end of the first year was $4800. How much money is invested at each rate?

11. Six grams of a tranquilizer in powdered form and of normal strength is mixed with a double-strength tranquilizer to form a mixture that is 1.5 times normal strength. How many grams of the double-strength powder must be used to yield this new mixture?

12. Ten liters of a 30% alcohol solution are mixed with an 80% alcohol solution to produce a 60% solution. How many liters of the 80% alcohol solution are used?

13. A corporation invested \$100,000 in 12% bonds, in 16% stocks, and in 18% mortgages. The amount invested in bonds is the same as the amount invested in stocks. The total income per year from all three investments is \$15,000. How much is invested in each of the three types of investment?

14. A married couple invested \$60,000 as follows. They put \$20,000 into a tax-free account earning 10% interest. They put some of the balance in Money Market Certificates earning 15% interest. The rest they left in their checking-savings account earning 8% interest. The yearly income from these investments was \$7300. How much was invested in Money Market Certificates?

15. How many liters of water must be added to a 75% acid solution (75% acid and 25% water by volume) to obtain 100 liters of a solution that is 30% acid?

16. How much water must be evaporated from a 1% salt solution weighing 60 kilograms to obtain a 3% solution?

17. How many pints of alcohol must be evaporated from a solution containing 40% alcohol to obtain 100 pints of a solution containing 50% alcohol?

18. How many ounces of pure silver should be combined with an alloy containing 75% silver to produce 20 ounces of an alloy containing 90% silver?

19. A caterer buys beef at \$3 a pound, ham at \$2 a pound, and chicken at \$1 a pound. He buys twice as much beef as ham and 10 pounds less chicken than beef. The total bill is \$290. How many pounds of each type of meat does the caterer buy?

20. A home gardener buys pepper plants at 20¢ a plant, melon plants at 40¢ a plant, and tomato plants at 30¢ a plant. She buys 2 more melon plants than pepper plants and twice as many tomato plants as melon plants. She pays \$6.80 for all the plants. How many plants of each type does she buy?

Number Problems

Number and other word problems have a long history. They are found in ancient Egyptian manuscripts such as the Rhind papyrus written by the Egyptian scribe Ahmes around 1650 B.C. They occur on Babylonian clay tablets dated from 2100 B.C. to 1600 B.C.

The Greek Anthology, written around A.D. 500 by Metrodorus, is one of the best sources for ancient Greek algebraic problems.

Hindu problems are found in the works of Aryabhata (ca. A.D. 510), Brahmagupta (ca. A.D. 628), Mahavira (ca. A.D. 850), Bhaskara (ca. A.D. 1150), and in the Bakhshali manuscript, believed to have been written somewhere between A.D. 300 and A.D. 1000.

Algebraic knowledge passed from India to the Arabs and then to the Europeans. The famous *Liber abaci*, written in 1202 by the Italian Leonardo Fibonacci, was an important source by which the Hindu-Arabic numerals, arithmetic, and algebra were introduced into Europe.

2.9 Applications: Uniform Motion

If an object moves with a constant rate of speed, then the motion of the object is called **uniform motion.** The basic principle involved in such motion is the formula

$$\text{rate} \times \text{time} = \text{distance}$$
$$r \cdot t = d$$

In solving problems of this type, it is useful to make a sketch illustrating the facts of the problem and a chart organizing the facts.

Uniform motion problems fall into three basic categories: opposite directions, round trip, and pursuit.

Opposite Directions

When two objects start at the same point and travel in opposite directions, then the sum of the distances of the two objects is equal to the distance between them. (See Example 1.)

When two objects travel in opposite directions toward each other, then the sum of the distances of the two objects is equal to the distance between their starting points. (See Example 2.)

Example 1 At a certain time, two trains start from the same depot and travel in opposite directions. If one travels 35 mph and the other travels 60 mph, in how many hours will they be 285 miles apart?

SOLUTION

Since they start at the same time, the time that one train travels is the same as the time that the other train travels.

Let x = the time each travels.

See Fig. 2.11.

Formula:	r •	t =	d
One train	35	x	$35x$
Other	60	x	$60x$

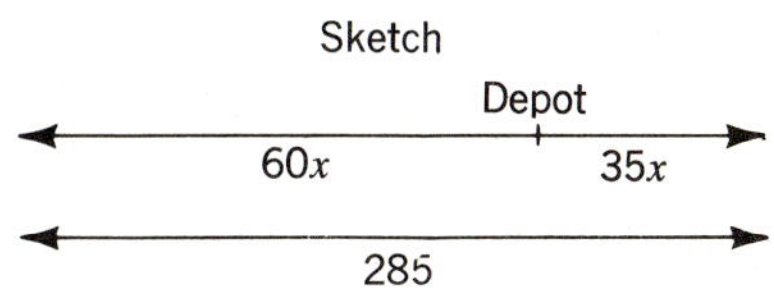

Figure 2.11

EQUATION

$$\begin{aligned} \text{distance of one train} + \text{distance of other} &= \text{total distance} \\ 35x + 60x &= 285 \\ x &= 3 \text{ hours} \end{aligned}$$

Example 2 City A and City B are 336 miles apart. A car left City A at 9 A.M. and traveled toward City B. One hour later, a helicopter left City B and traveled toward City A. At 12 noon it was directly over the car. If the rate of the helicopter was twice the rate of the car, how far did the helicopter travel?

SOLUTION

1. Make a sketch such as Fig. 2.12.

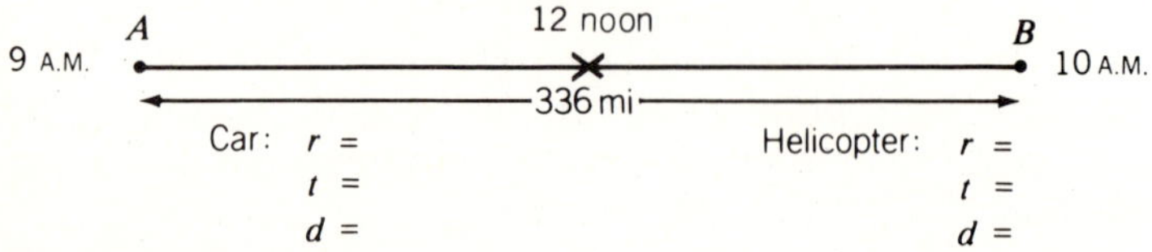

Figure 2.12

2. Let x = rate of car. Then $2x$ = rate of helicopter.
3. Put the information in a chart.

Formula:	r •	t =	d
Car	x	3	$3x$
Helicopter	$2x$	2	$4x$

4. Form an equation, using the sketch, and solve.

$$3x + 4x = 336$$
$$x = 48 \text{ mph}$$
$$4x = 4(48) = 192 \text{ miles, distance of helicopter}$$

Round Trip

In a round trip problem, an object travels to a certain point and then returns along the same route to its starting point. There are two important facts regarding this type of problem.

1. The distance going equals the distance returning.
2. The total time of the trip is the sum of the time going and the time returning. In general, these times are not the same.

Example 3 A boat traveled downstream at 24 mph and then immediately returned to its starting point. Its rate upstream was 18 mph. The total time for the trip was 7 hours.

a. Find the time it took the boat to travel downstream.
b. Find the distance the boat traveled downstream.

SOLUTION Let x = the time the boat traveled downstream.
Make a sketch such as Fig. 2.13.

continued

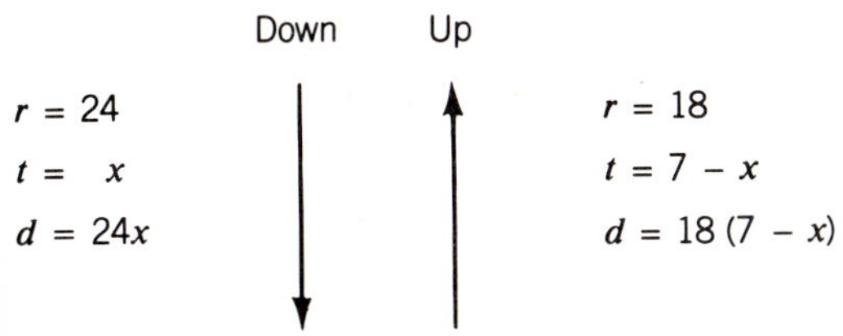

Figure 2.13

Form an equation and solve.

$$\text{distance downstream} = \text{distance upstream}$$
$$24x = 18(7 - x)$$
a. $x = 3$ hours

b. $d = 24(3) = 72$ miles

Pursuit

In a pursuit problem, two objects travel in the same direction until one overtakes the other.

If they **start** at the **same point,** their distances are equal. In general, their times will be different.

If they **start** at **different points,** the distance the slower object travels equals the sum of the distance of the faster object and the distance between their starting points.

Example 4 A truck leaves a station and travels along a highway at a rate of 40 mph. Two hours later, a police car leaves the same station and travels along the same highway until it overtakes the truck. The police car travels at a rate of 80 mph. How long does it take the police car to overtake the truck?

SOLUTION Let x = the time the police car travels.
$x + 2$ = the time the truck travels (it traveled 2 hours longer).

Sketch

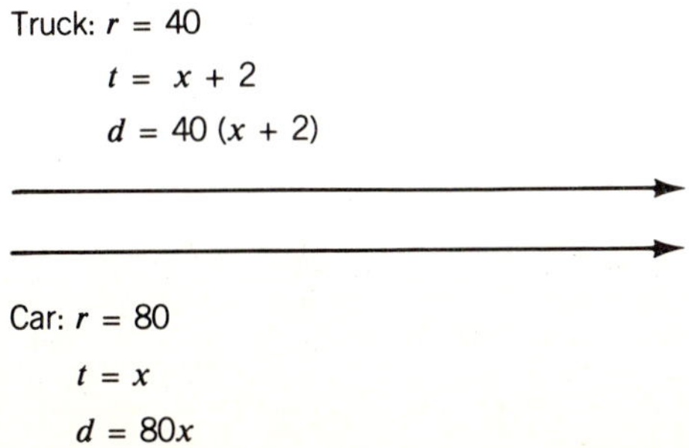

Equation

$$80x = 40(x + 2)$$
$$x = 2 \text{ hours}$$

Example 5 A ship is traveling at 20 knots (nautical miles per hour). When the ship is 15 nautical miles from a dock, a Coast Guard boat traveling at 50 knots leaves the dock in pursuit of the ship. What distance does the Coast Guard boat travel to overtake the ship?

SOLUTION Let $x =$ the time the boat travels.
Then $x =$ the time the ship travels.

Sketch

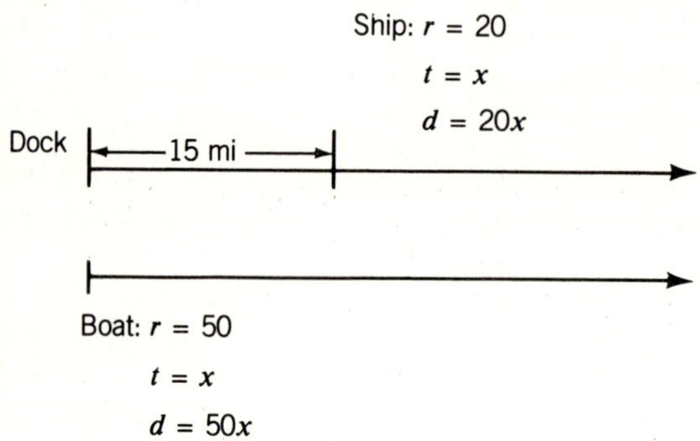

Equation

$$50x = 20x + 15$$
$$x = \frac{1}{2} \text{ hour}$$
$$d = 50\left(\frac{1}{2}\right)$$
$$= 25 \text{ nautical miles}$$

Exercises 2.9

Solve and check Exercises 1–20.

1. A car and a motorcycle stop at a rest stop on the highway. The car travels north at 55 mph, and the motorcycle goes south at 50 mph. If the two vehicles leave the rest stop at the same time, how long will it take them to be 210 miles apart?

2. Bill's home town is 315 miles from his college. At 8 A.M. Bill left college and drove home for Christmas vacation, traveling at a rate of 65 mph. Three hours after Bill started his trip, his girlfriend Abby left his home town and drove to meet him. Abby traveled at a rate of 55 mph. What time did Bill and Abby meet?

3. Two trains, 465 miles apart, travel toward each other, one traveling 15 mph faster than the other. They pass each other in 3 hours. What is the rate of the slower train?

4. At noon two cars leave Sacramento. One car, southbound for Fresno, travels 22 kph faster than the other car, northbound for Portland, Oregon. If the two cars are 712 kilometers apart at the end of 4 hours,

find the rate of speed of each car.

5. A plane took 4 hours to fly to Chicago from San Francisco. The return trip from Chicago to San Francisco took 1 hour longer because strong headwinds reduced the plane's speed by 100 kph. Find the rate of the plane going to Chicago.

6. A fisherman drove from his home to a mountain lake in 4 hours. Returning by the same route, he took 1 hour longer because heavy traffic forced him to drive 10 kph slower. Find his average rate going to the lake.

7. How far can a boat travel out to sea at 30 mph and return at 20 mph if the boat has just enough fuel for a 5-hour trip?

8. A man cycles from his home into the countryside at 12 mph. He visits with friends for an hour. Then his friends drive him home at a rate of 60 mph. The man returned home 4 hours after he left his home. How far is his home from his friends' house?

9. At 8 A.M. a truck leaves a depot, traveling west at 36 mph. Two hours later a car starts from the same depot and travels at 60 mph until it overtakes the truck. At what clock time does the car overtake the truck? (It is assumed that the truck driver does not stop for lunch.)

10. A Coast Guard boat averaging 30 mph starts one hour later from the same dock to overtake another boat, moving at 10 mph. How long does it take the Coast Guard boat?

11. A police car moving at 60 mph pursues a traffic offender driving at 40 mph. If the offender was 1 mile ahead when the police car started, how long does it take for the police car to overtake the offender?

12. A car speeding at 80 mph is 3 miles beyond a police car when the police car starts in pursuit at 90 mph. How long does it take the police car to overtake the speeding car?

13. Two planes leave an airport at the same time and travel in opposite directions. One plane travels 80 mph faster than the other plane. What is the rate of each plane if the planes are 2190 miles apart at the end of 3 hours?

14. At 6 A.M. two trains leave the same depot and travel in opposite directions. They are 972 kilometers apart at the end of 6 hours. Find the rate of each train if one travels 18 kph slower than the other.

15. A ship is heading due west out of San Francisco at a rate of 32 knots. A passenger on board realizes he has forgotten to bring his medicine. A helicopter, flying at a rate of 96 knots, is sent to deliver the medicine to the ship. If the helicopter starts out when the ship has been underway for 6 hours, how far from San Francisco is the ship when the helicopter reaches it?

16. At 1 P.M. an airplane left Airport A and flew due east to Airport B. At 2 P.M. another airplane left an airport 700 kilometers due east of Airport A and flew due east to Airport B. Both planes reached Airport B at 6 P.M. The plane that left at 1 P.M. flew 50 kilometers per hour faster than the other plane. Find the rate of each plane.

17. A couple attends an out-of-town wedding. The trip to the wedding takes 3 hours, the wedding and reception last 2 hours, and as a result of the celebration, the trip home takes 4 hours, at a rate of 15 mph less than the going rate. What was the couple's rate of travel on the way home?

18. An airplane flew due south with a tailwind at a rate of 570 kph. On the return trip, a headwind decreased its speed to 475 kph. The total flying time for both trips was 11 hours. Find the time of the southbound trip.

19. A salesman has a business appointment in a city 240 kilometers away. He starts driving at a rate of 75 kph. Suddenly he realizes he will be late, so he increases his rate to 90 kph. If his total driving time is 3 hours, how far did he travel at 75 kph?

20. A saleswoman drove 4 hours from her home to another city. The return trip on a new highway was 15 miles longer. However, it took 1 hour less time because she could drive 20 mph faster. Find the distance traveled on the shorter route.

Review Exercises

Solve and check the equations in Exercises 1–40.

1. $x + 7 = 12$
2. $3x - 2 = 13$
3. $2 - 3x = 23$
4. $x + 7 = x - 3$
5. $x + 7 = 3 - x$
6. $5x = 2x + 6$
7. $3x = 8x - 20$
8. $5x - (x - 2) = 18$
9. $6x + 7 - 5x = 2$
10. $2x - 5 = 7x - 5$
11. $x + 5 = -7$
12. $3x + 2 = x + 4$
13. $5 - 2y = 15$
14. $3 + 4x - 2 = 4 + 2x$
15. $2(x - 4) = 3x - 16$
16. $3(y + 2) + 4 = 5y - 2$
17. $4 - (3x + 8) = 17$
18. $5(z - 1) - (1 - 4z) = 30$
19. $5t + 7t - 2 = 5 + 7 - 2t$
20. $x + 2 = x + 3$
21. $2(x - 5) = 2x - 10$
22. $6 - \dfrac{x + 2}{3} = 4$
23. $4 = 2x + 3$
24. $4x - (x - 5) = 4 - (x + 7)$
25. $0 = \dfrac{x + 5}{2}$
26. $7x - 5(x - 2) = 20$
27. $9 - 6(2 - x) = 7x$
28. $x + 4 = 2x - 3$
29. $z + 1 = z$
30. $3 - x = x - 3$
31. $4(x + 3) = 12$
32. $2(x - 3) - 3(x + 1) = 0$
33. $y - 2 = 2 - y$
34. $4 - 3(x + 2) = 7$
35. $3(x + 2) = 3x + 6$
36. $0 = 2(x + 6)$
37. $\dfrac{t + 7}{2} = 5$
38. $5 - \dfrac{x + 3}{2} = 6$
39. $2(x - 3) + 3(x + 2) = x + 8$
40. $x - 3[x - 3(x - 3)] = 1$

Solve the equations in Exercises 41–60 for the specified variables.

41. $3x - y = 5;\ y$
42. $3x - y = 5;\ x$
43. $2x + 5 + y = 0;\ y$
44. $2x + 5 + y = 0;\ x$
45. $5x - 2y = 10;\ x$
46. $5x - 2y = 10;\ y$
47. $y = 3x + 2;\ x$
48. $x = 3y + 2;\ y$
49. $ax + by = c;\ y$
50. $ax + by = c;\ x$

51. $C = \pi d$; d

52. $A = P + Prt$; t

53. $P = x + y + z$; y

54. $S = 2a + 2b$; b

55. $R = \dfrac{g}{T - t}$; t

56. $M = n(n + 2d)$; d

57. $B = 15 - \dfrac{A}{2}$; A

58. $W - 5H + 190 = 0$; H

59. $A + 2B - 3C = 20$; B

60. $A + 2B - 3C = 20$; C

Solve and check Exercises 61–69.

61. Six more than twice a number is equal to 5 times the difference obtained when 3 is subtracted from the number. Find the number.

62. A bus and a train leave the same station at the same time, and both travel north. The rate of the train is 15 kph less than twice the rate of the bus. At the end of 3 hours, the train is 135 kilometers farther north than the bus. Find the rate of each.

63. How many grams of copper must be added to 60 grams of an alloy composed of 40% copper to make an alloy composed of 50% copper?

64. How much paint is required to paint 3 walls of a room if each wall is a rectangle with a length 2 feet less than twice the height? The perimeter of each wall is 56 feet, and 1 gallon of paint covers about 360 square feet.

65. A woman paid $7.65 for some fruit consisting of peaches costing 40¢ a pound, plums at 45¢ a pound, and bananas at 35¢ a pound. If she bought twice as many pounds of plums as peaches and 1 pound less of bananas than plums, how many pounds of each kind of fruit did she buy?

66. A corner lot has the shape of a right triangle (one angle is 90 degrees). The sides forming the 90-degree angle are along the streets, and one of these sides is 11 feet longer than the other side. To widen the street along the longer side, a strip 10 feet wide is removed from the lot, thus reducing the length of this side by 11 feet. The area of the lot was decreased by 2310 square feet. Find the original length of the shorter side along the street.

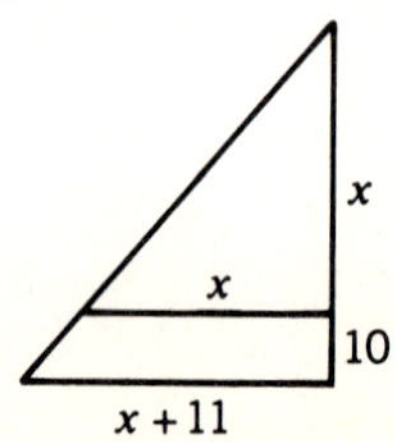

67. During an average life of 70 years, it is estimated that a person sleeps 3 years more than he or she works; that one half the years he or she works are spent on recreation and church functions; that 7 years are spent on eating and drinking; 5 years on traveling; 2 years on dressing; and 3 years on being ill. How many years does he or she work?

68. A person traveling on a plane is allowed 20 kilograms of baggage free but must pay $1.50 per kilogram for all excess baggage. A wife had 5 kilograms of baggage more than her husband had. Together they paid $28.50 for excess baggage. How much baggage did each have?

69. A rush order is received in Boomtown for some machine parts that must be delivered from Bolttown, 280 miles away. To speed the delivery, a station wagon traveling 65 mph leaves Boomtown at 11 A.M. to intercept the delivery truck traveling 50 mph. If the delivery truck left Bolttown at 10 A.M., at what clock time is the delivery truck intercepted?

In Exercises 70–75, solve and graph on a number line.

70. $3x + 5 > x - 3$

71. $5 - x \leq x - 5$

72. $0 > 4x - 12$

73. $2x + 3 - x < 5 + x$

74. $2x + 3 > 5 + 2x$

75. $12 \leq -3(2x - 4)$

3 Operations on Polynomials

Objectives

OBJECTIVES	TYPICAL PROBLEMS	LOCATIONS
Find the product of monomials.	1. Multiply: $(-3x^3y)(4x^2y^2)$	*Section 3.1* *Examples 1–5* *Exercises 1–60*
Add and subtract polynomials.	2. Add: $(5x^2 - 4x + 7) + (x^2 + 5x - 9)$ 3. Subtract: $(6x - 10y + 12) - (x - 10y - 2)$	*Section 3.2* *Examples 2–7* *Exercises 1–50*
Multiply polynomials.	4. Multiply: $-3x^2(2x^2 - 4x + 5)$ 5. Multiply: $(x - 4)(x^2 + 2x - 2)$ 6. Multiply: $(3x - 2)(2x + 5)$	*Section 3.3* *Examples 1–8* *Exercises 1–90*
Find the greatest common monomial factor of a polynomial.	7. Factor: $16x^3 - 40x^2 + 8x$	*Section 3.4* *Examples 1–6* *Exercises 1–50*
Factor a difference of squares.	8. Factor: $9x^2 - 16y^2$	*Section 3.5* *Examples 5–9* *Exercises 21–46 and 61–76*
Factor a simple trinomial.	9. Factor: $x^2 - 2x - 48$	*Section 3.6* *Examples 4–9* *Exercises 21–100*
Factor a perfect square trinomial.	10. Factor: $16x^2 - 56xy + 49y^2$	*Section 3.7* *Examples 3–9* *Exercises 1–100*
Factor a general trinomial.	11. Factor: $5x^2 + 7x - 6$	*Section 3.8* *Examples 1–7* *Exercises 1–100*
Completely factor special polynomials over the integers.	12–15. Factor, if possible: 12. $x^4 - 24x^2 - 25$ 13. $12x^3 - 14x^2 - 10x$ 14. $4x^4 - 64$ 15. $81x^4 - 72x^2 + 16$	*Section 3.9* *Examples 1–7* *Exercises 1–60*

Solve special polynomial equations by factoring.	16–18. Solve: 16. $x^2 = 36$ 17. $4x^2 - 28x = 0$ 18. $x^2 + x = 30$	*Section 3.10* *Examples 10–15* *Exercises 1–60*

Answers to Typical Problems

1. $-12x^5y^3$
2. $6x^2 + x - 2$
3. $5x + 14$
4. $-6x^4 + 12x^3 - 15x^2$
5. $x^3 - 2x^2 - 10x + 8$
6. $6x^2 + 11x - 10$
7. $8x(2x^2 - 5x + 1)$
8. $(3x - 4y)(3x + 4y)$
9. $(x + 6)(x - 8)$
10. $(4x - 7y)^2$
11. $(5x - 3)(x + 2)$
12. $(x - 5)(x + 5)(x^2 + 1)$
13. $2x(2x + 1)(3x - 5)$
14. $4(x - 2)(x + 2)(x^2 + 4)$
15. $(3x - 2)^2(3x + 2)^2$
16. $x = \pm 6$
17. $x = 0$ or $x = 7$
18. $x = 5$ or $x = -6$

A fundamental concept in the study of algebra is the class of algebraic expressions known as polynomials. The applications of polynomials to real-world situations are too numerous to list. For instance, the distance a missile travels when thrown into the air and allowed to fall back to earth can be expressed in terms of a polynomial—so can the velocity of the same missile and its acceleration. In business, revenue obtained from the sale of a commodity is often expressed by a polynomial and so are marginal revenue, supply and demand, efficiency of investment, and so forth. A study of any of the fields of application of mathematics will surely convince you of the importance of polynomials.

It is the object of this chapter to introduce polynomials, to define operations on polynomials, to present some basic factoring techniques, and to show the solution of factorable quadratic equations in one variable.

3.1 Products of Monomials

In Chapter 1 the operation "raising to a power" was discussed. For example,

$$x^2 = x \cdot x$$
$$x^3 = x \cdot x \cdot x$$

Also,

$$x^4 = x \cdot x \cdot x \cdot x \quad \text{and} \quad x^5 = x \cdot x \cdot x \cdot x \cdot x$$

A definition is needed here to explain what is meant by x^n, where n is a natural number.

Definition of x^n

If x is a real number and n is a natural number, then

$$x^1 = x \text{ and } x^n = \overbrace{x \cdot x \cdot x \cdot \;\cdots\; \cdot x}^{n \text{ factors}} \text{ if } n > 1$$

The variable x is called the **base,** the natural number n is called the **exponent,** and x^n is called a **power.**

Each of the following is an example of a monomial:

$$3 \quad , \quad x^2 \quad , \quad 5x^3 \quad , \quad \sqrt{2}xy \quad , \quad x^2y^9 \quad , \quad \frac{1}{2}xyz^5$$

In general, we have the following definition.

Definition

A **monomial** is one of the following:

1. A constant;
2. A term of the form cx^n,
   where c is a constant, x is a variable, and n is a natural number; or
3. A product of terms as described in (2).

Note that $\sqrt{x}$ is *not* a monomial and neither is $\frac{1}{x}$, $\frac{5}{x^2}$, or any other term that does not fit the description of being a constant or of the form cx^n, where c is constant and n is a natural number.

A monomial that has only one variable is called a **monomial in one variable.** For example, $3x$, $15x^3$, and $-2y^2$ are monomials in one variable.

The monomial $4x^2y^3$ is not a monomial in one variable but a **monomial in two variables.**

The **degree of a monomial in one variable** is equal to the exponent on the variable.

As examples,

The degree of $4x^5$ is 5 and the degree of $-6x$ is 1, since $x = x^1$.

The **degree of a monomial in two or more variables** is equal to the sum of the exponents on the variables.

As examples,

The degree of $7x^2y^4$ is 6 , since $2 + 4 = 6$
The degree of $5xy$ is 2 , since $xy = x^1y^1$ and $1 + 1 = 2$
The degree of $-3x^2y^3z^4$ is 9 , since $2 + 3 + 4 = 9$

The constant factor of a monomial is called the **numerical coefficient** of the monomial. Usually, the numerical coefficient is just called the coefficient, for convenience.

As examples,

The coefficient of $4x^5$ is 4
The coefficient of $-6x$ is -6
The coefficient of x^2y^4 is 1

Monomials are multiplied by rearranging the factors so that the coefficients are written first (at the left) and the literal factors are written from left to right in alphabetical order, using the exponential form so that no letter is repeated.

Example 1

$$\begin{aligned} x^2 \cdot x^3 &= (x \cdot x) \cdot (x \cdot x \cdot x) \\ &= x \cdot x \cdot x \cdot x \cdot x \\ &= x^5 \end{aligned}$$

Example 2

$$\begin{aligned} 3x^2 \cdot 2x^3 &= (3 \cdot x \cdot x) \cdot (2 \cdot x \cdot x \cdot x) \\ &= (3 \cdot 2) \cdot (x \cdot x \cdot x \cdot x \cdot x) && \text{(Associative and commutative axioms, multiplication)} \\ &= 6x^5 \end{aligned}$$

Example 3

$$\begin{aligned} (5xy^2)(-3x^2y^3) &= 5(-3)x(xx)(yy)(yyy) && \text{(Rearranging factors)} \\ &= -15x^3y^5 && \text{(Writing in exponential form)} \end{aligned}$$

Example 4

$$\begin{aligned} (-3x^3)^2 &= (-3x^3)(-3x^3) \\ &= (-3)(-3)(xxx)(xxx) \\ &= 9x^6 \end{aligned}$$

Example 5

$$(-2y)(-3y^2)(-5y^4) = (-2)(-3)(-5)y(yy)(yyyy)$$
$$= -30y^7$$

Exercises 3.1

1–10. Rewrite each of the following in exponential form. State the degree of each monomial.

1. $2xxx$
2. $5yy$
3. $-2tttt$
4. $-8x$
5. y
6. $4ttttt$
7. xyy
8. $xxxy$
9. $-3xxyyy$
10. $-7xxxyyy$

11–60. Do the indicated multiplications.

11. $(2x)(3x)$
12. $(4y)(7y)$
13. $(-5y)(-y)$
14. $(-x)(-6x)$
15. $4x^2(-6x^2)$
16. $(-3y^2)(4y^2)$
17. $(-3xy)(5xy)$
18. $(2xy)(-5xy)$
19. $(-x^2y^2)(-x^2y^2)$
20. $(-2x^2y^2)(-3x^2y^2)$
21. x^3x^2
22. y^2y^4
23. $3y^2(4y^2)$
24. $2x^3(4x^3)$
25. $5x(x^3)$
26. $4y(y^4)$
27. $7z(5z^5)$
28. $2z(3z^3)$
29. $(-2x^4)(3x^2)$
30. $(-5x^2)(x^6)$
31. $(-y^3)(-2y^5)$
32. $(-y^5)(-3y^2)$
33. $(6t^2)(-4t^2)$
34. $(9t^3)(-4t^3)$
35. $(-3x)(-5x)$
36. $(-4y)(-2y)$
37. $(4x^2)^2$
38. $(-3y^2)^2$
39. $(-5t^3)^2$
40. $(6t^3)^2$
41. $(-x)(-x^2)(-x^3)$
42. $(-z)(-z^2)(-z^2)$
43. $(-2y)(-3y)(4y^2)$
44. $5y(-5y^2)(-2y^3)$
45. $-2(3x)^2$
46. $-3(-5y)^2$
47. $5t(-2t^2)^2$
48. $-4t(5t^3)^2$
49. $(-2x^2)^3$
50. $(-10x^3)^2$
51. $(xy)(-x^2y)$
52. $(-x^2y)(2xy^2)$
53. $(-xy^2)(-3xy)$
54. $(-5xy)(-4x^2y)$
55. $(-rs)(rs)$
56. $(uv)(-uv)$
57. $(-xy)(-xy)(-xy)$
58. $(rt)(-rt)(-rt)$
59. $uv(-vw)(-uvw)$
60. $(-rst)(-rs)(-st)$

3.2 Sums and Differences of Polynomials

Definition
A **polynomial** is a monomial or an algebraic sum of monomials.

Example 1 Each of the following is an example of a polynomial:

a. $x^2 + 3x + 2$ is a polynomial in one variable.
b. $15y^9 - y$ is a polynomial in one variable.
c. $x^2 - 3xy + 2y^2$ is a polynomial in two variables.
d. $9x^5 - 3x^4 + 2x^2 + 5x - 1$ is a polynomial in one variable.
e. $x^2 + y^2 + z^2 - 4xyz$ is a polynomial in three variables.

Some common types of polynomials have special names. A **monomial** is a polynomial of one term, a **binomial** is a polynomial of two terms, and a **trinomial** is a polynomial of three terms.

Each of the following is an example of a binomial:

$$x + 1 \quad , \quad 3y^2 + y \quad , \quad -5x^2 - 2 \quad , \quad a + b \quad , \quad x^2 - y^2$$

Each of the following is an example of a trinomial:

$$x^2 + 2x + 1 \quad , \quad 5y^3 - 3y + 2 \quad , \quad a + b + c \quad , \quad x^2 - 6xy + 9y^2$$

The **degree of a polynomial** is the greatest degree of any of its terms.

As examples,

The degree of $3x^4 + 5x^3 - 2x^2 + 8$ is 4, the degree of $3x^4$
The degree of $6x^4 - 4x^2y^3 + 7y^4$ is 5, the degree of $-4x^2y^3$

It is conventional to express a polynomial in **descending powers** of a variable. This means that the term containing the highest power of the variable is written first, at the left; the next highest power is written second; and so on, with the lowest power written last, at the right. For example,

$$5x^4 - 6x^3 - 2x^2 + 8x - 4$$

is written in descending powers of the variable x. As another example,

$$y^3 + 3y^2z - z^4$$

is written in descending powers of the variable y.

If the order just described is reversed, then the polynomial is said to be written in **ascending powers** of the variable. For example,

$$1 - 2x - x^2$$

is written in ascending powers of the variable x.

Some problems are more easily done if each polynomial in the problem is arranged in either descending powers of a variable or ascending powers of a variable.

Polynomials are added or subtracted by using the commutative and associative axioms to rearrange the terms and by using the distributive axiom to combine like terms.

Example 2 Add $8x^2 - 5x + 7$ to $x^2 + 3x - 4$.

SOLUTION

$$\begin{aligned}&(8x^2 - 5x + 7) + (x^2 + 3x - 4)\\ &= (8x^2 + x^2) + (-5x + 3x) + (7 - 4) && \text{(Rearrange terms)}\\ &= (8 + 1)x^2 + (-5 + 3)x + (7 - 4) && \text{(Combine like terms)}\\ &= 9x^2 - 2x + 3\end{aligned}$$

Most of these steps can be done mentally, and, as a rule, the final result can be written immediately after inspection of the problem.

Example 3 Simplify by doing the indicated operations:

$$(4x + 8y + 9) + (3x - 9y - 2).$$

SOLUTION

$$(4x + 8y + 9) + (3x - 9y - 2) = 7x - y + 7$$

To subtract one polynomial from another, the subtraction problem must be changed to an addition problem by using the definition of subtraction: $r - s = r + (-s)$.

It is important to remember that the sign of each term of the subtrahend must be changed, as shown in the following example.

Example 4 Subtract $4x^2 + xy - 6y^2$ from $2x^2 + 3xy - 9y^2$.

SOLUTION

$(2x^2 + 3xy - 9y^2) - (4x^2 + xy - 6y^2)$ (Recalling that "subtract *a* from *b*" is translated as "*b* − *a*")

$= (2x^2 + 3xy - 9y^2) + (-4x^2 - xy + 6y^2)$ (Using the definition of subtraction)

$= (2x^2 - 4x^2) + (3xy - xy) + (-9y^2 + 6y^2)$ (Using the associative and commutative axioms several times)

$= (2 - 4)x^2 + (3 - 1)xy + (-9 + 6)y^2$ (Using the distributive axiom)

$= -2x^2 + 2xy - 3y^2$

Again, it is usually possible to write the final result immediately after inspection of the problem. Thus, by performing much of the work mentally, we would write the solution as illustrated in Example 5.

Example 5 Subtract $5x - 2y + 4$ from $-3x + 2y + 4$.

SOLUTION

$$\begin{aligned} &(-3x + 2y + 4) - (5x - 2y + 4) \\ &= (-3x + 2y + 4) + (-5x + 2y - 4) \\ &= -8x + 4y \end{aligned}$$

A subtraction problem can be checked by adding the result to the subtrahend. If $a - b = c$, then $a = b + c$.

Example 6 Check that $(-3x + 2y + 4) - (5x - 2y + 4) = -8x + 4y$

SOLUTION Add $-8x + 4y$ to $5x - 2y + 4$

$$\begin{array}{r} 5x - 2y + 4 \\ -8x + 4y \phantom{{}+4} \\ \hline -3x + 2y + 4 \end{array}$$

Note that the addition in Example 6 is done vertically. An addition or subtraction problem can be done either vertically or horizontally.

Example 7 Simplify $(3y - 6 - 2y^2) + (5y^2 + 9 - 2y)$.

SOLUTION First, rearrange each polynomial in descending powers.

Horizontal Format:

$$\begin{aligned} &(-2y^2 + 3y - 6) + (5y^2 - 2y + 9) \\ &= (-2y^2 + 5y^2) + (3y - 2y) + (-6 + 9) \\ &= 3y^2 + y + 3 \end{aligned}$$

Vertical Format:

$$\begin{array}{r} -2y^2 + 3y - 6 \\ 5y^2 - 2y + 9 \\ \hline 3y^2 + y + 3 \end{array}$$

Exercises 3.2

1–10. Add the given polynomials vertically.

1. $\begin{array}{r} 2x^2 - 5x + 7 \\ \underline{3x^2 + 8x - 9} \end{array}$

2. $\begin{array}{r} -x^2 + 6x - 4 \\ \underline{2x^2 - 7x - 3} \end{array}$

3. $\begin{array}{r} 4x - 6y - 14 \\ \underline{3x + 6y + 10} \end{array}$

4. $\begin{array}{r} 10x + 2y - 16 \\ \underline{-10x + 6y - 20} \end{array}$

5. $\begin{array}{r} y^3 + 2y^2 - 3y \quad \\ \underline{- \; y^2 - 2y + 3} \end{array}$

6. $\begin{array}{r} y^3 - 2y^2 + \; y \quad \\ \underline{- \; y^2 + 2y - 1} \end{array}$

7. $\begin{array}{r} 3x^2 + 4xy - y^2 \\ \underline{5x^2 - 4xy - y^2} \end{array}$

8. $\begin{array}{r} 2x^2 - 5xy + 3y^2 \\ \underline{7x^2 + 5xy - 3y^2} \end{array}$

9. $\begin{array}{r} 2x + 3y - z \\ \underline{-x - 2y + z} \end{array}$

10. $\begin{array}{r} 3x - 4y + 2z \\ \underline{x + \; y - 5z} \end{array}$

11–30. Simplify.

11. $(3x^2 + 2x) + (4x^2 + 3x + 1)$
12. $(y^3 + y + 1) + (2y^2 - 2y - 4)$
13. $(y^2 - 3y + 4) + (5y^2 + 6y - 9)$
14. $(7x^2 - 5x + 2) + (x^2 + x - 6)$
15. $(3x + 5y - 7) + (2x - 5y - 3)$
16. $(2x + y - z) + (x - 2y + z)$
17. $(t^3 - 4t^2 + 4t) + (-2t^2 + 8t - 8)$
18. $(t^3 - 6t^2 + 9t) + (-3t^2 + 18t - 27)$
19. $(8z - 4x - 2y) + (4x + 4y - 12z)$
20. $(a + 2c - 2b) + (3b - 2c - 4a)$
21. $(2x - 3y - 4) - (2x + 3y + 4)$
22. $(2a - b - c) - (a + b + c)$
23. $(r + s - t) - (r - s + t)$
24. $(3x + 4y - 6) - (2x - y + 6)$
25. $(x^2 + 6x + 9) - (x^2 - 4x + 4)$
26. $(25x^2 - 10xy + y^2) - (9x^2 + 6xy + y^2)$
27. $(1 + x - x^2) - (2 - x^2 + 2x)$
28. $(y - 3y^2 - 6) - (8 - 2y - 4y^2)$
29. $(y^2 + 2xy - x^2) - (2xy - x^2 - y^2)$
30. $(x^2 - x^3 + 5x^4) - (1 - x^3 - x^2)$

31–50. Do the operations as directed.

31. Add $3x^2 + 5 - x$, $x - 8$, $x^2 + 4$.

32. Subtract $2a^2 - 5b^2 - 4$ from $a^2 - b^2 + 1$.

33. From $-t^2 - 3t - 5$ subtract $-t^2 - 5t + 4$.

34. Find the sum of $c^3 - 3c^2 - c - 7$ and $4c - 3c^2 - 6c^3 + 6$.

35. Find the sum of $x^3 - 3x^2 + x - 5$ and $x^2 - 3x + 4$.

36. From the sum of $3x^2 + 5x + 9$ and $x^2 - 7x + 14$ subtract $2x^2 - 3x + 5$.

37. Subtract $3a^2 + 5ab - 6b^2$ from the sum of $2a^2 - 3ab + 4b^2$ and $5ab - 6b^2$.

38. Add $4 - 3x^2$, $2 + 3x + 8x^2$, $3 - 2x^2 - 5x$.

39. From $4a + 2b - 4c$ subtract $3a - 3b - 3c$.

40. Subtract $2y^3 - 5y^2 - 3y + 9$ from $-4y^3 - 3y^2 - 4$.

41. Subtract $2x^2 - 6$ from the sum of $-3x^2 + 5x$ and $4 - 7x$.

42. Find the number that must be added to $2x^2 - 3xy - y^2$ to obtain $y^2 + xy + x^2$.

43. Subtract $x^3 - 17x + 1$ from the sum of $x^4 + 3x^3 - x^2 + 2$ and $5x^4 - x^3 + 4x - 7$.

44. From the sum of $a^2 + 3ab + 4b^2$ and $2a^2 - 4ab + 5b^2$ subtract $3a^2 - 7ab + 8b^2$.

45. Subtract the sum of $x^4 + 3x^2 + 2$ and $2x^4 - 5x^3 + x^2 - 1$ from $6x^3 - 7x^2 + x$.

46. From $2x^3 - 4x^2y + 8xy^2 + 11y^3$ subtract the sum of $x^3 + y^3$ and $3x^2y - 5xy^2 - 7y^3$.

47. Find the sum of $a^3 + b^3 - a^2b + ab^2$, $a^2b + b^3 - 3a^3$, and $b^3 - 5ab^2 + a^2b$.

48. From $x^3 - x^2 - 3x + 2$ subtract the sum of $x^4 + 5x - 1$ and $x^3 + 3x^2 - 6x + 2$.

49. Subtract the sum of $x^3 + 2x^2y - xy^2 + y^3$ and $5x^2y + 6xy^2 - 2y^3$ from $4x^3 - 3x^2y + 3xy^2 + 4$.

50. Find the sum of $a^3 - 3a^2b + 5ab^2 - b^3$, $2a^3 + 5a^2b - 2ab^2$, and $6a^2b + 15ab^2 - 9b^3$.

3.3 Products of Polynomials

In order to find the products of polynomials, the distributive axiom is used. It is restated below for emphasis.

The Distributive Axiom
If r, s, and t are real numbers, then

$$r(s + t) = rs + rt$$

This axiom can be applied for multiplying a monomial by a binomial, and it can be extended to multiplying a binomial by a binomial, a monomial by a trinomial, and so on.

When parentheses are removed in the multiplication of a polynomial by a monomial, it is important to remember that each term of the polynomial must be multiplied by the monomial.

Example 1 Multiply $x(2x^2 + 1)$.

SOLUTION

$$\begin{aligned} x(2x^2 + 1) &= x \cdot 2x^2 + x \cdot 1 && \text{(Distributive axiom)} \\ &= 2x^3 + x \end{aligned}$$

Therefore,

$$x(2x^2 + 1) = 2x^3 + x$$

Example 2 Multiply $3x^3(2x^2 + 4x - 2)$.

SOLUTION

$$\begin{aligned} 3x^3(2x^2 + 4x - 2) &= 3x^3(2x^2) + 3x^3(4x) + 3x^3(-2) \\ &= 6x^5 + 12x^4 - 6x^3 \end{aligned}$$

The product of a polynomial multiplied by a polynomial is obtained by applying the distributive axiom several times.

Example 3 Multiply $(2x + 5)(3x + 2)$.

SOLUTION

$$\begin{aligned} &(2x + 5)(3x + 2) \\ &= (2x + 5)(3x) + (2x + 5)(2) && \text{(Distributive axiom)} \\ &= 3x(2x + 5) + 2(2x + 5) && \text{(Commutative axiom)} \\ &= 6x^2 + 15x + 4x + 10 && \text{(Distributive axiom)} \\ &= 6x^2 + 19x + 10 && \text{(Distributive axiom)} \end{aligned}$$

Example 4 Write $(x^2 + 2)(3x^2 + x + 1)$ as a single polynomial (multiply).

SOLUTION

$$\begin{aligned} &(x^2 + 2)(3x^2 + x + 1) \\ &= (x^2 + 2)(3x^2) + (x^2 + 2)(x) + (x^2 + 2)(1) \\ &= 3x^4 + 6x^2 + x^3 + 2x + x^2 + 2 \\ &= 3x^4 + x^3 + 7x^2 + 2x + 2 \end{aligned}$$

Note that the answer was stated in descending powers of the variable—that is, the variable was arranged so that the first term had the largest exponent, the second term the next largest exponent, and so on, with the constant term last. This rearrangement can be done because of the commutative and associative properties of addition of real numbers, and it makes the answer easier to check.

Sometimes it is convenient to arrange the polynomials to be multiplied in vertical order rather than horizontal order, as shown in the following example.

Example 5 Multiply $3x^2 + 2x + 4$ by $4x^2 - x + 1$.

SOLUTION

$$\begin{array}{rrrrrl} & 3x^2 & + 2x & + 4 & & \\ & 4x^2 & - x & + 1 & & \\ \hline 12x^4 & + 8x^3 & + 16x^2 & & & \text{(Multiply top row by } 4x^2\text{)} \\ & - 3x^3 & - 2x^2 & - 4x & & \text{(Multiply top row by } -x\text{)} \\ & & + 3x^2 & + 2x & + 4 & \text{(Multiply top row by } +1\text{)} \\ \hline 12x^4 & + 5x^3 & + 17x^2 & - 2x & + 4 & \text{(Combine like terms)} \end{array}$$

There is a special method, called the FOIL method, that is very useful for finding the product of two binomials.

Multiplying, by using the distributive axiom,

$$\begin{aligned} (A + B)(C + D) &= A(C + D) + B(C + D) \\ &= AC + AD + BC + BD \end{aligned}$$

Note the following about the four terms of the product polynomial:

1. The **first** term, AC, is the product of the **first** terms of the factors.
2. The **last** term, BD, is the product of the **last** terms of the factors.
3. The **two middle terms** form the **sum** of the **inner product** and the **outer product** as shown below.

$(A + B)(C + D)$

BC Inner product

AD Outer product

The word FOIL is a useful device for remembering the terms of this special product: F means first terms product, O means outer product, I means inner product, and L means last terms product.

First term	Last term	First term	Last term		First terms product	Outer product	Inner product	Last terms product
↓	↓	↓	↓		↓	↓	↓	↓
$(A$	$+\ B)$	$(C$	$+\ D)$	$=$	AC	$+\ AD$	$+\ BC$	$+\ BD$
					F	O	I	L

Examples 6, 7, and 8 show products of binomials obtained by using the FOIL method.

Example 6 Use the FOIL method to find $(3x + 4)(5x - 2)$.

SOLUTION

$$\begin{aligned} & \qquad\quad \text{F} \qquad\quad \text{O} \qquad\quad \text{I} \qquad\quad \text{L} \\ (3x + 4)(5x - 2) &= 3x(5x) + 3x(-2) + 4(5x) + 4(-2) \\ &= 15x^2 - 6x + 20x - 8 \\ &= 15x^2 + 14x - 8 \end{aligned}$$

Example 7 Use the FOIL method to find $(2x - 5)(y + 6)$.

SOLUTION

$$\begin{aligned} & \qquad\quad \text{F} \qquad\quad \text{O} \qquad\quad \text{I} \qquad\quad \text{L} \\ (2x - 5)(y + 6) &= 2x(y) + 2x(6) + (-5)y + (-5)(6) \\ &= 2xy + 12x - 5y - 30 \end{aligned}$$

Example 8 Use the FOIL method to find $(x^2 - 9)(x - 4)$.

SOLUTION

$$\begin{aligned} & \qquad\quad \text{F} \qquad\quad \text{O} \qquad\quad \text{I} \qquad\quad \text{L} \\ (x^2 - 9)(x - 4) &= x^2(x) + x^2(-4) + (-9)x + (-9)(-4) \\ &= x^3 - 4x^2 - 9x + 36 \end{aligned}$$

Exercises 3.3

1–20. Multiply.

1. $x(x - 4)$
2. $x(x + 3)$
3. $2x(3x + 5)$
4. $5x(2x - 4)$
5. $-3y(4y - 1)$
6. $-4y(2y + 3)$
7. $-x^2(x^2 + 4)$
8. $-x^3(x^2 - 16)$
9. $4x^2(2x^2 - y^2)$
10. $6x^3(3x - 5y)$
11. $5(2x - 3y + 6)$
12. $4(3x + 2y - 5)$
13. $-x(x^2 + 2x + 1)$
14. $-y(y^2 - 4y + 4)$
15. $-6y(2y^2 - y - 3)$
16. $-2x(9x^2 - x - 1)$
17. $5xy(x^2 + 2xy - y^2)$
18. $3xy(x^2 - 3xy + 6)$
19. $-2x^2(x^2 - 3x - 4)$
20. $-5x^2(x^2 - 2x - 8)$

21–34. Multiply in vertical order.

21. $\begin{array}{l} x^2 - 2x + 1 \\ \underline{x - 1} \end{array}$
22. $\begin{array}{l} x^2 + 4x + 4 \\ \underline{x + 2} \end{array}$
23. $\begin{array}{l} y^2 + 5y + 25 \\ \underline{y - 5} \end{array}$
24. $\begin{array}{l} y^2 - 6y + 36 \\ \underline{y + 6} \end{array}$
25. $\begin{array}{l} 4x^2 + 8xy + y^2 \\ \underline{4x + y} \end{array}$
26. $\begin{array}{l} 9x^2 - 12xy + 4y^2 \\ \underline{3x - 2y} \end{array}$
27. $\begin{array}{l} 49r^2 - 14rs + 4s^2 \\ \underline{7r + 2s} \end{array}$
28. $\begin{array}{l} 100r^2 + 90rt + 81t^2 \\ \underline{10r - 9t} \end{array}$
29. $\begin{array}{l} x^2 - x + 1 \\ \underline{x^2 + x - 1} \end{array}$
30. $\begin{array}{l} x^2 + 3x + 2 \\ \underline{x^2 - 3x + 2} \end{array}$
31. $\begin{array}{l} x + y - 5 \\ \underline{x + y + 5} \end{array}$
32. $\begin{array}{l} x - y - 4 \\ \underline{x - y + 4} \end{array}$
33. $\begin{array}{l} a + b + c \\ \underline{a + b + c} \end{array}$
34. $\begin{array}{l} a - b - c \\ \underline{a - b - c} \end{array}$

35–70. Multiply by using the FOIL method.

35. $(x + 3)(x + 5)$
36. $(y + 2)(y + 7)$
37. $(x - 3)(x - 5)$
38. $(y - 2)(y - 7)$
39. $(x + 3)(x - 5)$
40. $(y + 2)(y - 7)$
41. $(x - 3)(x + 5)$
42. $(y - 2)(y + 7)$
43. $(2x + y)(x - 3y)$
44. $(3x - 2y)(x + 4y)$
45. $(3x - 4y)(3x - 2y)$
46. $(5x - 3y)(5x - y)$
47. $(x^2 - 4)(x^2 + 9)$
48. $(x^2 - 25)(x^2 + 1)$

49. $(x^2 - 36)(x^2 - 1)$
50. $(x^2 - 16)(x^2 - 100)$
51. $(y^2 - 25)(y^2 + 25)$
52. $(y^2 - 9)(y^2 + 9)$
53. $(3x + 2y)(3x - 2y)$
54. $(4x + 5y)(4x - 5y)$
55. $(3x + 2y)(2x - 3y)$
56. $(4x + 5y)(5x - 4y)$
57. $(x + 4)(y + 6)$
58. $(x + 3)(y + 8)$
59. $(x - 5)(2y - 1)$
60. $(x - 7)(3y - 2)$
61. $(x^2 - 1)(x + 1)$
62. $(x + 4)(x^2 - 9)$
63. $(y - 6)(y^2 + 4)$
64. $(y - 8)(y^2 + 1)$
65. $(4r - 1)(4s - 1)$
66. $(8u - 3)(8v + 3)$
67. $(x^2 + 2)(x^3 - 2)$
68. $(y^3 + 8)(y^2 - 8)$
69. $(8x + 3y)^2$
70. $(7x - 2y)^2$

71–90. Multiply and arrange in descending powers of x.

71. $(2 - x)(1 - x - 2x^2)$
72. $(x + 3)(x - 2 + x^2)$
73. $(3x - 1)(3x - 1 - x^2)$
74. $(1 - 4x)(1 - 2x - 4x^2)$
75. $(x^2 - 4)(x + x^2 - 2)$
76. $(x^2 + 9)(9 - x - x^2)$
77. $(x^2 + 1 - 3x)(x^2 - 1)$
78. $(2x - x^2 - 4)(1 - x^2)$
79. $(x - 1)(1 + x + x^2 + x^3)$
80. $(1 + x)(1 - x + x^2 - x^3)$
81. $(1 - x)^2$
82. $(3 - x)^2$
83. $(4 - 5x)^2$
84. $(1 - 2x)^2$
85. $(1 + x + x^2)^2$
86. $(1 - x - x^2)^2$
87. $(1 - x)(1 + x + x^2)$
88. $(1 + x)(1 - x + x^2)$
89. $(1 + x)(1 + 2x + x^2)$
90. $(1 - x)(1 - 2x + x^2)$

3.4 Monomial Factors

Since the sides of an equation may be exchanged without changing the truth (or falsity) of the equation, the distributive axiom may also be written in the following form:

$$AB + AC = A(B + C)$$

The product $A(B + C)$ is the factored form of $AB + AC$. *To factor an expression means to write it as a product of factors.*

Example 1 Factor $7x + 7y$.

SOLUTION Using the form $AB + AC = A(B + C)$ with $A = 7$, $B = x$, and $C = y$, then

$$7x + 7y = 7(x + y)$$

Example 2 Factor $2x^2 + 5x$.

SOLUTION Note that each term has x as a factor.

$$2x^2 + 5x = x(2x) + x(5) = x(2x + 5)$$

It is customary to write the common monomial factor, such as x in Example 2, at the left in the factored form.

It is desirable to express a polynomial as a product of the *greatest* common monomial factor and another polynomial. Thus, although

$$6x^4 + 12x^2 = 2x(3x^3 + 6x)$$

$2x$ is not the greatest common factor.
Note that

$$3x^3 + 6x = 3x(x^2 + 2)$$

and

$$\begin{aligned} 2x(3x^3 + 6x) &= 2x(3x)(x^2 + 2) \\ &= 6x^2(x^2 + 2) \end{aligned}$$

The greatest common factor is $6x^2$, so the desired factored form is

$$6x^4 + 12x^2 = 6x^2(x^2 + 2)$$

Factoring polynomials can be thought of as the inverse or opposite process of multiplying polynomials. For example, the problem

Multiply $6x^2(x^2 + 2)$

$$6x^2(x^2 + 2) = 6x^4 + 12x^2$$

is the opposite of the problem

Factor $6x^4 + 12x^2$

$$6x^4 + 12x^2 = 6x^2(x^2 + 2)$$

The following example illustrates the procedure for finding the greatest common monomial factor of a polynomial.

Example 3 Factor $15x^4 - 10x^3 + 5x^2$.

SOLUTION
Writing each term of the polynomial in factored form, it reads

$$3(5)xxxx - 2(5)xxx + 5xx$$

Now it is easy to identify the greatest common factor as $5xx = 5x^2$. Writing each term as the product of the greatest common factor and another factor,

$$\begin{aligned} 15x^4 - 10x^3 + 5x^2 &= \mathbf{5x^2}(3x^2) + \mathbf{5x^2}(-2x) + \mathbf{5x^2}(1) \\ &= \mathbf{5x^2}(3x^2 - 2x + 1) \end{aligned}$$

Note the necessity for writing 1 as the third term of the second factor. If this term had been omitted, the product of these two expressions would not have yielded the original polynomial.

A factoring problem can be checked by multiplication.

Example 4 Check the result of Example 3 by multiplication.

SOLUTION

$$\begin{aligned} 5x^2(3x^2 - 2x + 1) &= 5x^2(3x^2) + 5x^2(-2x) + 5x^2(1) \\ &= 15x^4 - 10x^3 + 5x^2 \end{aligned}$$

When the leftmost term of a polynomial written in descending powers of a variable begins with a minus sign, it is customary to include -1 as one of the factors of the greatest common monomial factor. The next example demonstrates this convention.

Example 5 Find the greatest common monomial factor of $-x^3 - x^2 + x$.

SOLUTION

$$\begin{aligned} -x^3 - x^2 + x &= (-1)x^3 + (-1)x^2 + (-1)(-x) \\ &= \mathbf{(-x)}x^2 + \mathbf{(-x)}x + \mathbf{(-x)}(-1) \\ &= \mathbf{(-x)}(x^2 + x - 1) \\ &= -x(x^2 + x - 1) \end{aligned}$$

CHECK

$$\begin{aligned} -x(x^2 + x - 1) &= (-x)(x^2) + (-x)(x) + (-x)(-1) \\ &= -x^3 - x^2 + x \end{aligned}$$

A factoring problem can also be checked by assigning a value to each variable and showing that the value of the expanded form is equal to the value of the factored form.

Example 6 Factor $8ax^3 - 6ax^2 + 12ax$ and check by letting $a = 5$ and $x = 2$.

SOLUTION

$$\begin{aligned} &8ax^3 - 6ax^2 + 12ax \\ &= 2ax(4x^2) - 2ax(3x) + 2ax(6) \\ &= 2ax(4x^2 - 3x + 6) \end{aligned}$$

CHECK For $a = 5$ and $x = 2$,

$$\begin{aligned} 8ax^3 - 6ax^2 + 12ax &= 8(5)(2^3) - 6(5)(2^2) + 12(5)(2) \\ &= 40(8) - 30(4) + 12(10) \\ &= 320 - 120 + 120 = 320 \\ 2ax(4x^2 - 3x + 6) &= 2(5)(2)(4 \cdot 2^2 - 3 \cdot 2 + 6) \\ &= 2(10)(16 - 6 + 6) \\ &= 20(16) = 320 \end{aligned}$$

It is so easy to make a mistake in algebraic work that good algebraic techniques must involve some type of checking.

Exercises 3.4

In Exercises 1–40 factor by finding the greatest common monomial factor.

1. $4x + 12$
2. $5x - 20$
3. $6x - 3$
4. $14x + 7$
5. $3x^2 + 6x + 9$
6. $5x^2 + 15xy + 5y^2$
7. $8ab - 12a + 4$
8. $6x + 3y - 3$
9. $4y^3 - 100y^2$
10. $4x^2 + 2x$
11. $ay - a$
12. $by + by^2$
13. $x^4 - x^3$
14. $x^3 - x^4$
15. $x^4 - x^5$
16. $x^5 - x^4$
17. $24x^3 - 30x^2 + x$
18. $n^3 - 3n^2 + n$
19. $c^2n^2 + cn^3$
20. $x^2y^2 - 3xy^2$
21. $ax + ay - az$
22. $x^3 - x^2 + x$
23. $-5x^2 - 25x$
24. $-36x^2 - 9x$
25. $24x^2 + 12x + 6$
26. $20y^2 - 8y + 12$
27. $c^3 - c^2 - c$
28. $x^2 + xy - x$

29. $6y^3 - 2y^2 + 10y$
30. $3t^3 + 3t^2 - 12t$
31. $-25x^3 - 15x$
32. $-8x^3 - 20x^2$
33. $-xy^3 - xy^2 - xy$
34. $-x^3 - x^2y - x^3y$
35. $-x^4 + 2x^3 - 6x^2$
36. $-5x^4 - 10x^3 + 25x^2$
37. $4x^2y - 4xy^2 - 4xy$
38. $9x^2y - 18xy^2 - 18xy$
39. $-6a^3b - 6ab^3 + 6a^2b^2$
40. $-8an^3 - 40an^2 + 32an$

In Exercises 41–50 factor by first replacing the binomial in parentheses by a single letter, as follows:

$$\begin{aligned} x(x-3) - y(x-3) &= xN - yN \text{ where } N = x - 3 \\ &= N(x-y) \\ &= (x-3)(x-y) \end{aligned}$$

41. $x(x + 6) + 3(x + 6)$
42. $x(x - 4) + 5(x - 4)$
43. $2x(x - 7) - 3(x - 7)$
44. $3x(x + 2) - 4(x + 2)$
45. $5x(x + 1) - (x + 1)$
46. $4x(x - 5) - (x - 5)$
47. $x^2(x - 4) + 4(x - 4)$
48. $x^2(x + 8) - 2(x + 8)$
49. $t(t^2 - 5) + (t^2 - 5)$
50. $t(t^2 + 6) - (t^2 + 6)$

3.5 Product of Sum and Difference; Factoring Difference of Squares

Some products occur so often that they deserve special attention. In this section we will consider the product of the sum and difference of the same terms.

As an example,

$(3x + 4y)(3x - 4y)$ is a product of a sum and difference of the same terms.

Multiplying by the FOIL method,

$$\begin{aligned} (3x + 4y)(3x - 4y) &= 3x(3x) + 3x(-4y) + 4y(3x) + 4y(-4y) \\ &= 9x^2 - 12xy + 12xy - 16y^2 \\ &= 9x^2 - 16y^2 \end{aligned}$$

Note that the outer and inner products are alike except for sign. This means they are additive inverses with a sum of zero.

In general,

$$\begin{aligned} (A + B)(A - B) &= A^2 + A(-B) + BA + B(-B) \\ &= A^2 - AB + AB - B^2 \\ &= A^2 - B^2 \end{aligned}$$

Since the product is the difference of the square of A and the square of B, this special product is called a **Difference of Squares.**

Theorem 1. Product of Sum and Difference

$$(A + B)(A - B) = A^2 - B^2$$

Theorem 1 is useful for finding quickly the product of a sum and difference of the same terms. We simply write the square of the first term minus the square of the second term.

Example 1 Multiply $(x - 3)(x + 3)$.

SOLUTION We see a product of a sum and difference of the same terms. Thus, Theorem 1 applies.

Note that $(x - 3)(x + 3) = (x + 3)(x - 3)$ by the commutative axiom for multiplication.

$$(A - B)(A + B) = A^2 - B^2$$
$$(x - 3)(x + 3) = x^2 - 3^2 = x^2 - 9$$

Example 2 Find the product $(2y + 5)(2y - 5)$.

SOLUTION Apply the theorem for the product of a sum and difference with $A = 2y$ and $B = 5$.

$$(2y + 5)(2y - 5) = (2y)^2 - 5^2 = 4y^2 - 25$$

Example 3 Multiply $(x^2 + 4y)(x^2 - 4y)$.

SOLUTION

$$\begin{aligned}(x^2 + 4y)(x^2 - 4y) &= (x^2)^2 - (4y)^2 \\ &= x^4 - 16y^2\end{aligned}$$

Example 4 Multiply $(9x^2 - 1)(9x^2 + 1)$.

SOLUTION

$$\begin{aligned}(9x^2 - 1)(9x^2 + 1) &= (9x^2)^2 - 1^2 \\ &= 81x^4 - 1\end{aligned}$$

Since $(A + B)(A - B) = A^2 - B^2$, it follows that

$$A^2 - B^2 = (A + B)(A - B)$$

This form of the equation is useful for factoring an expression that is the difference of squares.

Theorem 2. Difference of Squares

$$A^2 - B^2 = (A - B)(A + B)$$

Example 5 Factor $x^2 - 25$.

SOLUTION Since x^2 and 25 are both perfect squares and the binomial is of the form $A^2 - B^2$, it factors as $(A + B)(A - B)$, where $A^2 = x^2$ and $B^2 = 25 = 5^2$.

$$A^2 - B^2 = (A + B)(A - B)$$
$$x^2 - 25 = x^2 - 5^2 = (x + 5)(x - 5)$$

Example 6 Write $9x^2 - 4$ in factored form.

SOLUTION $9x^2$ is the square of $3x$, since $(3x)(3x) = 9x^2$; 4 is the square of 2. Therefore

$$\begin{aligned} 9x^2 - 4 &= (3x)^2 - (2)^2 \\ &= (3x + 2)(3x - 2) \end{aligned}$$

Example 7 Factor $4x^2 - 25y^2$.

SOLUTION $4x^2$ is the square of $2x$; $25y^2$ is the square of $5y$. Therefore

$$\begin{aligned} 4x^2 - 25y^2 &= (2x)^2 - (5y)^2 \\ &= (2x + 5y)(2x - 5y) \end{aligned}$$

Example 8 Factor $2x^2 - 50$.

SOLUTION The two terms have a common factor, 2, so find this common factor first:

$$\begin{aligned} 2x^2 - 50 &= 2(x^2 - 25) \\ &= 2(x - 5)(x + 5) \end{aligned}$$

Example 9 Factor $x^2 + 1$, if possible.

SOLUTION x^2 is the square of x; 1 is the square of 1. However, these squares are *not* separated by a minus sign, and they are therefore *not the difference* of two squares but their sum. Note that *none* of the following possibilities work:

$$\begin{aligned} (x + 1)(x + 1) &= x^2 + 2x + 1 \\ (x - 1)(x - 1) &= x^2 - 2x + 1 \\ (x + 1)(x - 1) &= x^2 - 1 \end{aligned}$$

Therefore, $x^2 + 1$ is prime and cannot be factored.

In general, a binomial of the form $x^2 + a^2$ cannot be expressed as the product of two linear factors. However, it may have a monomial factor. $x^2 + 4$ is prime, $x^2 + 9$ is prime, and so on. $4x^2 + 16 = 4(x^2 + 4)$, but $x^2 + 4$ is prime.

If all the numerical coefficients and constants of a factored expression are integers, the polynomial is said to be **factored over the integers.** All the examples worked thus far have been factored over the integers.

A polynomial that cannot be factored over the integers is said to be **prime** over the integers.

Exercises 3.5

1–20. Multiply by using Theorem 1 on the product of a sum and difference.

$$(A + B)(A - B) = A^2 - B^2$$

See Examples 1–4.

1. $(x + 4)(x - 4)$
2. $(x + 7)(x - 7)$
3. $(y - 5)(y + 5)$
4. $(y - 6)(y + 6)$
5. $(x + 3y)(x - 3y)$
6. $(x - 2y)(x + 2y)$
7. $(8x - 1)(8x + 1)$
8. $(9x + 1)(9x - 1)$
9. $(10x + 9y)(10x - 9y)$
10. $(8x + 5y)(8x - 5y)$
11. $(x^2 - 1)(x^2 + 1)$
12. $(x^2 - 5)(x^2 + 5)$
13. $(9x^2 + 1)(9x^2 - 1)$
14. $(4x^2 + 1)(4x^2 - 1)$
15. $(x^2 - y^2)(x^2 + y^2)$
16. $(x^2 - 3y^2)(x^2 + 3y^2)$
17. $(25x^2 - 4)(25x^2 + 4)$
18. $(10x^2 - 9)(10x^2 + 9)$
19. $(7a^2 + 6b^2)(7a^2 - 6b^2)$
20. $(20a^2 - 7b^2)(20a^2 + 7b^2)$

21–46. Factor over the integers, if possible. Check by multiplication. See Examples 5, 6, 7, and 9.

21. $x^2 - 81$
22. $x^2 - 9$
23. $y^2 - 4$
24. $y^2 - 16$
25. $25x^2 - 1$
26. $64x^2 - 1$
27. $49y^2 - 1$
28. $36y^2 - 1$
29. $x^2 - 5$
30. $x^2 + 4$
31. $x^2 + 9$
32. $x^2 - 2$
33. $36x^2 - 25$
34. $4x^2 - 49$
35. $64x^2 - 49y^2$
36. $81x^2 - 64y^2$
37. $x^4 - 9$
38. $x^4 - 25$
39. $25y^4 - 16$
40. $16y^4 - 121$
41. $49x^4 - 144y^4$
42. $100x^4 - 9y^4$
43. $4x^4 + 9y^4$
44. $25x^4 + 16$
45. $y^2 - 8$
46. $y^2 - 6$

47–60. First multiply the binomials. Then multiply the product by the monomial factor.

47. $3(x - 6)(x + 6)$
48. $5(x + 4)(x - 4)$
49. $x(x + 7)(x - 7)$
50. $x(x - 5)(x + 5)$
51. $2x(x + 9)(x - 9)$
52. $4x(x - 3)(x + 3)$

53. $y^2(y - 8)(y + 8)$
54. $y^2(y + 10)(y - 10)$
55. $4x^2(x - 5)(x + 5)$
56. $9x^2(x + 1)(x - 1)$
57. $-3y^2(y + 4)(y - 4)$
58. $-5y^2(y - 8)(y + 8)$
59. $16(x^2 - 1)(x^2 + 1)$
60. $25(x^2 - 2)(x^2 + 2)$

61–76. Factor over the integers, if possible. First remove any common monomial factor. Check by multiplication. See Example 8.

61. $4x^2 - 36$
62. $5x^2 - 80$
63. $25x^3 - 4x$
64. $64x^3 - 9x$
65. $4x^2 + 36$
66. $4x^2 + 16$
67. $45x^3 - 5x$
68. $100x^3 - 4x$
69. $36x^4 - 49x^2y^2$
70. $4x^2y^2 - 81y^4$
71. $25x^4 - 100$
72. $9x^4 - 81$
73. $25x^4 - 100x^2$
74. $9x^4 - 81x^2$
75. $25x^4 + 100x^2$
76. $9x^4 + 81x^2$

3.6 Product of Related Binomials; Factoring Simple Trinomials

When the first terms of two binomials are the same, there is a special form for the product of these binomials.

Consider $(x + 2)(x + 3)$. Note that the first term of each of the binomial factors is x.

Multiplying by the FOIL method,

$$\begin{aligned}(x + 2)(x + 3) &= x^2 + 3x + 2x + 2(3)\\ &= x^2 + (3 + 2)x + 3(2)\\ &= x^2 + 5x + 6\end{aligned}$$

Note that 5, the coefficient of x in the product, is the **sum** of 2 and 3.

Note also that 6, the constant term of the product, is the **product** of 2 and 3.

In general, we have the following special product.

Product of Related Binomials

$$(X + A)(X + B) = X^2 + (A + B)X + AB$$

By using this special product, when it applies, we can quickly find the product of two related binomials.

Example 1 Find the product $(x + 7)(x + 5)$.

SOLUTION

$$\begin{aligned}(X + A)(X + B) &= X^2 + (A + B)X + AB \\ (x + 7)(x + 5) &= x^2 + (7 + 5)x + 7(5) \\ &= x^2 + 12x + 35\end{aligned}$$

Example 2 Multiply $(x^2 - 4)(x^2 - 9)$.

SOLUTION The special product form applies, with $X = x^2$ and $A = -4$ and $B = -9$.

$$\begin{aligned}(x^2 - 4)(x^2 - 9) &= (x^2)^2 + (-4 - 9)x^2 + (-4)(-9) \\ &= x^4 - 13x^2 + 36\end{aligned}$$

Example 3 Multiply $(x + 4y)(x - 6y)$.

SOLUTION This time $X = x$ and $A = 4y$ and $B = -6y$.

$$\begin{aligned}(x + 4y)(x - 6y) &= x^2 + (4y - 6y)x + (4y)(-6y) \\ &= x^2 + (-2y)x + (-24y^2) \\ &= x^2 - 2xy - 24y^2\end{aligned}$$

A trinomial having the form $X^2 + (A + B)X + AB$ is called a simple trinomial. Its factors are given by the special product for related binomials.

Simple Trinomial Formula

$$X^2 + (A + B)X + AB = (X + A)(X + B)$$

To factor a simple trinomial, we have to find two numbers whose product is the last term and whose sum is the coefficient of X.

Example 4 Factor $x^2 + 7x + 10$.

SOLUTION

1. We start by knowing that the factors have the form $(x + a)(x + b)$.
2. We need to find two integers whose product is 10 and whose sum is 7.

   $$ab = 10 \quad \text{and} \quad a + b = 7$$

   $$10 = 1(10) \quad \text{or} \quad 10 = 2(5)$$

   Thus ($a = 1$ and $b = 10$) or ($a = 2$ and $b = 5$).
3. We test each pair to see which, if any, will work.

   $$(x + 1)(x + 10) = x^2 + 11x + 10 \quad \text{(No)}$$

   $$(x + 2)(x + 5) = x^2 + 7x + 10 \quad \text{(Yes)}$$

4. Therefore, $x^2 + 7x + 10 = (x + 2)(x + 5)$.

Example 5 Factor $x^2 - 8x + 15$.

SOLUTION

1. The factored form is $(x + a)(x + b)$, where $ab = +15$ and $a + b = -8$.
2. Finding the pairs of factors of $+15$,

$$ab = (-1)(-15) \quad \text{or} \quad ab = (-3)(-5)$$

   Note that both factors must be negative, since the sum is negative and the product is positive.
3. Forming each possibility and multiplying to find the correct set of factors,

$$(x - 1)(x - 15) = x^2 - 16x + 15$$
$$(x - 3)(x - 5) = x^2 - 8x + 15$$

4. Therefore, $x^2 - 8x + 15 = (x - 3)(x - 5)$.

Note, for Example 5, that also $x^2 - 8x + 15 = (x - 5)(x - 3)$. This is so because $(x - 3)(x - 5) = (x - 5)(x - 3)$ by the commutative axiom for multiplication. In other words, the order in which the factors are written does not change the product.

Example 6 Factor $x^2 + 4x - 5$.

SOLUTION

1. The factored form is $(x + a)(x + b)$, where the product $ab = -5$ and the sum $a + b = 4$.
2. Since the product is negative, one factor will be positive and the other negative.

$$ab = (-1)(+5) \quad \text{or} \quad ab = (+1)(-5)$$

3. Testing the possible binomial factors,

$$(x - 1)(x + 5) = x^2 + 4x - 5$$
$$(x + 1)(x - 5) = x^2 - 4x - 5$$

4. Thus, $x^2 + 4x - 5 = (x - 1)(x + 5)$.

Example 7 Factor $x^2 - 10x - 24$ over the integers and check by the FOIL method.

SOLUTION

The integral factors of -24 are $-1(24)$, $1(-24)$, $-2(12)$, $2(-12)$, $-3(8)$, $3(-8)$, $-4(6)$, $4(-6)$. The only pair of factors whose sum is -10 is 2 and -12. Therefore

continued

$$x^2 - 10x - 24 = (x + 2)(x - 12)$$

Also

$$x^2 - 10x - 24 = (x - 12)(x + 2)$$

CHECK

$$\begin{aligned}(x + 2)(x - 12) &= \overset{F}{x^2} + \overset{O}{(-12x)} + \overset{I}{2x} + \overset{L}{2(-12)} \\ &= x^2 - 10x - 24\end{aligned}$$

$$x^2 - 10x - 24 = (x + 2)(x - 12)$$

Example 8 Factor $x^2 + 3xy - 10y^2$ over the integers and check by the FOIL method.

SOLUTION

1. Comparing the forms:

$$X^2 + (A + B)X + AB = (X + A)(X + B)$$
$$x^2 + 3xy - 10y^2 = (x\ ?\ y)(x\ ?\ y)$$

   It is seen that A and B must have the forms py and qy where $p + q = 3$ and $pq = -10$.
2. The pairs of factors of -10 are $(-1)(10)$, $(10)(-1)$, $(-2)(5)$, and $(2)(-5)$. Since $(-2) + (5) = 3$, select $p = -2$ and $q = 5$.
3. Then $x^2 + 3xy - 10y^2 = (x - 2y)(x + 5y)$.

   Also, by the commutative axiom for multiplication,

$$x^2 + 3xy - 10y^2 = (x + 5y)(x - 2y)$$

CHECK

$$\begin{aligned}(x - 2y)(x + 5y) &= \overset{F}{x^2} + \overset{O}{5xy} - \overset{I}{2xy} - \overset{L}{10y^2} \\ &= x^2 + 3xy - 10y^2\end{aligned}$$

Example 9 Factor $x^2 + x + 1$ over the integers, if possible.

SOLUTION Comparing the forms:

$$X^2 + (A + B)X + AB = (X + A)(X + B)$$

it is seen that $A + B = 1$ and $AB = 1$. The only pairs of factors of 1 that are integers are $(1)(1)$ and $(-1)(-1)$; in other words, $A = 1$ and $B = 1$ or $A = -1$ and $B = -1$. But

$$A + B = 1 + 1 = 2 \quad \text{or} \quad A + B = -1 + (-1) = -2$$

and according to the problem, $A + B = 1$. Therefore, $x^2 + x + 1$ cannot be factored over the integers.
$x^2 + x + 1$ is prime over the integers.

Exercises 3.6

1–20. Multiply by using the special product formula for related binomials. See Examples 1–3.

1. $(x + 6)(x + 3)$
2. $(x + 2)(x + 7)$
3. $(x - 3)(x - 6)$
4. $(x - 7)(x - 2)$
5. $(x - 3)(x + 6)$
6. $(x - 7)(x + 2)$
7. $(x + 3)(x - 6)$
8. $(x + 7)(x - 2)$
9. $(x + 5y)(x - y)$
10. $(x + y)(x - 6y)$
11. $(x - 3y)(x - 8y)$
12. $(x - 5y)(x - 4y)$
13. $(x^2 - 4)(x^2 - 25)$
14. $(x^2 - 1)(x^2 - 9)$
15. $(x^2 + 4)(x^2 - 25)$
16. $(x^2 + 1)(x^2 - 9)$
17. $(y^2 - 9)(y^2 + 16)$
18. $(y^2 - 4)(y^2 + 8)$
19. $(x^2 + y^2)(x^2 - 2y^2)$
20. $(x^2 - y^2)(x^2 + 3y^2)$

21–30. Supply the missing factor.

21. $x^2 + 8x + 15 = (x + 5)(\quad\quad)$
22. $x^2 + 10x + 24 = (x + 4)(\quad\quad)$
23. $x^2 - 8x + 15 = (x - 5)(\quad\quad)$
24. $x^2 - 10x + 24 = (x - 4)(\quad\quad)$
25. $x^2 - 2x - 15 = (x - 5)(\quad\quad)$
26. $x^2 + 2x - 24 = (x - 4)(\quad\quad)$
27. $x^2 + 7x - 8 = (x + 8)(\quad\quad)$
28. $x^2 + 8x - 9 = (x + 9)(\quad\quad)$
29. $x^2 - 9xy - 10y^2 = (x - 10y)(\quad\quad)$
30. $x^2 - xy - 2y^2 = (x - 2y)(\quad\quad)$

31–80. Factor over the integers, if possible. Check by the FOIL method. See Examples 4–9.

31. $x^2 + 6x + 5$
32. $x^2 + 8x + 7$
33. $x^2 - 6x + 5$
34. $x^2 - 8x + 7$
35. $x^2 - 4x - 5$
36. $x^2 - 6x - 7$
37. $x^2 + 4x - 5$
38. $x^2 + 6x - 7$
39. $x^2 + 5x - 4$
40. $x^2 - 7x - 6$
41. $y^2 + 5y - 6$
42. $y^2 + 8y + 15$
43. $y^2 - 7y + 6$
44. $y^2 - 16y + 15$
45. $y^2 - y - 6$
46. $y^2 + 2y - 15$
47. $y^2 - 5y + 6$
48. $y^2 - 14y - 15$

49. $x^2 - 2x + 3$
50. $x^2 + 3x - 2$
51. $x^2 - 3x + 2$
52. $x^2 + 2x - 3$
53. $x^2 - 2xy - 3y^2$
54. $x^2 - 3xy + 2y^2$
55. $x^2 + 3xy - 10y^2$
56. $x^2 - 2xy - 35y^2$
57. $x^2 - 10xy + 21y^2$
58. $x^2 + 2xy - 8y^2$
59. $x^2 + 13xy - 14y^2$
60. $x^2 + 6xy - 27y^2$
61. $x^2 - 2xy + 4y^2$
62. $x^2 + 2xy + y^2$
63. $x^2 - 4xy + 4y^2$
64. $x^2 + xy + y^2$
65. $x^2 - 6x + 9$
66. $x^2 - 10x - 25$
67. $x^2 + 6x - 9$
68. $x^2 - 10x + 25$
69. $2 + x - x^2$
70. $4 - 3x - x^2$
71. $x^4 + 3x^2 + 2$
72. $x^4 + 8x^2 + 7$
73. $x^4 + 5x^2 - 14$
74. $x^4 - 20x^2 - 21$
75. $x^4 - 7x^2 - 8$
76. $x^4 + 4x^2 - 12$
77. $x^4 - 12x^2 + 36$
78. $x^4 - 16x^2 + 64$
79. $y^4 + y^2 - 12$
80. $y^4 - y^2 - 30$

In Exercises 81–100, factor over the integers by first removing the greatest common monomial factor and then factoring the general trinomial factor, if possible.

81. $3x^2 - 9x + 6$
82. $10x^2 - 20x - 350$
83. $150x^4 - 1050x^3 + 1500x^2$
84. $4a^3y + 4a^2y - 48ay$
85. $20y^2 + 140y - 120$
86. $35b^2 - 30b - 5b^3$
87. $4x^2 + 32x + 60$
88. $5x^2 + 30x - 35$
89. $100x^3 - 200x^2 - 800x$
90. $6y^4 - 24y^3 - 72y^2$
91. $15y^2 + 30y + 45$
92. $84 + 7a - 7a^2$
93. $20cd - cd^2 - cd^3$
94. $30u^2 + 150u - 120$
95. $2x^3 + 8x^2y - 120xy^2$
96. $5uv - uv^3 - u^3v$
97. $28nk^2 - 3nk^3 - nk^4$
98. $x^3y - 2x^2y^2 - 35xy^3$
99. $u^4 + 2u^3v - 3u^2v^2$
100. $18d^2 + 18dx + 36x^2$

3.7 Perfect Square Trinomials

Square of Binomial

The square of a binomial is a polynomial having three terms, called a **perfect square trinomial.**

Using the distributive axiom,

$$\begin{aligned}(A + B)^2 &= (A + B)(A + B)\\ &= (A + B)A + (A + B)B\\ &= A^2 + BA + AB + B^2\\ &= A^2 + 2AB + B^2\end{aligned}$$

The **Perfect Square Trinomial formula** is

$$(A + B)^2 = A^2 + 2AB + B^2$$

This formula shows that the square of a binomial is obtained by squaring the first term, doubling the product of the two terms, and squaring the last term, and then adding these three products.

Example 1 Use the formula to obtain the product $(6x + 7)^2$.

SOLUTION Use $A = 6x$ and $B = 7$.

$$\begin{aligned}(A + B)^2 &= A^2 + 2AB + B^2\\ (6x + 7)^2 &= (6x)^2 + 2(6x)(7) + 7^2\\ &= 36x^2 + 84x + 49\end{aligned}$$

Example 2 Use the formula to obtain the product $(3x - 5y)^2$.

SOLUTION Use $A = 3x$ and $B = -5y$.

$$\begin{aligned}(A + B)^2 &= A^2 + 2AB + B^2\\ (3x - 5y)^2 &= (3x)^2 + 2(3x)(-5y) + (-5y)^2\\ &= 9x^2 - 30xy + 25y^2\end{aligned}$$

When a trinomial can be recognized as a perfect square trinomial, then it can be factored by using the Perfect Square Trinomial form.

Perfect Square Trinomial Form

$$A^2 + 2AB + B^2 = (A + B)^2$$
$$A^2 - 2AB + B^2 = (A - B)^2$$

To recognize a perfect square trinomial

1. There must be three terms.
2. Two terms must be perfect square monomials: $(A)^2$ and $(B)^2$.
3. There must not be a minus sign before either A^2 or B^2.

4. The third term must be *twice* the product of A and B, where A^2 and B^2 are the perfect square monomials.

The following table may be useful. Also see the Table of Powers and Roots inside the front cover.

N	x	x^2	x^3	1	2	3	4	5	6	7	8	9	10
N^2	x^2	x^4	x^6	1	4	9	16	25	36	49	64	81	100

Example 3 Is $x^2 + 25$ a perfect square trinomial?

SOLUTION No, because it is not a trinomial. It has only two terms instead of three.

Example 4 Is $x^2 + 14x - 49$ a perfect square?

SOLUTION No, because the third term is *not* positive.

Example 5 Is $y^2 + 3y + 9$ a perfect square?

SOLUTION The only possibility is $(y + 3)^2$. But

$$(y + 3)^2 = y^2 + 2(3)y + 9 = y^2 + 6y + 9$$

Thus $y^2 + 3y + 9$ is not a perfect square. The middle term is not **twice** the product of y and 3.

Example 6 Is $x^2 + 16 - 8x$ a perfect square?

SOLUTION First, rearranging the polynomial in descending powers of x,

$$x^2 - 8x + 16 = x^2 - 8x + 4^2 = (x - 4)^2$$

The answer is yes because $2x(-4) = -8x$.

Example 7 Fill in the missing term so that $x^2 + (\quad) + 9y^2$ is a perfect square trinomial.

SOLUTION Using the form

$$A^2 + 2AB + B^2 = (A + B)^2$$
$$x^2 + (\quad) + 9y^2 = x^2 + (\quad) + (3y)^2 = (x + 3y)^2$$
$$A = x \quad \text{and} \quad B = 3y. \quad \text{Thus} \quad 2AB = 2x(3y) = 6xy$$

The missing term is $6xy$ and

$$x^2 + 6xy + 9y^2 = (x + 3y)^2$$

Example 8 Factor $x^2 + 10x + 25$

SOLUTION Using the form

$$A^2 + 2AB + B^2 = (A + B)^2$$
$$x^2 + 10x + 25 = x^2 + 2(5x) + 5^2 = (x + 5)^2$$

Example 9 Factor $4y^2 - 12y + 9$

SOLUTION

$$\begin{aligned} & \quad\; A^2 \quad + 2AB \qquad\quad + B^2 \\ 4y^2 - 12y + 9 &= (2y)^2 + 2(-6y) \qquad + 3^2 \\ &= (2y)^2 + 2(-3)(2y) + (-3)^2 = (2y + [-3])^2 \\ &= (2y - 3)^2 \end{aligned}$$

Exercises 3.7

1–30. Use the Perfect Square Trinomial formula to square each binomial. See Examples 1 and 2.

$$(A + B)^2 = A^2 + 2AB + B^2$$

1. $(x + 3)^2$
2. $(x + 4)^2$
3. $(x - 2)^2$
4. $(x - 5)^2$
5. $(y + 6)^2$
6. $(y - 7)^2$
7. $(y - 9)^2$
8. $(y + 8)^2$
9. $(4x - 1)^2$
10. $(3x + 1)^2$
11. $(5x + 1)^2$
12. $(2x - 1)^2$
13. $(2x + 7)^2$
14. $(6x + 5)^2$
15. $(3x - 8)^2$
16. $(9x - 4)^2$
17. $(x + y)^2$
18. $(x - y)^2$
19. $(10x - 7y)^2$
20. $(4x - 3y)^2$
21. $(x^2 - 10)^2$
22. $(x^2 + 3)^2$
23. $(x^2 + 1)^2$
24. $(x^2 - 4)^2$
25. $(x^3 - 8)^2$
26. $(x^3 - 1)^2$
27. $(4x^2 - 9y^2)^2$
28. $(25x^2 - 4y^2)^2$
29. $(12x^2 - y^2)^2$
30. $(100x^2 - y^2)^2$

31–50. Fill in the missing term so that the result is a perfect square trinomial. See Examples 4–7.

31. $x^2 + (\qquad) + 36$
32. $x^2 + (\qquad) + 100$
33. $t^2 - (\qquad) + 49$
34. $t^2 - (\qquad) + 25$
35. $y^2 - (\qquad) + 16$
36. $y^2 - (\qquad) + 400$
37. $y^2 + (\qquad) + 81$
38. $y^2 + (\qquad) + 121$
39. $x^4 + (\qquad) + 4$
40. $x^4 + (\qquad) + 1$
41. $y^4 - (\qquad) + 64$
42. $y^4 - (\qquad) + 9$
43. $x^2 + (\qquad) + 900y^2$
44. $x^2 + (\qquad) + 1600y^2$

45. $x^2 - (\quad) + y^2$
46. $x^2 - (\quad) + 36y^2$
47. $4x^2 + (\quad) + 9y^2$
48. $16x^2 - (\quad) + 25y^2$
49. $100t^2 - (\quad) + 49$
50. $81t^2 + (\quad) + 64$

51–60. Which of the following are perfect square trinomials?

51. $x^2 + 1$
52. $y^2 + 4$
53. $y^2 + 8y + 16$
54. $x^2 + 14x - 49$
55. $t^2 + 16t - 64$
56. $t^2 - 18t + 36$
57. $x^2 + 4x + 8$
58. $a^2 + 18a + 81$
59. $x^2 + 25 - 10x$
60. $x^2 + 9 - 6x$

61–100. Factor over the integers, if possible. Check by multiplication. See Examples 4–9.

61. $x^2 + 4x + 4$
62. $x^2 + 18x + 81$
63. $x^2 - 2x + 1$
64. $x^2 - 10x + 25$
65. $9y^2 + 6y + 1$
66. $16y^2 + 8y + 1$
67. $36t^2 - 12t + 1$
68. $49t^2 - 14t + 1$
69. $16x^2 - 40x + 25$
70. $9x^2 - 48x + 64$
71. $y^2 - 17y + 16$
72. $y^2 - 10y + 9$
73. $x^2 + 49$
74. $x^2 - 4$
75. $x^2 - 49$
76. $x^2 + 4$
77. $4x^2 + 12xy + 9y^2$
78. $25x^2 + 20xy + 4y^2$
79. $16 - 8t + t^2$
80. $36 - 12t + t^2$
81. $25y^2 - 80xy + 64x^2$
82. $9y^2 - 42xy + 49x^2$
83. $u^2 + 3u + 9$
84. $u^2 + 10u + 16$
85. $u^2 + 10u + 9$
86. $u^2 + 4u + 16$
87. $x^4 - 20x^2 + 100$
88. $x^4 + 60x^2 + 900$
89. $x^4 + 12x^2 + 36$
90. $x^4 - 16x^2 + 64$
91. $x^4 + 13x^2 + 36$
92. $x^4 - 34x^2 + 64$
93. $121y^4 - 22y^2 + 1$
94. $144y^4 - 24y^2 + 1$
95. $t^2 + 81$
96. $t^2 - 52t + 100$
97. $t^2 - 30t + 81$
98. $t^2 + 15t - 100$
99. $t^2 + 80t - 81$
100. $t^2 - 10t + 100$

3.8 Factoring General Trinomials

The trinomials factored so far either were perfect square trinomials or the coefficient of the term of greatest degree was 1. This section deals with factoring general types of trinomials. We begin by examining the product of two binomials.

$$\begin{aligned}(2x + 1)(5x + 3) &= 2x(5x) + 2x(3) + 1(5x) + 1(3)\\ &= 10x^2 + (6 + 5)x + 3\\ &= 10x^2 + 11x + 3\end{aligned}$$

Factoring $10x^2 + 11x + 3$ is the reverse of this process. We need to find two binomial factors whose product is $10x^2 + 11x + 3$. We note the following:

1. The x^2 term is the product of the x terms of the factors.

   $$10x^2 = 2x(5x)$$

2. The constant term is the product of the constant terms.

   $$3 = 1(3)$$

3. The x term is the sum of the inner and outer products.

   $$11x = 6x + 5x$$

We have the following general statement.

General Trinomial Form

$AX^2 + BX + C = (aX + b)(cX + d)$ if and only if

$$A = ac \quad \text{and} \quad C = bd \quad \text{and} \quad B = ad + bc$$

Factoring a general trinomial is a trial-and-error process, as follows.

1. Find the pairs of factors of A.
2. Find the pairs of factors of C.
3. List the possible pairs of binomial factors.
4. Multiply each pair of binomial factors to see which, if any, has a product equal to the given trinomial.

Example 1 Factor $5x^2 + 17x + 6$.

SOLUTION The factored form is to be $(ax + b)(cx + d)$.

1. $ac = 5 = (1)(5)$ (Product of the first terms' coefficients)
2. $bd = 6 = (1)(6)$ (Product of the last terms' coefficients)
   $= (2)(3)$

continued

3. The possible pairs of binomial factors are

$$(x+1)(5x+6) \qquad (x+2)(5x+3)$$
$$(x+6)(5x+1) \qquad (x+3)(5x+2)$$

4. Multiplying each possible pair of factors,

$$(x+1)(5x+6) = 5x^2 + 11x + 6 \quad \text{(No, wrong middle term)}$$
$$(x+6)(5x+1) = 5x^2 + 31x + 6 \quad \text{(No, wrong middle term)}$$
$$(x+2)(5x+3) = 5x^2 + 13x + 6 \quad \text{(No, wrong middle term)}$$
$$(x+3)(5x+2) = 5x^2 + 17x + 6 \quad \text{(Yes, right middle term)}$$

Therefore,

$$5x^2 + 17x + 6 = (x+3)(5x+2)$$

It is very important to **check that all possible binomial factors are listed.** Overlooking a pair may mean losing the factored form.

It is also important to **select the correct possibility.** Note that the product of each pair of binomials has the same first terms product and the same last terms product. The middle term of each is different. One and only one set of prime factors is correct (except for the order in which they are written). For example, $(x + 3)(5x + 2)$ is the same set of factors as $(5x + 2)(x + 3)$, since $(x + 3)(5x + 2) = (5x + 2)(x + 3)$ by the commutative axiom for multiplication.

Example 2 Factor $6x^2 - 17x + 7$.

SOLUTION The factored form is to be $(ax + b)(cx + d)$.

1. The pairs of factors of 6 are (1)(6) or (2)(3).
   The possible forms are $(x + b)(6x + d)$ or $(2x + b)(3x + d)$.
2. The pairs of factors of 7 are (1)(7) and $(-1)(-7)$.
3. The possible binomial factors are

$$(x-1)(6x-7) \qquad (2x-1)(3x-7)$$
$$(x-7)(6x-1) \qquad (2x-7)(3x-1)$$

   Note the minus signs. This must be the case because the product bd is positive and the sum of the inner and outer products is negative.

4. Multiplying,

$$(x-1)(6x-7) = 6x^2 - 13x + 7 \quad \text{(No)}$$
$$(x-7)(6x-1) = 6x^2 - 43x + 7 \quad \text{(No)}$$
$$(2x-1)(3x-7) = 6x^2 - 17x + 7 \quad \text{(Yes, right middle term)}$$
$$(2x-7)(3x-1) = 6x^2 - 23x + 7 \quad \text{(No)}$$

Therefore,

$$6x^2 - 17x + 7 = (2x-1)(3x-7)$$

and $6x^2 - 17x + 7 = (3x-7)(2x-1)$

Note in Example 2 that the last multiplication listed was not necessary because the correct set of factors had been found. If we are lucky, we find the correct set on the first few trials. If we are unlucky, we find the correct set on the last trial.

Example 3 Factor $8y^2 - 10y - 7$.

SOLUTION The factored form is to be $(ax + b)(cx + d)$.

1. The pairs of factors of 8 are (1)(8) and (2)(4).
2. The pairs of factors of -7 are $(-1)(7)$ and $(1)(-7)$.
3. The possible binomial factors are

$$\begin{array}{ll} (y-1)(8y+7) & (2y-1)(4y+7) \\ (y+7)(8y-1) & (2y+7)(4y-1) \\ (y+1)(8y-7) & (2y+1)(4y-7) \\ (y-7)(8y+1) & (2y-7)(4y+1) \end{array}$$

4. After several trial multiplications, we find that

$$\begin{aligned}(2y+1)(4y-7) &= 2y(4y) + 2y(-7) + 1(4y) + 1(-7) \\ &= 8y^2 - 10y - 7\end{aligned}$$

Therefore,

$$8y^2 - 10y - 7 = (2y + 1)(4y - 7)$$

Example 4 Factor $6x^2 + x - 15$.

SOLUTION

1. The factored form is $(2x \quad)(3x \quad)$ or $(x \quad)(6x \quad)$.
2. The pairs of factors of -15 are $(-1)(15)$, $(1)(-15)$, $(-3)(5)$, or $(3)(-5)$.
3. The possible binomial factors are

$$\begin{array}{ll} (2x+5)(3x-3) & (6x+5)(x-3) \\ (2x-5)(3x+3) & (6x-5)(x+3) \\ (2x+3)(3x-5) & (6x+3)(x-5) \\ (2x-3)(3x+5) & (6x-3)(x+5) \\ (2x+15)(3x-1) & (6x+15)(x-1) \\ (2x-15)(3x+1) & (6x-15)(x+1) \\ (2x+1)(3x-15) & (6x+1)(x-15) \\ (2x-1)(3x+15) & (6x-1)(x+15) \end{array}$$

4. After several multiplications, we find the factored form.

$$6x^2 + x - 15 = (2x - 3)(3x + 5)$$

Example 5 Factor $5x^2 + 6x - 1$ over the integers, if possible.

SOLUTION If the trinomial can be factored, the possibilities are

$$(x + 1)(5x - 1) \quad \text{and} \quad (x - 1)(5x + 1)$$

continued

Trying these,

$$\begin{aligned}(x+1)(5x-1) &= x(5x-1)+(1)(5x-1)\\ &= 5x^2-x+5x-1\\ &= 5x^2+4x-1\\ (x-1)(5x+1) &= x(5x+1)+(-1)(5x+1)\\ &= 5x^2+x-5x-1\\ &= 5x^2-4x-1\end{aligned}$$

Since none of these possibilities works and since they are the only possibilities, $5x^2+6x-1$ *cannot* be factored over the integers; it is **prime.**

Exercises 3.8

Supply the missing factor in Exercises 1–20.

Example 6

$6x^2 - 7x - 20 = (3x + 4)(\quad\quad)$

1. Check first terms product. $3x(\quad) = 6x^2$
   $3x(\mathbf{2x}) = 6x^2$
   $2x$ is the first term.
2. Check last terms product. $4(\quad) = -20$
   $4(\mathbf{-5}) = -20$
   -5 is the last term.
3. Supply the missing factor. $(3x + 4)(\mathbf{2x - 5})$

1. $2x^2 - 9x - 110 = (x - 10)(\quad\quad)$
2. $6x^2 - 7x - 20 = (3x + 4)(\quad\quad)$
3. $6x^2 - x - 2 = (2x + 1)(\quad\quad)$
4. $3y^2 + 25y - 50 = (y + 10)(\quad\quad)$
5. $12y^2 - 17y - 5 = (4y + 1)(\quad\quad)$
6. $8a^2 - 26a + 21 = (2a - 3)(\quad\quad)$
7. $10x^2 + 23x - 5 = (2x + 5)(\quad\quad)$
8. $10x^2 + 27x + 5 = (2x + 5)(\quad\quad)$
9. $10x^2 - 23x - 5 = (2x - 5)(\quad\quad)$
10. $10x^2 - 27x + 5 = (2x - 5)(\quad\quad)$
11. $6x^2 + 7x - 20 = (2x + 5)(\quad\quad)$
12. $8a^2 - 10a - 3 = (2a - 3)(\quad\quad)$
13. $3y^2 + 10y - 88 = (3y + 22)(\quad\quad)$
14. $6x^2 + x - 2 = (2x - 1)(\quad\quad)$
15. $10p^2 + 21p - 10 = (2p + 5)(\quad\quad)$

16. $6x^2 + x - 40 = (3x + 8)(\quad)$
17. $4x^2 - 21x - 18 = (4x + 3)(\quad)$
18. $4x^2 + 27x + 18 = (4x + 3)(\quad)$
19. $4x^2 + 21x - 18 = (4x - 3)(\quad)$
20. $4x^2 - 27x + 18 = (4x - 3)(\quad)$

21–30. Supply the missing signs.

Example 7

$$8x^2 + 2x - 15 = (4x \quad 5)(2x \quad 3)$$

1. Form outer product. $4x(3x) = 12x$
2. Form inner product. $5(2x) = 10x$
3. Select signs for products, so sum is given middle term. $+12x$, $-10x$, sum $+2x$
4. Insert signs to produce products of step 3. $8x^2 + 2x - 15 = (4x - 5)(2x + 3)$

21. $2x^2 - 11x + 15 = (2x \quad 5)(x \quad 3)$
22. $5x^2 - 28x + 15 = (5x \quad 3)(x \quad 5)$
23. $6x^2 - 11x - 10 = (3x \quad 2)(2x \quad 5)$
24. $8x^2 + 6x - 9 = (4x \quad 3)(2x \quad 3)$
25. $30x^2 + 7x - 2 = (6x \quad 1)(5x \quad 2)$
26. $18x^2 - 3x - 10 = (6x \quad 5)(3x \quad 2)$
27. $24x^2 + x - 10 = (8x \quad 5)(3x \quad 2)$
28. $6x^2 + x - 15 = (3x \quad 5)(2x \quad 3)$
29. $35x^2 - x - 12 = (7x \quad 4)(5x \quad 3)$
30. $12x^2 - x - 6 = (3x \quad 2)(4x \quad 3)$

31–80. Factor over the integers, if possible. See Examples 1–5.

31. $2x^2 + 3x + 1$
32. $3x^2 + 4x + 1$
33. $3x^2 - 5x + 2$
34. $2x^2 - 7x + 3$
35. $5x^2 + 3x - 2$
36. $7x^2 + 6x - 1$
37. $7x^2 - 13x - 2$
38. $5x^2 - 14x - 3$
39. $2x^2 + 3x - 1$
40. $3x^2 + 2x + 1$
41. $3y^2 + 4y - 7$
42. $5y^2 + 9y - 2$
43. $3y^2 - 20y - 7$
44. $5y^2 - 3y - 2$
45. $3y^2 + 4y + 7$
46. $5y^2 + 11y + 2$
47. $3y^2 - 22y + 7$
48. $5y^2 - 9y + 2$
49. $4t^2 - 21t + 5$
50. $6t^2 + 11t - 7$
51. $4t^2 + 12t + 9$
52. $6t^2 - t - 7$

53. $4t^2 - 8t - 5$
54. $6t^2 + t + 7$
55. $4t^2 - 6t + 9$
56. $9t^2 - 24t + 16$
57. $2u^2 - u - 1$
58. $5u^2 + u - 4$
59. $9x^2 + 21xy - 8y^2$
60. $4x^2 - 11xy + 6y^2$
61. $10x^2 + 11x - 6$
62. $15x^2 - 2x - 8$
63. $21x^2 - 2x - 8$
64. $14x^2 + x - 4$
65. $12x^2 - 23x + 5$
66. $12x^2 - 16x + 5$
67. $25x^2 - 30x + 9$
68. $16x^2 - 40x + 25$
69. $25x^2 - 34xy + 9y^2$
70. $16x^2 - 41xy + 25y^2$
71. $4x^4 + 17x^2 + 4$
72. $4x^4 + 37x^2 + 9$
73. $6x^4 - 7x^2 - 5$
74. $2x^4 + x^2 - 1$
75. $3x^4 - 4x^2 - 7$
76. $6x^4 + 7x^2 - 5$
77. $36t^4 - 12t^2 + 1$
78. $64t^4 - 80t^2 + 25$
79. $2t^4 + t^2 - 45$
80. $8t^4 - 22t^2 - 21$

81–96. Factor over the integers, if possible. First find any common monomial factor.

81. $6x^2 + 8x + 2$
82. $12y^2 - 12y + 3$
83. $20y^2 + 5y - 25$
84. $16x^2 + 8x - 24$
85. $81t^2 - 108t + 36$
86. $16x^2 - 120x + 56$
87. $10x^4 - 7x^3 + x^2$
88. $6x^4 + 12x^3 - 21x^2$
89. $30x^2 - 12x^3 - 6x^4$
90. $2x^2 + 2x^3 - 60x^4$
91. $16x^2 - 40xy - 24y^2$
92. $50x^2 - 260xy + 50y^2$
93. $2x^3y + 6x^2y^2 - 20xy^3$
94. $4x^3y + 4x^2y^2 - 80xy^3$
95. $4a^4 + 8a^2b^2 - 4a^3b$
96. $25a^4 + 25a^2b^2 - 100a^3b$
97. For what integers k is $x^2 + kx + 4$ factorable over the integers?
98. For what integers k is $x^2 + kx - 12$ factorable over the integers?
99. For what integers k is $2x^2 + kx - 5$ factorable over the integers?
100. For what integers k is $5x^2 + kx + 9$ factorable over the integers?

3.9 Complete Factoring

A polynomial is said to be **completely factored over the integers** when all common factors have been removed and when no further factoring over the integers is possible. For example, $2x^2 - 8 = 2(x^2 - 4)$. However, $x^2 - 4$ can be factored, since it is the difference of squares. Therefore, in completely factored form,

$$2x^2 - 8 = 2(x + 2)(x - 2)$$

It is often necessary to do several factoring steps before a polynomial is completely factored. (The instruction **completely factor** means **completely factor over the integers.**)

Example 1 Factor completely $3ax^4 - 3ay^4$.

SOLUTION

$$3ax^4 - 3ay^4$$
$$3a(x^4 - y^4)$$
$$3a(x^2 + y^2)(x^2 - y^2)$$
$$3a(x^2 + y^2)(x + y)(x - y)$$

Summarizing these steps:

1. Remove the greatest common monomial factor by the distributive axiom:

$$3ax^4 - 3ay^4 = 3a(x^4 - y^4)$$

2. Factor the difference of two squares:

$$= 3a(x^2 + y^2)(x^2 - y^2)$$

3. Factor the difference of two squares:

$$= 3a(x^2 + y^2)(x + y)(x - y)$$

Since the factors of a product can be written in any order, it also follows that

$$3ax^4 - 3ay^4 = 3a(x + y)(x - y)(x^2 + y^2)$$

It is desirable to arrange the factors in such an order that monomials are written first, then linear factors, followed by quadratic factors, and so on.

(A **linear** factor is a factor whose degree is 1, such as $x + y$. A **quadratic** factor is one whose degree is 2, such as x^2 or $x^2 + y^2$. A **cubic** factor is one whose degree is 3, such as $x^3 - 2x^2$ and so on.)

Procedure for Complete Factorization

1. Remove the greatest common monomial factor first.
2. Factor any binomial, if present, using the Difference of Squares form.
3. Factor any trinomial, if present, using the Perfect Square Trinomial form, the Simple Trinomial form, or the General Trinomial form.

Factoring Forms

1. Common Monomial Factor	$AB + AC = A(B + C)$ $AB + AC + AD = A(B + C + D)$
2. Difference of Squares	$A^2 - B^2 = (A - B)(A + B)$
3. Perfect Square Trinomial	$A^2 + 2AB + B^2 = (A + B)^2$ $A^2 - 2AB + B^2 = (A - B)^2$
4. Simple Trinomial	$X^2 + (a + b)X + ab = (X + a)(X + b)$
5. General Trinomial	$AX^2 + BX + C = (aX + b)(cX + d)$ where $A = ac$ and $C = bd$ and $B = ad + bc$

Example 2 Completely factor $6x^2 + 30x - 84$.

SOLUTION

$$6x^2 + 30x - 84 = 6(x^2 + 5x - 14) \quad \text{(Common Monomial Factor)}$$
$$= 6(x + 7)(x - 2) \quad \text{(Simple Trinomial)}$$

Example 3 Factor completely over the integers $x^4 - 5x^2 - 36$.

SOLUTION

$$x^4 - 5x^2 - 36 = (x^2 - 9)(x^2 + 4) \quad \text{(Simple Trinomial)}$$
$$= (x + 3)(x - 3)(x^2 + 4) \quad \text{(Difference of Squares)}$$

Example 4 Check Example 2 by multiplication.

SOLUTION

$$6(x + 7)(x - 2) = 6(x^2 + 5x - 14)$$
$$= 6x^2 + 30x - 84$$

Example 5 Check Example 3.

SOLUTION

$$(x + 3)(x - 3)(x^2 + 4) = (x^2 - 9)(x^2 + 4)$$
$$= x^4 - 5x^2 - 36$$

Example 6 Completely factor $8x^5 - 74x^3 + 18x$.

SOLUTION

$8x^5 - 74x^3 + 18x$

$2x(4x^4 - 37x^2 + 9)$ (Common Monomial Factor)

$2x(x^2 - 9)(4x^2 - 1)$ (General Trinomial form)

$2x(x - 3)(x + 3)(2x - 1)(2x + 1)$ (Difference of Squares)

Example 7 Completely factor $16x^4 - 200x^2 + 625$.

SOLUTION

$16x^4 - 200x^2 + 625$

$(4x^2 - 25)(4x^2 - 25)$ (Perfect Square Trinomial)

$(2x - 5)(2x + 5)(2x - 5)(2x + 5)$ (Difference of Squares)

$(2x - 5)^2(2x + 5)^2$ (Conventional style)

Exercises 3.9

Factor completely over the integers.

1. $3x^2 - 12$
2. $5x^2 - 80$
3. $9x^2 - 9x - 18$
4. $4x^2 - 6x + 2$
5. $36y^2 - 36y + 9$
6. $36y^2 + 24y + 4$
7. $15t^2 - 35t + 20$
8. $6t^2 - 30t - 84$
9. $18x^2 - 8x^4$
10. $-100x^2 - 4x^4$
11. $-36y^2 - 9y^4$
12. $9y^2 - 36y^4$
13. $x^3 - 5x^2 - 36x$
14. $5x^3 + 20x^2 - 105x$
15. $36x^4 - 180x^3 + 81x^2$
16. $36x^4 - 120x^3 + 64x^2$
17. $12y^3 + 14y^2 - 10y$
18. $6y^3 + y^2 - 12y$
19. $100t^3 + 160t^2 + 64t$
20. $27t^3 + 36t^2 + 12t$
21. $x^4 - 16y^4$
22. $x^4 - 81y^4$
23. $x^4 - 8x^2 + 16$
24. $x^4 - 18x^2 + 81$
25. $x^4 - 15x^2 - 16$
26. $x^4 + 80x^2 - 81$
27. $25x^4 + 399x^2 - 16$
28. $4x^4 - 36x^2 + 81$
29. $25x^4 - 40x^2 + 16$
30. $4x^4 + 27x^2 - 81$
31. $7x^4 - 70{,}000$
32. $25x^2 + x^4$
33. $x^2 + x^4$
34. $100x^4 - 62{,}500$
35. $2x^4 - 26x^2 + 72$
36. $x^5 - 32x^3 + 256x$
37. $49x^4 - 98x^2 + 49$
38. $x^5 - 37x^3 + 36x$
39. $36t^2 + 72t - 36$
40. $49t^2 - 49t + 49$
41. $2y^2 - 20y^3 + 50y^4$
42. $144y^2 - 48y^3 + 4y^4$

43. $27t - 3t^3$
44. $56t^2 - 8t - 98t^3$
45. $240y^2 - 320y - 45y^3$
46. $64y^2 - y^4$
47. $-8x^4 - 8x^3 - 2x^2$
48. $-48x^4 - 24x^3 - 3x^2$
49. $18x^4 + 66x^2 - 120$
50. $30x^4 + 15x^2 - 420$
51. $(x^2 + 1)^2 - x^2$
52. $(x^2 + 2)^2 - 4x^2$
53. $x(x^2 - 4) + 5(x^2 - 4)$
54. $x(x^2 - 1) - 8(x^2 - 1)$
55. $(a - b)^2 - c^2$
56. $(x - y)^2 - z^2$
57. $(x^2 - 6)^2 - x^2$
58. $(x^2 - 8)^2 - 4x^2$
59. $(x^2 + 4x)^2 - 25$
60. $(x^2 + 2x)^2 - 1$

3.10 Applications

Rapid Calculations (Optional)

A very useful application of factoring is to obtain shortcuts in arithmetic. The following examples illustrate the use of several factoring forms.

SUM AND DIFFERENCE FORM:

$$(A + B)(A - B) = A^2 - B^2$$

Example 1

$$(29)(31) = (30 + 1)(30 - 1) = (30)^2 - 1 = 900 - 1 = 899$$

Example 2

$$(52)(48) = (50 + 2)(50 - 2) = 2500 - 4 = 2496$$

DIFFERENCE OF SQUARES FORM:

$$A^2 - B^2 = (A - B)(A + B)$$

Example 3

$$(35)^2 - (34)^2 = (35 - 34)(35 + 34) = (1)(69) = 69$$

Example 4

$$(17)^2 - (15)^2 = (17 - 15)(17 + 15) = (2)(32) = 64$$

PERFECT SQUARE TRINOMIAL FORM:

$$(A + B)^2 = A^2 + 2AB + B^2$$

Example 5

$$(32)^2 = (30 + 2)^2 = 900 + 2(60) + 4 = 1024$$

Example 6

$$(21)^2 = (20 + 1)^2 = 400 + 40 + 1 = 441$$

FIRST AND TEN FORM:

$$(10a + b)(10a + [10 - b]) = a(a + 1) \oplus b(10 - b)$$

where $\oplus$ means *tack on*, as shown in the following examples.

The First and Ten form is used for numbers that are two-digit numbers having the same tens digit and with the sum of the units digits equal to 10.

Example 7

$$(32)(38) = (3)(4) \oplus (2)(8) = 12 \oplus 16 = 1216$$

Example 8

$$(65)(65) = (6)(7) \oplus (5)(5) = 42 \oplus 25 = 4225$$

Example 9

$$(26)(24) = (2)(3) \oplus (6)(4) = 6 \oplus 24 = 624$$

Solutions of Equations by Factoring

If the trinomial $ax^2 + bx + c$ can be factored over the set of integers, then there is an easy way to solve the equation $ax^2 + bx + c = 0$. The method depends on the following theorem.

The Zero-Product Theorem
A product of two real numbers is zero if and only if one or both of the two numbers are zero.
In symbols, $rs = 0$ if and only if $r = 0$ or $s = 0$.

Example 10 Given $3x = 0$. Then by the zero-product theorem $3 = 0$ or $x = 0$. Since $3 \neq 0$, x must $= 0$.

Example 11 Solve $(x - 4)(x - 7) = 0$.

SOLUTION
Since the product $(x - 4)(x - 7)$ is zero, the left factor, $x - 4$, is zero, or the right factor, $x - 7$, is zero.

continued

$$x - 4 = 0 \quad \text{or} \quad x - 7 = 0$$
$$x = 4 \quad \text{or} \quad x = 7$$

The solution set is $\{4, 7\}$, since each of these numbers makes the original statement true.

CHECK

For $x = 4$, $(4 - 4)(4 - 7) = (0)(-3) = 0$
For $x = 7$, $(7 - 4)(7 - 7) = (3)(0) = 0$

Example 12 Solve the equation $x^2 + 15x + 54 = 0$.

SOLUTION
In factored form,

$$x^2 + 15x + 54 = (x + 6)(x + 9)$$

and the given equation is equivalent to

$$(x + 6)(x + 9) = 0$$

By the zero-product theorem,

$$x + 6 = 0 \quad \text{or} \quad x + 9 = 0$$
$$x = -6 \quad \text{or} \quad x = -9$$

CHECK

$$\begin{aligned} x^2 + 15x + 54 &= 0 \\ (-6)^2 + (15)(-6) + 54 &= 0 \\ 36 - 90 + 54 &= 0 \\ 0 &= 0 \end{aligned} \qquad \begin{aligned} x^2 + 15x + 54 &= 0 \\ (-9)^2 + (15)(-9) + 54 &= 0 \\ 81 - 135 + 54 &= 0 \\ 0 &= 0 \end{aligned}$$

Therefore, the solution set is $\{-6, -9\}$.

Example 13 Solve for x: $x^2 + 3x = 0$.

SOLUTION
The quadratic polynomial $x^2 + 3x$ has a common factor, x. An equivalent equation is

$$x(x + 3) = 0$$

By the zero-product theorem,

$$x = 0 \quad \text{or} \quad x + 3 = 0$$
$$x = 0 \quad \text{or} \quad x = -3$$

and the solution set is $\{0, -3\}$.

This method can be extended to other types of polynomial equations. We simply use the zero-product theorem two or more times.

Example 14 Solve $x^3 - 25x = 0$.

SOLUTION

Factoring	$x(x^2 - 25) = 0$
Using the zero-product theorem	$x = 0$ or $x^2 - 25 = 0$
Factoring	$(x - 5)(x + 5) = 0$
Using the zero-product theorem	$x - 5 = 0$ or $x + 5 = 0$
Solving	$x = 0$ or $x = 5$ or $x = -5$

The solution set is $\{0, 5, -5\}$.

Example 15 Solve $x^4 + 5x^2 = 36$.

SOLUTION To use the zero-product theorem, the right side of the equation must be zero.

First, we must find an equivalent equation whose right side is zero.

Equivalent equation	$x^4 + 5x^2 - 36 = 0$
Factoring	$(x^2 - 4)(x^2 + 9) = 0$
Zero-product theorem	$x^2 - 4 = 0$ or $x^2 + 9 = 0$
Factoring	$(x - 2)(x + 2) = 0$
Zero-product theorem	$x - 2 = 0$ or $x + 2 = 0$
Solving	$x = 2$ or $x = -2$

The equation $x^2 + 9 = 0$ has no integers for solutions. In fact, there is no real number for which $x^2 + 9 = 0$.

The solution set is $\{2, -2\}$ over the set of real numbers.

Stated Problems

The solutions of many problems stated in words can be found by solving an equation of the form $ax^2 + bx + c = 0$. Such problems are shown in Chapter 6, Section 6.7. If desired, this section may be studied at this time.

Exercises 3.10

Solve and check. See Examples 10–15.

1. $(x + 2)(x - 3) = 0$
2. $(x - 1)(x + 4) = 0$
3. $x(x - 5) = 0$
4. $(x + 6)(x - 6) = 0$
5. $(x - 4)(x + 4) = 0$
6. $x(x + 7) = 0$
7. $(2x - 1)(x + 8) = 0$
8. $(x - 3)(3x + 2) = 0$
9. $(y + 6)(y + 9) = 0$
10. $(y + 5)(y + 8) = 0$
11. $x^2 - 3x + 2 = 0$
12. $x^2 - 7x + 12 = 0$
13. $x^2 - 2x = 24$
14. $x^2 - 3x = 0$
15. $x^2 - 2x = 0$
16. $x^2 - 3x = 40$
17. $y^2 - 36 = 0$
18. $y^2 - 49 = 0$
19. $t^2 = 81$
20. $t^2 = 64$
21. $3t^2 - 6t - 45 = 0$
22. $6t^2 + 24t - 72 = 0$
23. $y^2 - 14y + 49 = 0$
24. $y^2 - 20y + 100 = 0$
25. $x^2 = x + 12$
26. $x^2 = x + 6$
27. $(x + 2)(x + 3) = 12$
28. $(x + 1)(x - 2) = 4$
29. $2u^2 - 32 = 0$
30. $3u^2 - 75 = 0$
31. $4x^3 = 36x$
32. $16x^3 = 16x$
33. $(x - 4)(x + 4) = 6x$
34. $(x + 3)(x - 3) = 8x$
35. $(x - 1)^2 = 25$
36. $(x - 2)^2 = 64$
37. $5y^2 + 9y = 2$
38. $4y^2 + 15y = 4$
39. $(z + 3)^2 = 2z^2 + 9$
40. $(z - 2)^2 = 2z^2 + 4$
41. $(z - 6)^2 = z^2 + 36$
42. $(z + 9)^2 = z^2 + 81$
43. $x(x - 2)(x - 3) = 0$
44. $x(x - 4)(x + 6) = 0$
45. $x^3 - x^2 - 20x = 0$
46. $x^3 + 9x^2 + 18x = 0$
47. $x^4 - 5x^2 + 4 = 0$
48. $x^4 - 82x^2 + 81 = 0$
49. $x^4 - 8x^2 + 16 = 0$
50. $x^4 - 80x^2 - 81 = 0$
51. $x^4 - 24x^2 = 25$
52. $x^4 - 18x^2 + 81 = 0$
53. $x^4 + 1 = 2x^2$
54. $x^4 + 10{,}000 = 200x^2$
55. $x^4 = 8x^2 + 9$
56. $x^4 - 48x^2 = 49$
57. $x^4 + 144 = 40x^2$
58. $4x^4 + 4 = 17x^2$
59. $(x - 3)(x - 5)(x - 7) = 0$
60. $(x + 4)(x - 6)(x - 8) = 0$

In Exercises 61–80, find the products by using the rapid calculation methods of this section.

61. (37)(43)
62. (84)(86)
63. (18)(22)
64. $(22)^2$

65. $(34)^2$
66. $(31)(39)$
67. $(41)^2 - (40)^2$
68. $(85)^2$
69. $(27)(23)$
70. $(44)^2 - (42)^2$
71. $(38)(42)$
72. $(55)(45)$
73. $(35)(45)$
74. $(87)(93)$
75. $(87)(83)$
76. $(61)^2 - (60)^2$
77. $(42)^2$
78. $(65)^2 - (63)^2$
79. $(46)(44)$
80. $(51)^2$

Review Exercises

Simplify the products in Exercises 1–30.

1. $x^2 \cdot x^3$
2. $y \cdot y^2$
3. $x^3 \cdot x$
4. $p^2 \cdot p^3 \cdot p^4$
5. $(-3x)(4x^3)$
6. $-a^3(2a^5)$
7. $(5n^3x)(-2x^3)$
8. $3(a^2)^3$
9. $3x(-2x)^2$
10. $-2x(5x)^2$
11. $x(x^2 + 3x + 2)$
12. $x(x^3 + 2x - 1)$
13. $2x(3x^2 - 4x - 1)$
14. $4x(ax^2 - bx + c)$
15. $(x - 2)(x - 3)$
16. $(y + 1)(y - 4)$
17. $(2a + 3)(3a - 1)$
18. $(a + 5)^2$
19. $(5x + 1)(5x - 1)$
20. $(2y - 3)^2$
21. $x(x + 2)(2x - 7)$
22. $x^3(2x^2 + x - 2)$
23. $-x^4(3x^3 + 2x - 4)$
24. $(2x + y)(3x - 16y)$
25. $x(x + 4)(2x - 5)$
26. $2x(3x + 2)^2$
27. $(x + 2)(x + 3)^2$
28. $(x^2 + 3x + 2)(x - 4)$
29. $(2x^2 - x + 3)(3x + 1)$
30. $(x^2 + 5x + 6)(x^2 - 3x + 1)$

In Exercises 31–50, factor completely.

31. $3x^2 - 6x + 24xy$
32. $t^2 + 11t + 24$
33. $x^2 + 10x + 25$
34. $49y^4 - 1$
35. $x^3y - 16xy^3$
36. $16x^2 + 64$
37. $-3x^3 + 27x$
38. $6a^2 - 7a - 3$

39. $36x + 3x^3 - 3x^5$

40. $100p^2 - 25q^2$

41. $64y^3 - 36y$

42. $x^4 - 16$

43. $12x^2y - 22xy - 20y$

44. $8a^2 + 5ab - 3b^2$

45. $2y^5 - 162y$

46. $15ax^6 + 42ax^5 - 9ax^4$

47. $25r^2 + 10r + 1$

48. $36x^2 - 12x + 1$

49. $4p^2 - 20pq + 25q^2$

50. $x^4 - 18x^2 + 81$

Solve the equations in Exercises 51–60.

51. $2x^2 - 3x - 14 = 0$

52. $x^2 + 7x = 0$

53. $x^2 + 7x = 60$

54. $45y - 15y^2 = 0$

55. $t(t - 11) = -18$

56. $12z^2 + 24z + 12 = 0$

57. $2x - (x + 2) = 3x$

58. $x^2 + 6 = 7x$

59. $x^2 = 3x$

60. $x^2 = 64$

4 Fractions

Objectives

OBJECTIVES	TYPICAL PROBLEMS	LOCATIONS
Simplify a fraction.	1–2. Simplify: 1. $\dfrac{x^2 - 9x + 20}{x^2 + x - 20}$ 2. $\dfrac{36 + 6x}{36 - x^2}$	*Section 4.1* *Examples 1–9* *Exercises 1–90*
Multiply fractions.	3. Multiply: $\dfrac{3x^2 + 6x - 72}{2x^2 - 4x - 48} \cdot \dfrac{2x - 12}{3x - 12}$	*Section 4.2* *Examples 1–6* *Exercises 1–52*
Divide fractions.	4. Divide: $\dfrac{4x^2 - 8x}{2x^2 - 3x - 2} \div \dfrac{4x^2 - 2x}{4x^2 - 1}$	*Section 4.2* *Examples 7–11* *Exercises 53–80*
Raise a fraction to higher terms.	5. Raise to higher terms with the indicated denominator. $\dfrac{x + 1}{x + 5} = \dfrac{\quad}{2x^2 - 50}$	*Section 4.3* *Examples 1–9* *Exercises 1–62*
Add and subtract fractions.	6–7. Rewrite as a single fraction. 6. $\dfrac{x - 5}{4x - 16} + \dfrac{x - 2}{x^2 - 16}$ 7. $\dfrac{1}{x^2 - x} - \dfrac{1}{2x^2 - 3x + 1}$	*Section 4.4* *Examples 1–7* *Exercises 1–76*
Simplify a complex fraction.	8. Simplify: $\dfrac{1 + \dfrac{4}{x}}{1 - \dfrac{16}{x^2}}$	*Section 4.5* *Examples 1–7* *Exercises 1–30*
Divide polynomials using long division.	9. Divide by using the long division algorithm. $\dfrac{x^3 - 3x^2 + 5x - 6}{x - 3}$	*Section 4.6* *Examples 1–6* *Exercises 1–42*

Solve a fractional equation.	10. Solve: $\frac{3}{x-4} - \frac{2}{x+6} = \frac{24}{x^2+2x-24}$	*Section 4.7* *Examples 1–5* *Exercises 1–60*
Solve a ratio and proportion problem.	11. For an alloy rated 14K gold, the ratio of pure gold to the other metal in the alloy is 7 to 5. How many grams of pure gold are in 60 grams of an object rated 14K gold?	*Section 4.8* *Examples 1–6* *Exercises 1–30*
Solve a work problem.	12. An old machine can process a certain amount of data in 40 minutes. It takes a new machine 24 minutes to do the same job. How long would it take both machines, working together, to process these data?	*Section 4.9* *Examples 1, 2* *Exercises 1–16*

Answers to Typical Problems

1. $\frac{x-5}{x+5}$
2. $\frac{-6}{x-6}$
3. $\frac{x+6}{x+4}$
4. 2
5. $\frac{2x^2-8x-10}{2x^2-50}$
6. $\frac{x+7}{4(x+4)}$
7. $\frac{1}{x(2x-1)}$
8. $\frac{x}{x-4}$
9. $x^2+5+\frac{9}{x-3}$
10. -2
11. 35 grams
12. $\frac{x}{40}+\frac{x}{24}=1$; 15 minutes

A **rational algebraic expression** is an expression that is obtained by adding, subtracting, multiplying, or dividing polynomials. For example, the expressions

$$t - \frac{1}{t} + 5, \quad \frac{y^2+9}{y+3}, \quad \text{and} \quad \frac{x-2}{x^2-5x+6} + \frac{3x+5}{x^2-9}$$

are rational expressions.

The choice of the word **rational** to describe these expressions was influenced by the fact that these expressions become rational numbers—numbers that are the "ratio" of two integers, a and b, $b \neq 0$—whenever each variable is replaced by a rational number.

A rational algebraic expression is also called an **algebraic fraction** or, for simplicity, a **fraction.**

This chapter deals with the addition, subtraction, multiplication, and division of algebraic fractions. Next the solution of equations involving fractions is discussed. Finally, word problems are presented—word problems that can be solved by fractional equations.

4.1 Simplification of Fractions

From experience with the arithmetic of fractions, one learns that a fraction can be named in more than one way.

For example, by examining the geometric models shown in Figures 4.1 and 4.2, it can be seen that $\frac{3}{4}$ and $\frac{9}{12}$ are names for the same number.

Figures 4.1 and 4.2 show that further subdivisions of a basic unit cause a subdivision of the original parts of the unit used to represent the fraction.

Before stating a "fundamental theorem," the following definition is in order.

Definition of Equal Quotients

If a, b, c, and d are real numbers and if $bd \neq 0$, then

$$\frac{a}{b} = \frac{c}{d} \quad \text{if and only if} \quad ad = bc$$

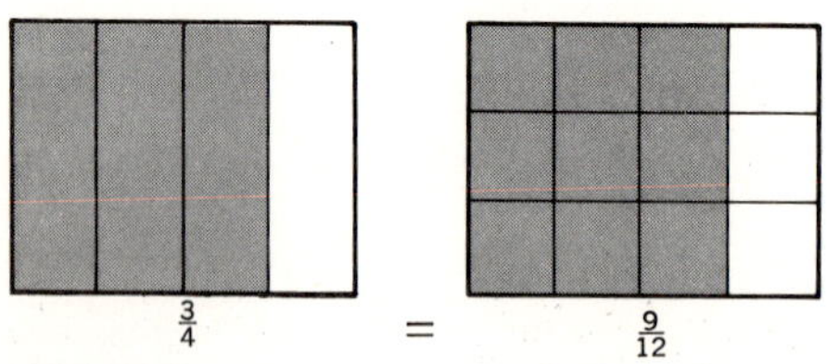

Figure 4.1

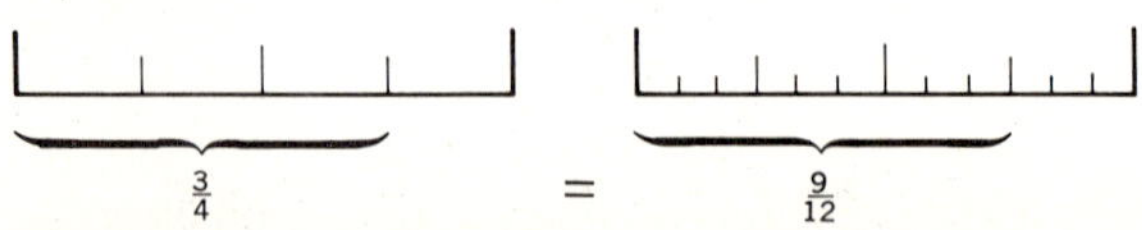

Figure 4.2

Thus $\frac{3}{4} = \frac{9}{12}$ because $3(12) = 4(9)$ and $\frac{2}{3} = \frac{4}{x}$ if and only if $2x = 12$ and $x = 6$.

Note that $\frac{3}{4} = \frac{3 \cdot 3}{4 \cdot 3} = \frac{9}{12}$.

The following theorem is so important in connection with everything pertaining to fractions that it is called the fundamental theorem of fractions.

The Fundamental Theorem of Fractions

If n, d, and k are real numbers and $d \neq 0$ and $k \neq 0$, then

$$\frac{nk}{dk} = \frac{n}{d}$$

This theorem is valid, since $(nk)d = (dk)n$ by the associative and commutative axioms for multiplication.

The fundamental theorem of fractions is used to simplify a fraction—that is, to reduce a fraction to lowest terms.

The simplification of a fraction, or the reduction of a fraction to lowest terms, means the renaming of a fraction so that the numerator and denominator do not have a factor in common.

Since all common factors of the numerator and denominator must be removed, simplification of a fraction requires that the numerator and denominator be completely factored.

Example 1 Simplify $\frac{60}{84}$

SOLUTION

$$\frac{60}{84} = \frac{\mathbf{2 \cdot 2 \cdot 3} \cdot 5}{\mathbf{2 \cdot 2 \cdot 3} \cdot 7} = \frac{5 \cdot 12}{7 \cdot 12} = \frac{5}{7}$$

Example 2 Reduce $\frac{3x^2}{15x^5}$ to lowest terms.

SOLUTION

$$\frac{3x^2}{15x^5} = \frac{\mathbf{3 \cdot x \cdot x}}{\mathbf{3} \cdot 5 \cdot \mathbf{x \cdot x} \cdot x \cdot x \cdot x} = \frac{1 \cdot 3x^2}{5x^3 \cdot 3x^2} = \frac{1}{5x^3}$$

Example 3 Simplify $\dfrac{35x^2y^3}{7xy}$

SOLUTION

$$\frac{35x^2y^3}{7xy} = \frac{5 \cdot \mathbf{7} \cdot \mathbf{x}xy\mathbf{y}y}{\mathbf{7xy}} = \frac{5xy^2 \cdot 7xy}{1 \cdot 7xy} = \frac{5xy^2}{1} = 5xy^2$$

Note in the preceding example that $\dfrac{5xy^2}{1}$ names the same number as $5xy^2$, just as $\dfrac{5}{1}$ names the same number as 5. The integral forms $5xy^2$ and 5 are considered to be the simplified forms of $\dfrac{5xy^2}{1}$ and $\dfrac{5}{1}$, respectively.

The factoring technique is also used to reduce a fraction with a polynomial numerator or a polynomial denominator.

Example 4 Simplify $\dfrac{x^2 - 9}{x^2 - 3x}$

SOLUTION Factoring numerator and denominator,

$$\frac{x^2 - 9}{x^2 - 3x} = \frac{(x + 3)\mathbf{(x - 3)}}{x\mathbf{(x - 3)}} = \frac{x + 3}{x}$$

Example 5 Reduce $\dfrac{x^2 - 1}{x - 1}$ to lowest terms.

SOLUTION

$$\frac{x^2 - 1}{x - 1} = \frac{(x + 1)(x - 1)}{x - 1} = \frac{(x + 1)\mathbf{(x - 1)}}{1\mathbf{(x - 1)}} = \frac{x + 1}{1} = x + 1$$

Example 6 Simplify $\dfrac{3x^2 - 75y^2}{3x^2 - 21xy + 30y^2}$

SOLUTION

$$\begin{aligned}\frac{3x^2 - 75y^2}{3x^2 - 21xy + 30y^2} &= \frac{3(x^2 - 25y^2)}{3(x^2 - 7xy + 10y^2)} \\ &= \frac{\mathbf{3(x - 5y)}(x + 5y)}{\mathbf{3(x - 5y)}(x - 2y)} \\ &= \frac{x + 5y}{x - 2y}\end{aligned}$$

Since the number $\frac{n}{d}$ can be considered as the quotient obtained when the numerator, n, is divided by the denominator, d, there are three numbers involved in this operation; accordingly, there are three signs associated with the number $\frac{n}{d}$:

1. The sign of the numerator n
2. The sign of the denominator d
3. The sign of the quotient $\frac{n}{d}$

As a result, there are four different ways to name a negative number such as $\frac{-3}{7}$:

$$\frac{-3}{7} = -\frac{3}{7} = \frac{3}{-7} = -\frac{-3}{-7}$$

The preferred form is $\frac{-3}{7}$, that is, the minus sign is placed in the numerator. It is very easy to "drop" a minus sign in a calculation, and this convention helps to prevent this error.

Example 7 Simplify $\frac{21xy}{-14xz}$

SOLUTION

$$\frac{21xy}{-14xz} = \frac{-21xy}{14xz} = \frac{(-3y)(7x)}{2z(7x)} = \frac{-3y}{2z}$$

Special attention is called to the forms $a - b$ and $b - a$, which are additive inverses of each other.

$$\begin{aligned} -(b - a) &= -(b + (-a)) = -1(b) + (-1(-a)) \\ &= -b + a = a + (-b) = a - b \end{aligned}$$

Thus

$$-(b - a) = a - b \quad \text{and} \quad a - b = -(b - a).$$

This property is needed sometimes in the process of renaming fractions. Its use is illustrated in the following examples.

Example 8 Simplify $\frac{x - 1}{1 - x}$

SOLUTION Since $1 - x = -(x - 1)$,

$$\frac{x - 1}{1 - x} = \frac{x - 1}{-(x - 1)} = \frac{1(x - 1)}{(-1)(x - 1)} = \frac{1}{-1} = -1$$

Example 9 Simplify $\dfrac{15 - 5y}{y^2 - 3y}$

SOLUTION

$$\frac{15 - 5y}{y^2 - 3y} = \frac{5(3 - y)}{y(y - 3)} = \frac{5 \cdot -(y - 3)}{y \cdot (y - 3)} = \frac{-5 \cdot (y - 3)}{y \cdot (y - 3)} = \frac{-5}{y}$$

Exercises 4.1

In Exercises 1–90 reduce each fraction to lowest terms.

1. $\dfrac{-4}{-12}$
2. $\dfrac{12}{-36}$
3. $\dfrac{15}{-25}$
4. $\dfrac{-42}{-105}$
5. $-\dfrac{150}{200}$
6. $-\dfrac{60}{90}$
7. $\dfrac{6xy^2}{6xy}$
8. $\dfrac{7a^2b}{7ab}$
9. $\dfrac{25x^3}{75x^5}$
10. $\dfrac{75x^5}{25x^3}$
11. $\dfrac{m^3}{m}$
12. $\dfrac{3x}{6xy}$
13. $\dfrac{-300x^3y^2}{-75xy}$
14. $\dfrac{-80a^2b^2}{-16ab}$
15. $\dfrac{a^3b}{-a^3b}$
16. $\dfrac{3x^2y^2}{-xy}$
17. $\dfrac{xy}{3x^2y^2}$
18. $\dfrac{xy^2z}{-x^2y^2z^2}$
19. $\dfrac{-ab^2c}{-a^2b^2c^2}$
20. $\dfrac{3x(a + b)}{12x^2}$
21. $\dfrac{7x + 21}{7x}$
22. $\dfrac{4y}{4y^2 - 8y}$
23. $\dfrac{x^2 - 4x}{x^2 - 2x}$
24. $\dfrac{y^2 - 3y}{y^2 - 9}$
25. $\dfrac{x^2 - 4}{x^2 - 2x}$
26. $\dfrac{y^2 - 3y}{y^2 - 9y}$
27. $\dfrac{5y - 40}{5y}$
28. $\dfrac{6xy}{2x^2y + 6xy^2}$

29. $\frac{5x - 5y}{5x + 5y}$

30. $\frac{y^2 + 5y}{25 + 5y}$

31. $\frac{5x - 5y}{x^2 - xy}$

32. $\frac{y^2 + 5y}{y^2 - 5y}$

33. $\frac{4r + r^2}{4r + 16}$

34. $\frac{3a + 3b}{9a + 9b}$

35. $\frac{(c - d)^2}{6c - 6d}$

36. $\frac{u^2v + uv^2}{(u + v)^2}$

37. $\frac{4t^2 - 16}{t - 2}$

38. $\frac{9t^2 - 81}{t + 3}$

39. $\frac{8a + 8b}{8a - 8b}$

40. $\frac{7x - 7y}{7y - 7x}$

41. $\frac{8a - 8b}{8b - 8a}$

42. $\frac{7x - 7y}{7x + 7y}$

43. $\frac{8y^2 - 12y}{4y}$

44. $\frac{3x^2 + 6x}{3x}$

45. $\frac{4m^2 + 4}{4m + 4}$

46. $\frac{3a^2 + 3}{3a + 3}$

47. $\frac{y + y^2}{3y + 3}$

48. $\frac{3y + y^2}{3y + 9}$

49. $\frac{(a - b)^3}{(a - b)^2}$

50. $\frac{(r + 2s)^2}{(r + 2s)^5}$

51. $\frac{8a^2 + 8ab}{4b^2 + 4ab}$

52. $\frac{2uv + 2v^2}{6uv + 6v^2}$

53. $\frac{75n^2 - 3}{5n + 1}$

54. $\frac{3t + 1}{4 - 36t^2}$

55. $\frac{x^2 - 1}{x^2 + 3x - 4}$

56. $\frac{x^2 + 2x - 8}{x^2 - 4}$

57. $\frac{z^2 - 10z + 25}{z^2 - z - 20}$

58. $\frac{4t^2 - 4t - 24}{6t^2 - 36t + 54}$

59. $\frac{t - 3}{2t^2 - 5t - 3}$

60. $\frac{3x + 15}{x^2 + 3x - 10}$

61. $\frac{x^2 - 9}{9 - x^2}$

62. $\frac{25 - t^2}{t^2 - 25}$

63. $\frac{9x^2 - 4y^2}{2y - 3x}$

64. $\frac{x + 6}{36 - x^2}$

65. $\frac{75n^2 - 3}{1 + 5n}$

66. $\frac{8y - 5x}{25x^2 - 64y^2}$

67. $\dfrac{6x^2 - 6}{6x^2 + 36x - 42}$

68. $\dfrac{3n^3 - 21n^2 - 24n}{3n^3 - 3n}$

69. $\dfrac{2y + 4}{y^2 - 2y - 8}$

70. $\dfrac{6y^4 + 15y^3 - 9y^2}{9y^2 + 27y}$

71. $\dfrac{a^2 - b^2}{(a - b)^2}$

72. $\dfrac{(x - y)^2}{x^2 - y^2}$

73. $\dfrac{4a^2 - 4ab}{a^2 - b^2}$

74. $\dfrac{x^2 - y^2}{x^2y - xy^2}$

75. $\dfrac{2n^3 + 4n^2 + 2n}{n^3 - n^2 - 2n}$

76. $\dfrac{5a^2 + 15a - 140}{10a^2 + 110a + 280}$

77. $\dfrac{7 - x}{x^2 - 49}$

78. $\dfrac{5 - x}{x^2 - 25}$

79. $\dfrac{49n^2 - 21n - 10}{49n^2 - 25}$

80. $\dfrac{5n^2 - 17n + 6}{5n^2 - 13n - 6}$

81. $\dfrac{m^2n - mn^2 + mn}{m^2n + mn^2 - mn}$

82. $\dfrac{x^2y - xy^2 + xyz}{yz^2 - y^2z + xyz}$

83. $\dfrac{a^2 + ab - ac}{b^2 + ab - bc}$

84. $\dfrac{r^2 + rs - rt}{r^2 - rs + rt}$

85. $\dfrac{4x - 20}{10 - 2x}$

86. $\dfrac{12 - 3x}{6x - 24}$

87. $\dfrac{(r - s)(r - t)}{(t - r)(t - s)}$

88. $\dfrac{(a - b)(a - c)(b - c)}{(c - a)(c - b)(b - a)}$

89. $\dfrac{x^2 - 3x - 10}{x^2 - 2x - 15}$

90. $\dfrac{8t^2 - 12t - 20}{8t^2 - 28t + 20}$

4.2 Multiplication and Division

Multiplication

Theorem: Product of Fractions

The product of two fractions is obtained by multiplying the numerators to obtain the new numerator and by multiplying the denominators to obtain the new denominator.

In symbols, if n, d, r, and s are any real numbers such that $d \neq 0$ and $s \neq 0$, then

$$\frac{n}{d} \cdot \frac{r}{s} = \frac{nr}{ds}$$

Example 1 Express $\frac{3}{5} \cdot \frac{2}{7}$ as a single fraction.

SOLUTION

$$\frac{3}{5} \cdot \frac{2}{7} = \frac{3 \cdot 2}{5 \cdot 7} = \frac{6}{35} \quad \begin{matrix}\text{(Product of numerators)} \\ \text{(Product of denominators)}\end{matrix}$$

Example 2 Simplify $\frac{5}{8} \cdot \frac{4}{15}$

SOLUTION

$$\frac{5}{8} \cdot \frac{4}{15} = \frac{5 \cdot 4}{8 \cdot 15} \quad \begin{matrix}\text{(Product of numerators)} \\ \text{(Product of denominators)}\end{matrix}$$

$$= \frac{5 \cdot 4}{(4 \cdot 2)(5 \cdot 3)} \quad \text{(Factoring)}$$

$$= \frac{1}{6} \quad \text{(Simplifying by the fundamental theorem)}$$

Example 3 Simplify $\frac{3}{10} \cdot \frac{5}{12} \cdot \frac{4}{24}$

SOLUTION

First method:

$$\frac{3}{10} \cdot \frac{5}{12} \cdot \frac{4}{24} = \frac{3 \cdot 5 \cdot 4}{10 \cdot 12 \cdot 24} = \frac{60}{2880}$$

$$= \frac{2 \cdot 2 \cdot 3 \cdot 5}{2 \cdot 2 \cdot 2 \cdot 2 \cdot 2 \cdot 2 \cdot 3 \cdot 3 \cdot 5} = \frac{1}{48}$$

continued

Second method:

$$\frac{3}{10}\cdot\frac{5}{12}\cdot\frac{4}{24}=\frac{3}{10}\cdot\frac{5}{12}\cdot\frac{1}{6}$$
$$=\frac{3\cdot 5}{(2\cdot 5)(3\cdot 4)\cdot 6}=\frac{1}{48}$$

Notice how the arithmetic in Example 3 was much simpler when the numerator and denominator were factored as soon as possible rather than multiplying everything first.

Example 4 Express $\frac{7x^2}{3y}\cdot\frac{12y^2}{35x^3}$ as a simplified single fraction.

SOLUTION

$$\frac{7x^2}{3y}\cdot\frac{12y^2}{35x^3}=\frac{7\cdot 12x^2y^2}{3\cdot 35x^3y}$$ (Using the commutative and associative axioms to rearrange the factors so that the numerals occur first and the letters are arranged in alphabetical order)

$$=\frac{4y(3\cdot 7x^2y)}{5x(3\cdot 7x^2y)}$$ (Rearranging the factors to form the pattern $\frac{ak}{bk}$)

$$=\frac{4y}{5x}$$ (Since $3\cdot 7x^2y$ is a common **factor,** the fraction can be reduced)

Example 5 Simplify $\frac{t^2+3t}{6t-2}\cdot\frac{9t^2-1}{t^2-9}$

SOLUTION

$$\frac{t^2+3t}{6t-2}\cdot\frac{9t^2-1}{t^2-9}=\frac{t(t+3)}{2(3t-1)}\cdot\frac{(3t-1)(3t+1)}{(t-3)(t+3)}$$ (Factoring)

$$=\frac{t(3t+1)(t+3)(3t-1)}{2(t-3)(t+3)(3t-1)}$$ (Rearranging factors)

$$=\frac{t(3t+1)}{2(t-3)}$$ (Eliminating the common factors)

The final simplified fraction in Example 5 is left in factored form. Note that each fraction was expressed in factored form, and the terms were rearranged to simplify reducing.

Example 6 Simplify $\frac{x^2+3x+2}{x^2-4x-12}\cdot\frac{x-6}{x+1}$

continued

SOLUTION First factor the trinomials, if possible:

$$\frac{x^2 + 3x + 2}{x^2 - 4x - 12} = \frac{(x + 2)(x + 1)}{(x + 2)(x - 6)}$$

Thus

$$\frac{x^2 + 3x + 2}{x^2 - 4x - 12} \cdot \frac{x - 6}{x + 1} = \frac{(x + 2)(x + 1)(x - 6)}{(x + 2)(x - 6)(x + 1)} \quad \text{(Factoring)}$$

$$= \frac{\overset{1}{\cancel{(x + 2)}}\overset{1}{\cancel{(x + 1)}}\overset{1}{\cancel{(x - 6)}}}{\underset{1}{\cancel{(x + 2)}}\underset{1}{\cancel{(x - 6)}}\underset{1}{\cancel{(x + 1)}}} \quad \text{(Lining out common factors)}$$

$$= \frac{1}{1} = 1 \quad \text{(Simplifying)}$$

Note in Example 6 that it is not necessary to rearrange the factors to get the form $\frac{ak}{bk}$. Once the use of the fundamental theorem is understood, common factors can be lined out. This reduces the amount of writing.

Warning! A common error is to line out common terms. Be very careful that **only factors** common to the numerator and denominator are lined out.

Caution! A lined out factor is replaced with the factor 1. The 1 may be written as shown in Example 6 or mentally understood.

Division

By the definition of division, $\frac{10}{2}$ represents the number that must be multiplied by 2 to produce 10. Thus $\frac{10}{2} = 5$, since $2 \cdot 5 = 10$.

Now let us use the fundamental theorem of fractions and multiply the numerator and denominator of $\frac{10}{2}$ by $\frac{1}{2}$.

$$\frac{10}{2} = \frac{10\left(\frac{1}{2}\right)}{2\left(\frac{1}{2}\right)} = \frac{5}{1} = 5$$

Since $\frac{1}{2}$ is the reciprocal (multiplication inverse of 2), we see that the division of 10 by 2 is the same as the multiplication of 10 by the reciprocal of the divisor 2.

Every real number x, except 0, has a reciprocal $\frac{1}{x}$ such that $x\left(\frac{1}{x}\right) = 1$.

Example 7 Find the reciprocal of

a. $\frac{5}{7}$ b. $\frac{-5}{7}$ c. $x - 2$

SOLUTION

a. The reciprocal of $\frac{5}{7}$ is $\frac{7}{5}$, since $\frac{5}{7}\left(\frac{7}{5}\right) = 1$.

b. The reciprocal of $\frac{-5}{7}$ is $\frac{-7}{5}$, since $\frac{-5}{7}\left(\frac{-7}{5}\right) = 1$.

c. The reciprocal of $x - 2$ is $\frac{1}{x-2}$, since $(x-2)\left(\frac{1}{x-2}\right) = 1$.

Using the fact that $\frac{7}{5}$ is the reciprocal of $\frac{5}{7}$,

$$\frac{2}{3} \div \frac{5}{7} = \frac{2}{3}\cdot\frac{7}{5} = \frac{14}{15}$$

In general, we have the following theorem.

Theorem: Quotient of Fractions

The quotient of one fraction divided by another fraction is obtained by multiplying the numerator by the reciprocal of the denominator.

In symbols, for n, d, r, and s any real numbers with $d \neq 0$, $s \neq 0$, and $r \neq 0$,

$$\frac{n}{d} \div \frac{r}{s} = \frac{n}{d}\cdot\frac{s}{r} = \frac{ns}{dr}$$

Example 8 Express $\frac{3}{5} \div \frac{9}{4}$ as a single fraction.

SOLUTION

$$\frac{3}{5} \div \frac{9}{4} = \frac{3}{5}\cdot\frac{4}{9} = \frac{3\cdot 4}{5\cdot 3\cdot 3} = \frac{4}{15}$$

Example 9 Simplify $\frac{3a^2}{5b^3} \div \frac{6a}{10b}$

SOLUTION

$$\frac{3a^2}{5b^3} \div \frac{6a}{10b} = \frac{3a^2}{5b^3}\cdot\frac{10b}{6a}$$

$$= \frac{(3a^2)(10b)}{(6a)(5b^3)} = \frac{(a)(2)}{(2)(b^2)} = \frac{a}{b^2}$$

Example 10 Simplify $\frac{x+1}{x} \div \frac{x^2+3x+2}{x^2+x}$

SOLUTION

$$\frac{x+1}{x} \cdot \frac{x^2+x}{x^2+3x+2}$$ (Multiplying by the reciprocal of the divisor)

$$= \frac{x+1}{x} \cdot \frac{x(x+1)}{(x+2)(x+1)}$$ (Factoring)

$$= \frac{\cancel{x}(x+1)\cancel{(x+1)}}{\cancel{x}(x+2)\cancel{(x+1)}}$$ (Using the fundamental theorem)

$$= \frac{x+1}{x+2}$$ (Simplifying)

Example 11 Simplify $\frac{x^2+4x}{x^3+4x^2+4x} \div \frac{5x+20}{x^3-4x^2-12x}$

SOLUTION

$$\frac{x^2+4x}{x^3+4x^2+4x} \cdot \frac{x^3-4x^2-12x}{5x+20}$$ (Multiplying by the reciprocal of the divisor)

$$= \frac{x(x+4)(x)(x+2)(x-6)}{x(x+2)(x+2)5(x+4)}$$ (Factoring)

$$= \frac{x(x-6)\cancel{(x)}\cancel{(x+2)}\cancel{(x+4)}}{5(x+2)\cancel{(x)}\cancel{(x+2)}\cancel{(x+4)}}$$ (Using the fundamental theorem)

$$= \frac{x(x-6)}{5(x+2)}$$ (Simplifying)

Exercises 4.2

Simplify Exercises 1–80.

1. $\frac{2}{3} \cdot \frac{4}{3}$

2. $\frac{3}{8} \cdot \frac{16}{27}$

3. $\frac{5}{8} \cdot \frac{2}{3} \cdot \frac{3}{5}$

4. $\frac{1}{10} \cdot \frac{4}{5} \cdot \frac{7}{12}$

5. $\frac{2}{3} \div \frac{4}{3}$

6. $\frac{3}{8} \div \frac{1}{4}$

7. $\frac{1}{3} \div \frac{1}{9}$

8. $\left(\frac{2}{3} \cdot \frac{3}{5}\right) \div \frac{5}{12}$

9. $\left(\frac{2}{3} \div \frac{3}{5}\right) \cdot \frac{5}{12}$

10. $\left(\frac{1}{4} \div \frac{2}{3}\right) \div \frac{3}{8}$

11. $\frac{2}{x} \cdot \frac{3}{y}$

12. $\frac{5}{r} \cdot \frac{4}{t}$

13. $\frac{2}{x} \cdot \frac{x^2}{4}$

14. $\frac{3}{y} \cdot \frac{y^2}{6xy}$

15. $\frac{7y}{3x} \cdot \frac{3x}{14y^2}$

16. $\frac{2}{5ab^2} \cdot \frac{15ab}{22}$

17. $\frac{14x^2}{9y^2} \cdot \frac{36y^2}{7x}$

18. $\frac{10t^2}{3r} \cdot \frac{9r}{2t}$

19. $\frac{15a^2}{4b^2} \cdot \frac{8b^3}{45ab^2}$

20. $\frac{5}{3xy} \cdot \frac{21y^3}{7x}$

21. $\frac{3x}{x-3} \cdot \frac{2y}{y-2}$

22. $\frac{x+y}{4} \cdot \frac{x-y}{4}$

23. $\frac{4ab}{-6xy^2} \cdot \frac{-8x^2y}{12a^2b}$

24. $\frac{-6xy}{4a^2b} \cdot \frac{10ab^2}{15xy^2}$

25. $\frac{11r^2}{36t^2} \cdot \frac{18t}{-55r^2}$

26. $\frac{-8x^3}{27y^3} \cdot \frac{3y}{-2x}$

27. $\frac{9x-18}{2x-6} \cdot \frac{8x-24}{9x+18}$

28. $\frac{x^2}{x^2-1} \cdot \frac{x^2+x}{x^2}$

29. $\frac{x-y}{2xy} \cdot \frac{8x^2y^2}{(x-y)^2}$

30. $\frac{(r+s)^2}{r^2s^2} \cdot \frac{rs}{r+s}$

31. $\frac{t^2-4}{t^2-4t} \cdot \frac{2t-8}{2t+4}$

32. $\frac{3y-9}{3y+27} \cdot \frac{y^2+9y}{y^2-9}$

33. $\frac{a^2-25}{5} \cdot \frac{10a}{a^2+4a-5}$

34. $\frac{x^2-4xy+4y^2}{9x^3} \cdot \frac{18x}{4x-8y}$

35. $\dfrac{x^2+6x+9}{3x^2+6x}\cdot\dfrac{x^2-4}{x^2+x-6}$

36. $\dfrac{3x^2+3}{x^2-x-6}\cdot\dfrac{x^2-5x+6}{x^2+1}$

37. $\dfrac{a^2-b^2}{3}\cdot\dfrac{3a-3b}{a^2+ab}$

38. $\dfrac{5y^2-5y}{4y-40}\cdot\dfrac{y^3}{3}\cdot\dfrac{y^2-9y-10}{2-2y}$

39. $\dfrac{2n^2-3n-2}{n^3+2n^2-3n}\cdot\dfrac{6n^2-6n}{6n^2+3n}$

40. $\dfrac{x^2-5x}{x^2-3x+2}\cdot\dfrac{x^2-x}{5x-25}$

41. $\dfrac{a^2-3a+2}{a-1}\cdot\dfrac{a+5}{a^2+3a-10}$

42. $\dfrac{t+2}{t^2+t-2}\cdot\dfrac{t^2-4t+3}{t-3}$

43. $\dfrac{x^2+6x+9}{x^2+4x+3}\cdot\dfrac{2x^2+2x}{x^2-9}$

44. $\dfrac{7a+14}{a^2+4a+4}\cdot\dfrac{a^2-4}{5a-10}$

45. $\dfrac{3x^2-3x}{x^2-81}\cdot\dfrac{x^2+4x-45}{3x-15}$

46. $\dfrac{10y-21}{y^2-4y+4}\cdot\dfrac{y^2+y-6}{10y^2+9y-63}$

47. $\dfrac{2x^2+13x+15}{2x^2-x-6}\cdot\dfrac{x^2-x-2}{x^2+6x+5}$

48. $\dfrac{10y^2-11y-6}{3y^2-6y}\cdot\dfrac{y^2-y-2}{6y^2-13y+6}$

49. $(9t^2-25)\dfrac{5}{3t+5}$

50. $(2y+5)^2\dfrac{2y-5}{2y+5}$

51. $(x+6)^2\left(\dfrac{x-6}{x+6}\right)$

52. $(t^2-16)\left(\dfrac{t}{t-4}\right)$

53. $\dfrac{4}{x}\div\dfrac{12}{x}$

54. $\dfrac{y}{9}\div\dfrac{y}{45}$

55. $\dfrac{12x}{55y^2}\div\dfrac{60x^2}{33y}$

56. $\dfrac{2a}{9b^2}\div\dfrac{8a^2}{3b}$

57. $\dfrac{10}{3t-6}\div\dfrac{15}{2t-4}$

58. $\dfrac{3x-3}{5x-10}\div\dfrac{5x-5}{6x-12}$

59. $\dfrac{x^2-81}{x-3}\div\dfrac{x-9}{x^2-9}$

60. $\dfrac{y^2-25}{y+4}\div\dfrac{y+5}{y^2-16}$

61. $\dfrac{xy+x^2}{2xy-y^2}\div\dfrac{xy+y^2}{2x^2-xy}$

62. $\dfrac{4x^2-8xy}{4x^2-9y^2}\div\dfrac{4x^2-16y^2}{4x^2-6xy}$

63. $\dfrac{7x}{x^2-49}\div\dfrac{x^2-14x+49}{x^2+14x+49}$

64. $\dfrac{y^2-y-12}{y^2-16}\div\dfrac{y^2-9}{12y^2-36y}$

65. $\dfrac{x^2+x-2}{x^2-x-12}\div\dfrac{x^2+3x+2}{x^2-7x+12}$

66. $\dfrac{x^2+12x+36}{x^2-12x+36}\div\dfrac{x^2+5x-6}{x^2-5x-6}$

67. $\dfrac{b^2-3b-18}{b^2+3b-54}\div(b+3)$

68. $\dfrac{a^2+6a-16}{a^2-64}\div(a-2)$

69. $\dfrac{x^2+x-42}{x^2-x-42}\div\dfrac{x^2-36}{x^2-49}$

70. $\dfrac{x^2+x-20}{x^2-x-20}\div\dfrac{x^2-25}{x^2-16}$

71. $\dfrac{x^2+xy}{x^2-xy}\div\dfrac{xy+x^2}{xy-y^2}$

72. $\dfrac{x^2+xy}{y^2+xy}\div\dfrac{xy-x^2}{xy-y^2}$

73. $\dfrac{x^2}{x + x^2} \div \dfrac{x - x^2}{x^2 - 1}$

74. $\dfrac{4 - 2t}{t^2 - 4} \div \dfrac{t^2 + 4}{4 + 2t}$

75. $(y^2 + 3y - 10) \div \dfrac{y - 2}{y + 5}$

76. $(y^2 - 3y - 28) \div \dfrac{y + 4}{y + 7}$

77. $\dfrac{4a - 4b}{9a + 9b} \div \dfrac{8b - 8a}{6b + 6a}$

78. $\dfrac{3ab - 3a^2}{5ab + 5b^2} \div \dfrac{5a^2 - 5ab}{3b^2 + 3ab}$

79. $\dfrac{3y - 1}{3y + 9} \div (3y^2 + 8y - 3)$

80. $\dfrac{10y - 20}{10y + 2} \div (25y^2 - 45y - 10)$

4.3 Raising a Fraction to Higher Terms

In order to add or subtract certain fractions, the fractions must be renamed so that they have the same denominator, called a **common denominator.** The process of renaming a fraction by multiplying the numerator and denominator by the same number is called **raising the fraction to higher terms.** This process is justified by the fundamental theorem of fractions.

$$\frac{n}{d} = \frac{nk}{dk}$$

In practice, it is useful to consider the renaming of a fraction as a multiplication of the original fraction by the number 1, with 1 renamed as $\frac{k}{k}$. Thus

$$\frac{n}{d} = \frac{n}{d} \cdot 1 = \frac{n}{d} \cdot \frac{k}{k} = \frac{nk}{dk}$$

Example 1 Express the fraction $\frac{3}{5}$ in higher terms with denominator 20.

SOLUTION

$$\frac{3}{5} = \frac{3k}{5k} = \frac{3k}{20}$$

Since $5k = 20$, then $k = 4$ and $3k = 3 \cdot 4 = 12$.
Thus

$$\frac{3}{5} = \frac{3 \cdot 4}{5 \cdot 4} = \frac{12}{20}$$

In other words, the question is "What number multiplied by 5 equals 20?" Clearly, the answer is 4. Thus 4 is the multiplier of the numerator and the denominator.

Example 2 Raise $\frac{7}{9y}$ to higher terms with denominator $18y^3$.

SOLUTION

$$\frac{7}{9y} = \frac{7 \cdot k}{9y \cdot k} = \frac{7k}{18y^3}$$

Since $18y^3 = 9y \cdot 2y^2$, the multiplier k is $2y^2$.

In other words, the multiplier can be found by dividing the new denominator by the original denominator:

$$\frac{18y^3}{9y} = 2y^2$$

Therefore

$$\frac{7}{9y} = \frac{7(2y^2)}{9y(2y^2)} = \frac{14y^2}{18y^3}$$

Example 3 Raise $\frac{8}{15xy^2}$ to higher terms with denominator $45x^3y^3$.

SOLUTION Find the multiplier:

$$\frac{45x^3y^3}{15xy^2} = 3x^2y$$

Multiply the numerator and denominator by $3x^2y$:

$$\frac{8}{15xy^2} = \frac{8(3x^2y)}{15xy^2(3x^2y)} = \frac{24x^2y}{45x^3y^3}$$

Another type of problem uses the property that $x = \frac{x}{1}$ to build up a fraction or to change a polynomial into a rational expression with a designated denominator.

Example 4 Rename 6 as a fraction with denominator 5.

SOLUTION

$$6 = \frac{6}{1} = \frac{6(5)}{1(5)} = \frac{30}{5}$$

Example 5 Express the polynomial $x - 2$ as a rational expression with denominator $x + 3$.

SOLUTION

$$x - 2 = \frac{x - 2}{1} = \frac{(x - 2)(x + 3)}{1(x + 3)} = \frac{x^2 + x - 6}{x + 3}$$

What is a Fraction?

Throughout history people have had difficulty in understanding fractions. The ancient Egyptians limited their fractions by requiring that the numerator be the number 1. They were unable to grasp the concept of a pair of natural numbers representing a single number, such as the fraction $\frac{5}{6}$, which is a single number made from the integers 5 and 6.

Before the time of Archimedes (287–212 B.C.), the Greeks disliked the idea of breaking unity into parts, so they worked with ratios of integers. Thus if Alpha had 50 coins and Beta had 60 coins, the ratio of their amounts would be 5 to 6.

The Romans avoided fractions by the use of subunits; feet were divided into inches and pounds into ounces. A twelfth part of the Roman unit was called *uncia*, from which is derived our modern "ounce" and "inch." Instead of regarding a measurement as $\frac{5}{6}$ of a unit, the Romans considered this measurement as 10 *uncias* $\left(\frac{5}{6} = \frac{5 \cdot 2}{6 \cdot 2} = \frac{10}{12}\right)$.

Now the rational number $\frac{5}{6}$ is interpreted as the quotient obtained when 5 is divided by 6. The number $\frac{5}{6}$ is also interpreted as the ratio of 5 to 6, and this interpretation influenced the choice of the word "rational" to describe numbers that are the quotient of two integers.

Some problems involve the signs of a fraction and the special property that $b - a = -(a - b)$.

Example 6 Express $\frac{3}{7 - x}$ with denominator $x - 7$.

SOLUTION Since $7 - x = -(x - 7)$, then

$$\frac{3}{7 - x} = \frac{3}{-(x - 7)} = \frac{-3}{x - 7}$$

Example 7 Rename $\frac{x}{x + 3}$ as an equal fraction with denominator $x^2 - 9$.

SOLUTION First find the multiplier:

$$\frac{x^2 - 9}{x + 3} = \frac{(x - 3)(x + 3)}{x + 3} = x - 3$$

Thus

$$\frac{x}{x + 3} = \frac{x(x - 3)}{(x + 3)(x - 3)} = \frac{x^2 - 3x}{x^2 - 9}$$

Example 8 Express $\frac{3}{x + 2}$ in higher terms having the denominator $x^2 - 5x - 14$.

SOLUTION First find the multiplier:

$$\frac{x^2 - 5x - 14}{x + 2} = \frac{(x - 7)(x + 2)}{x + 2} = x - 7$$

Then

$$\frac{3}{x + 2} = \frac{3(x - 7)}{(x + 2)(x - 7)} = \frac{3x - 21}{x^2 - 5x - 14}$$

Example 9 Find n so that $\frac{6}{6 - x} = \frac{n}{x^2 - 36}$

SOLUTION

$$\frac{x^2 - 36}{6 - x} = \frac{(x + 6)(x - 6)}{(-1)(x - 6)} = \frac{x + 6}{-1} = (-1)(x + 6)$$

$$\frac{6}{6 - x} = \frac{6(-1)(x + 6)}{(-1)(x - 6)(-1)(x + 6)} = \frac{-6(x + 6)}{x^2 - 36}$$

Thus $n = -6(x + 6)$.

Exercises 4.3

Express the fractions in Exercises 1–62 in terms of the denominators indicated.

1. $\frac{5}{9} = \frac{}{36}$

2. $\frac{3}{8} = \frac{}{56}$

3. $\frac{4}{7} = \frac{}{21}$

4. $\frac{3}{5} = \frac{}{25}$

5. $\frac{2}{-3} = \frac{}{42}$

6. $-5 = \frac{}{35}$

7. $\frac{4}{9} = \frac{}{36}$

8. $\frac{-2}{5} = \frac{}{20}$

9. $\frac{3}{-7} = \frac{}{42}$

10. $4 = \frac{}{-10}$

11. $\frac{x}{15} = \frac{}{45}$

12. $\frac{3}{4} = \frac{}{16x^2}$

13. $\frac{2}{5} = \frac{}{15xy}$

14. $\frac{3}{-5x} = \frac{}{45x^2}$

15. $\frac{2x}{15} = \frac{}{45x^2}$

16. $-\frac{5x}{7y^2} = \frac{}{21xy^3}$

17. $-\frac{2xy}{3z} = \frac{}{42x^2z^2}$

18. $\frac{a^2}{b^3} = \frac{}{4a^2b^4}$

19. $\frac{3x}{-8} = \frac{}{40x^2}$

20. $-\frac{2x}{5y^2} = \frac{}{30xy^3}$

21. $-\frac{3xy}{7z} = \frac{}{42xyz}$

22. $\frac{a}{b^2} = \frac{}{5ab^3}$

23. $\frac{5}{x+1} = \frac{}{2x+2}$

24. $\frac{x}{2x-5} = \frac{}{6x-15}$

25. $\frac{x-3}{x+4} = \frac{}{5x+20}$

26. $\frac{2x}{x-2} = \frac{}{x^2-2x}$

27. $\frac{x}{x-5} = \frac{}{2x^2-10x}$

28. $\frac{2}{y} = \frac{}{y^2+2y}$

29. $\frac{3y}{y+3} = \frac{}{y^2-9}$

30. $\frac{t+3}{t-2} = \frac{}{t^2+t-6}$

31. $x+2 = \frac{}{x+1}$

32. $\frac{1}{4x-3} = \frac{}{3-4x}$

33. $\frac{-x}{7-x} = \frac{}{x-7}$

34. $\frac{1}{5-2x} = \frac{}{2x^2-5x}$

35. $\frac{y}{4-y} = \frac{\qquad}{y^2-16}$

36. $\frac{x+3}{x+4} = \frac{\qquad}{x^2+6x+8}$

37. $\frac{a+1}{a-3} = \frac{\qquad}{a^2-2a-3}$

38. $x = \frac{\qquad}{6x^2-6x}$

39. $\frac{t-2}{t+2} = \frac{\qquad}{14+9t+t^2}$

40. $\frac{a-b}{a+b} = \frac{\qquad}{(a+b)^2}$

41. $\frac{2x}{2x-1} = \frac{\qquad}{4x^2-1}$

42. $\frac{1+t}{1+3t} = \frac{\qquad}{18t^2+6t}$

43. $\frac{r}{r+4} = \frac{\qquad}{r^2+r-12}$

44. $\frac{u-5}{u-6} = \frac{\qquad}{u^2-4u-12}$

45. $6x-5 = \frac{\qquad}{x-2}$

46. $\frac{1}{5-6y} = \frac{\qquad}{6y-5}$

47. $\frac{-t}{1-t} = \frac{\qquad}{t-1}$

48. $\frac{x}{5-x} = \frac{\qquad}{x^2-25}$

49. $\frac{-1}{7-2r} = \frac{\qquad}{4r^2-49}$

50. $\frac{x+2}{x+4} = \frac{\qquad}{2x^3+5x^2-12x}$

51. $\frac{y-1}{y-2} = \frac{\qquad}{12-8y+y^2}$

52. $\frac{6x}{x+6} = \frac{\qquad}{2x^2+2x-60}$

53. $x = \frac{\qquad}{5x+10}$

54. $\frac{2x+3y}{2x-3y} = \frac{\qquad}{(2x-3y)^2}$

55. $\frac{x}{3} = \frac{\qquad}{3x^2-75x}$

56. $\frac{3}{3y+1} = \frac{\qquad}{1-9y^2}$

57. $\frac{6a^2}{5a-1} = \frac{\qquad}{10a^2+13a-3}$

58. $\frac{r+1}{r+2} = \frac{\qquad}{7r^4-28r^2}$

59. $\frac{-a}{2} = \frac{\qquad}{4a^2-10a}$

60. $\frac{9}{x^2-9} = \frac{\qquad}{(x-3)(x+3)^2}$

61. $\frac{5n+5}{n-1} = \frac{\qquad}{3n^3-3n}$

62. $\frac{ab}{(a-b)^2} = \frac{\qquad}{(b-a)^2}$

4.4 Addition and Subtraction

The procedures for adding and subtracting algebraic fractions are based on the theorems for the addition and subtraction of quotients of real numbers and on the fundamental theorem of fractions.

Let n, m, and d be any real numbers with $d \neq 0$.

The Addition of Quotients Theorem

$$\frac{n}{d} + \frac{m}{d} = \frac{n + m}{d}$$

The Subtraction of Quotients Theorem

$$\frac{n}{d} - \frac{m}{d} = \frac{n - m}{d}$$

Examination of the addition and subtraction theorems for the quotients of real numbers reveals that fractions cannot be combined into a single fraction by the operations of addition and subtraction unless the denominators of the fractions being combined are the same.

The fundamental theorem of fractions provides a technique for renaming fractions so that they will have a common denominator. A common denominator can always be found by multiplying the denominators of the fractions that are to be added or subtracted.

To provide insight into this process, a visual model for the addition of two fractions of arithmetic is provided in Fig. 4.3.

Now consider the addition problem from arithmetic,

$$\frac{5}{12} + \frac{7}{18}$$

By multiplying the denominators, a common denominator can be found: $12 \cdot 18 = 216$

$$\frac{5}{12} = \frac{5 \cdot 18}{12 \cdot 18} = \frac{90}{216} \quad \text{and} \quad \frac{7}{18} = \frac{7 \cdot 12}{18 \cdot 12} = \frac{84}{216}$$

Thus

$$\frac{5}{12} + \frac{7}{18} = \frac{90}{216} + \frac{84}{216} = \frac{90 + 84}{216} = \frac{174}{216}$$

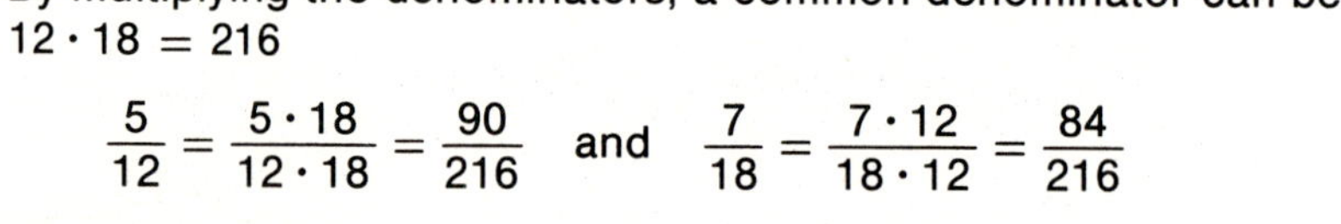

Figure 4.3

However, $\frac{174}{216}$ is not in its lowest terms and so must be simplified:

$$\frac{174}{216} = \frac{2 \cdot 87}{2 \cdot 108} = \frac{2 \cdot 3 \cdot 29}{2 \cdot 3 \cdot 36} = \frac{29}{36}$$

Thus

$$\frac{5}{12} + \frac{7}{18} = \frac{29}{36}$$

There is another procedure for finding a common denominator that simplifies the arithmetical calculations involved and sometimes, but not always, avoids the final simplification. This procedure is called finding the **least common denominator,** or **LCD.** The objective is to find the smallest common denominator—that is, the smallest number that has each of the original denominators for a factor. Thus, to find the least common denominator, the original denominators must be factored into prime factors.

The LCD of two or more numbers is the product of all the primes that occur in the prime factorization of each number, with each prime taken the greatest number of times it occurs in any denominator.

For example, to find the LCD of $\frac{5}{12}$ and $\frac{7}{18}$, factor 12 and 18:

$$12 = 2 \cdot 2 \cdot 3 \quad \text{and} \quad 18 = 2 \cdot 3 \cdot 3$$

Then the LCD is $2 \cdot 2 \cdot 3 \cdot 3 = 36$.

Now the sum $\frac{5}{12} + \frac{7}{18}$ is obtained as follows:

$$\frac{5}{12} + \frac{7}{18} = \frac{15}{36} + \frac{14}{36} = \frac{29}{36}$$

This procedure is similar to the one that is used in adding or subtracting algebraic fractions, with the objective of finding the smallest number for the common denominator replaced by the objective of **finding the polynomial of smallest degree for the common denominator.**

Example 1 Write $\frac{3}{16x} + \frac{5}{16x}$ as a single fraction.

SOLUTION The denominators of the two fractions are the same; therefore, the addition of quotients theorem can be applied directly:

$$\frac{3}{16x} + \frac{5}{16x} = \frac{3 + 5}{16x} = \frac{8}{16x}$$

But

$$\frac{8}{16x} = \frac{8 \cdot 1}{8 \cdot 2x} = \frac{1}{2x}$$

continued

Therefore, in lowest terms,

$$\frac{3}{16x} + \frac{5}{16x} = \frac{1}{2x}$$

Example 2 Write $\dfrac{5}{6x^2} + \dfrac{1}{9x}$ as a single fraction.

SOLUTION The two fractions have unlike denominators.

1. Find the LCD by factoring $6x^2$ and $9x$:

$$6x^2 = 2 \cdot 3 \cdot x \cdot x \quad \text{and} \quad 9x = 3 \cdot 3 \cdot x$$

   Thus the LCD is $2 \cdot 3 \cdot 3 \cdot x \cdot x = 18x^2$.

2. Rename the fractions by using the fundamental theorem of fractions:

$$\frac{5}{6x^2} = \frac{5 \cdot 3}{6x^2 \cdot 3} = \frac{15}{18x^2} \quad \text{and} \quad \frac{1}{9x} = \frac{1 \cdot 2x}{9x \cdot 2x} = \frac{2x}{18x^2}$$

3. Apply the theorem for the addition of quotients:

$$\frac{5}{6x^2} + \frac{1}{9x} = \frac{15}{18x^2} + \frac{2x}{18x^2} = \frac{15 + 2x}{18x^2}$$

$$= \frac{2x + 15}{18x^2} \quad \text{(Conventional form)}$$

Example 3 Write $\dfrac{3}{y} + \dfrac{2}{3} - \dfrac{y}{y+3}$ as a single fraction.

SOLUTION

1. The LCD is $3y(y + 3)$.
2. Rename the fractions so that each has the denominator $3y(y + 3)$:

$$\frac{3}{y} = \frac{3 \cdot 3(y+3)}{y \cdot 3(y+3)} = \frac{9y + 27}{3y(y+3)}$$

$$\frac{2}{3} = \frac{2 \cdot y(y+3)}{3 \cdot y(y+3)} = \frac{2y^2 + 6y}{3y(y+3)}$$

$$\frac{y}{y+3} = \frac{y \cdot 3y}{(y+3) \cdot 3y} = \frac{3y^2}{3y(y+3)}$$

3. Apply the theorem for the addition and subtraction of quotients and combine like terms:

$$\frac{9y + 27}{3y(y+3)} + \frac{2y^2 + 6y}{3y(y+3)} - \frac{3y^2}{3y(y+3)}$$

continued

$$= \frac{(9y + 27) + (2y^2 + 6y) - (3y^2)}{3y(y + 3)}$$

$$= \frac{-y^2 + 15y + 27}{3y(y + 3)}$$

Example 4 Simplify $\dfrac{1}{x^2 - 2x - 3} + \dfrac{1}{x^2 + 6x + 5}$

SOLUTION

1. Factor each denominator to find the LCD.

$$x^2 - 2x - 3 = (x + 1)(x - 3)$$
$$x^2 + 6x + 5 = (x + 1)(x + 5)$$

2. Find the LCD.

$$\text{LCD} = (x + 1)(x - 3)(x + 5)$$

3. Rename each fraction with the LCD as denominator.

$$\frac{1}{(x + 1)(x - 3)} = \frac{1(x + 5)}{(x + 1)(x - 3)(x + 5)}$$

$$\frac{1}{(x + 1)(x + 5)} = \frac{1(x - 3)}{(x + 1)(x + 5)(x - 3)}$$

4. Add the resulting numerators and place the sum over the LCD.

$$\frac{(x + 5) + (x - 3)}{(x + 1)(x - 3)(x + 5)} = \frac{2x + 2}{(x + 1)(x - 3)(x + 5)}$$

5. Factor the resulting numerator to see if the fraction can be simplified.

$$\frac{2x + 2}{(x + 1)(x - 3)(x + 5)} = \frac{2(x + 1)}{(x + 1)(x - 3)(x + 5)}$$

6. Simplify, when possible.

$$\frac{2(x + 1)}{(x + 1)(x - 3)(x + 5)} = \frac{2}{(x - 3)(x + 5)}$$

In subtracting fractions, it is especially important to remember that the bar, the horizontal line separating the numerator and denominator, is a grouping symbol indicating that the numerator is to be considered as a single number and that the denominator is to be considered as a single number.

In Example 5 the parentheses are used to enclose the numerator of each fraction in order to stress this fact.

Warning! When a minus sign precedes a fraction, each term of the numerator of the fraction must have its sign changed.

Example 5 Simplify $\frac{5}{x^2 - 6x + 9} - \frac{4}{x^2 - 9}$

SOLUTION

1. Find the LCD.

$$x^2 - 6x + 9 = (x - 3)(x - 3)$$
$$x^2 - 9 = (x - 3)(x + 3)$$

Thus the LCD $= (x - 3)^2(x + 3)$.

2. Rename fractions.

$$\frac{5}{(x-3)^2} = \frac{5(x+3)}{(x-3)^2(x+3)} = \frac{5x+15}{(x-3)^2(x+3)}$$
$$\frac{4}{x^2-9} = \frac{4(x-3)}{(x-3)(x+3)(x-3)} = \frac{4x-12}{(x-3)^2(x+3)}$$

3. Subtract fractions (numerators).

$$\frac{5}{(x-3)^2} - \frac{4}{x^2-9} = \frac{(5x+15)}{(x-3)^2(x+3)} - \frac{(4x-12)}{(x-3)^2(x+3)}$$
$$= \frac{(5x+15)-(4x-12)}{(x-3)^2(x+3)}$$

4. Simplify.

$$= \frac{x+27}{(x-3)^2(x+3)}$$

Example 6 Simplify $\frac{2}{x^2 - 4x + 4} + \frac{3x}{4 - x^2}$

SOLUTION

1. Find the LCD.

$$x^2 - 4x + 4 = (x - 2)(x - 2)$$
$$4 - x^2 = -(x^2 - 4) = -(x - 2)(x + 2)$$

It is advisable at this time to rewrite the second fraction.

$$\frac{3x}{4-x^2} = \frac{3x}{-(x^2-4)} = \frac{-3x}{x^2-4}$$

Thus the LCD $= (x - 2)(x - 2)(x + 2)$
$= (x - 2)^2(x + 2)$

2. Rename fractions.

$$\frac{2}{x^2-4x+4} = \frac{2(x+2)}{(x-2)(x-2)(x+2)} = \frac{2x+4}{(x-2)^2(x+2)}$$
$$\frac{-3x}{x^2-4} = \frac{-3x(x-2)}{(x-2)(x-2)(x+2)} = \frac{-3x^2+6x}{(x-2)^2(x+2)}$$

continued

3. Add fractions.

$$\frac{2}{x^2 - 4x + 4} + \frac{3x}{4 - x^2} = \frac{2}{x^2 - 4x + 4} + \frac{-3x}{x^2 - 4}$$

$$= \frac{(2x + 4) + (-3x^2 + 6x)}{(x - 2)^2(x + 2)}$$

4. Simplify.

$$= \frac{-3x^2 + 8x + 4}{(x - 2)^2(x + 2)}$$

Example 7 Simplify $\frac{3}{x - 2} - \frac{2}{3x} - \frac{18}{3x^2 - 6x}$

SOLUTION

1. Find the LCD.

$$x - 2 = x - 2$$
$$3x = 3x$$
$$3x^2 - 6x = 3x(x - 2)$$

Thus the LCD $= 3x(x - 2)$.

2. Rename fractions.

$$\frac{3}{x - 2} = \frac{3 \cdot 3x}{3x(x - 2)} = \frac{9x}{3x(x - 2)}$$

$$\frac{2}{3x} = \frac{2(x - 2)}{3x(x - 2)} = \frac{2x - 4}{3x(x - 2)}$$

$$\frac{18}{3x^2 - 6x} = \frac{18}{3x(x - 2)}$$

3. Combine fractions (numerators).

$$\frac{3}{x - 2} - \frac{2}{3x} - \frac{18}{3x^2 - 6x}$$

$$= \frac{9x}{3x(x - 2)} - \frac{2x - 4}{3x(x - 2)} - \frac{18}{3x(x - 2)}$$

$$= \frac{9x - (2x - 4) - 18}{3x(x - 2)}$$

4. Simplify.

$$= \frac{9x - 2x + 4 - 18}{3x(x - 2)}$$

$$= \frac{7x - 14}{3x(x - 2)}$$

continued

5. Factor numerator. Simplify, if possible.

$$= \frac{7(x-2)}{3x(x-2)}$$

$$= \frac{7}{3x}$$

Exercises 4.4

Rewrite as a single fraction in lowest terms.

1. $\frac{3}{7} + \frac{2}{7}$
2. $\frac{2}{9} + \frac{5}{9}$
3. $\frac{3}{x} + \frac{2}{x}$
4. $\frac{2}{y} + \frac{5}{y}$
5. $\frac{x+1}{7} + \frac{x-7}{7}$
6. $\frac{y-2}{9} + \frac{y+9}{9}$
7. $\frac{11}{12} - \frac{5}{12}$
8. $\frac{7}{8} - \frac{1}{8}$
9. $\frac{11}{3x} - \frac{5}{3x}$
10. $\frac{7}{2y} - \frac{1}{2y}$
11. $\frac{x-2}{5} - \frac{x-5}{5}$
12. $\frac{y+6}{6} - \frac{y+4}{6}$
13. $\frac{x-1}{x+1} + \frac{2x+4}{x+1}$
14. $\frac{2y-3}{y-2} + \frac{3y-7}{y-2}$
15. $\frac{5t-2}{6t} - \frac{2t-5}{6t}$
16. $\frac{7t-8}{10t} - \frac{2t-3}{10t}$
17. $\frac{2}{3} + \frac{5}{6}$
18. $\frac{5}{12} + \frac{3}{4}$
19. $\frac{2}{3t} + \frac{5}{6t}$
20. $\frac{5}{12t} + \frac{3}{4t}$
21. $\frac{x}{14y} + \frac{x}{35y}$
22. $\frac{a}{15b} + \frac{a}{10b}$
23. $\frac{1}{5y} - \frac{3y-5}{15y^2}$
24. $\frac{1}{3x} - \frac{4x+3}{12x^2}$
25. $\frac{2}{x} + \frac{3}{x^2}$
26. $\frac{5}{y^2} - \frac{2}{y}$
27. $\frac{x}{y} - \frac{x}{y+1}$
28. $\frac{3}{x} - \frac{3}{x+1}$
29. $\frac{5}{x+2} - \frac{5}{x}$
30. $\frac{3}{y+3} - \frac{3}{y}$

31. $x - \frac{1}{x}$

32. $y + \frac{1}{y}$

33. $\frac{1}{x} + \frac{1}{y}$

34. $\frac{1}{a} - \frac{1}{b}$

35. $\frac{2}{y+1} + \frac{1}{y+2}$

36. $\frac{1}{x-3} + \frac{2}{x+6}$

37. $2 - \frac{x-1}{x+1}$

38. $x - 1 - \frac{x-1}{x+1}$

39. $\frac{2}{x-2} + \frac{3}{2-x}$

40. $\frac{5}{y-5} + \frac{y}{5-y}$

41. $\frac{a-2}{3a+3} + \frac{a-3}{2a+2}$

42. $\frac{x}{5x-5} - \frac{2}{3-3x}$

43. $\frac{t-4}{(t+4)^2} + \frac{1}{t+4}$

44. $\frac{5x}{(3x-1)^2} + \frac{4}{3x-1}$

45. $\frac{4x}{(2x-3)^2} - \frac{2}{2x-3}$

46. $\frac{5t}{(t-5)^2} - \frac{5}{t-5}$

47. $\frac{1}{y^2+2y} + \frac{1}{y^2-2y}$

48. $\frac{3}{x^2-3x} + \frac{3}{x^2+3x}$

49. $\frac{x+1}{x-5} - \frac{x+3}{x+2}$

50. $\frac{x-6}{x+6} - \frac{x-2}{x+1}$

51. $\frac{2t}{9t+18} + \frac{t}{3t-6}$

52. $\frac{3}{2t-6} + \frac{1}{6t+2}$

53. $\frac{x}{3x-12} - \frac{x^2}{x^2-16}$

54. $\frac{x+4}{x^2-49} - \frac{x+1}{4x^2+28x}$

55. $\frac{x+7}{4-6x} + \frac{3x^2+14}{9x^2-4}$

56. $\frac{1}{2x-x^2} + \frac{2}{x^2-4}$

57. $\frac{x+3}{x+2} + \frac{7}{x^2-3x-10}$

58. $\frac{x-1}{x+6} - \frac{35}{x^2+7x+6}$

59. $\frac{t^2-2t-21}{t^2-4t-5} + \frac{t-6}{5-t}$

60. $\frac{x^2+3}{24-2x-x^2} + \frac{2x+1}{2x-8}$

61. $\frac{1}{a^2-1} - \frac{1}{a^2+2a+1}$

62. $\frac{1}{9x^2+6x+1} - \frac{2}{9x^2-1}$

63. $\frac{3}{x^2+5x+4} - \frac{2}{x^2+4x+3}$

64. $\frac{5}{x^2+x-6} + \frac{1}{x^2-5x+6}$

65. $\frac{y+9}{4y^2-5y-6} - \frac{y+7}{5y^2-11y+2}$

66. $\frac{7y}{2y^2+5y-3} - \frac{10y}{3y^2+8y-3}$

67. $\frac{2y}{y^2-5y+6} + \frac{3}{2+3y-2y^2}$

68. $\frac{x+2}{3x^2+5x+2} - \frac{x+3}{3x^2-16x-12}$

69. $\frac{3}{y^2-5y} + \frac{2}{y^2+5y} - \frac{4}{y^2-25}$

70. $\frac{1}{x^2-49} + \frac{1}{(x+7)^2} - \frac{1}{(x-7)^2}$

71. $1 + \frac{2}{x} - \frac{x - 1}{x^2 - x}$

72. $2a - 5 + \frac{25}{2a + 5}$

73. $\frac{5}{2x + 8} - \frac{7}{3x - 12} + \frac{20}{x^2 - 16}$

74. $\frac{2}{2x + x^2} - \frac{3}{2x - x^2} - \frac{4}{x^2 - 4}$

75. $\frac{x}{x + y} + \frac{y}{x - y} - \frac{2xy}{x^2 - y^2}$

76. $\frac{4a^2}{x^2 - a^2} - \frac{x - a}{x + a} - \frac{x + a}{x - a}$

4.5 Complex Fractions

A complex fraction is one that contains a fraction in the numerator or in the denominator or in both numerator and denominator.

A complex fraction is simplified first by expressing the numerator and denominator as single fractions and then by dividing the resulting fractions.

Example 1 Simplify $\dfrac{3 + \frac{1}{3}}{2 - \frac{3}{5}}$

SOLUTION This problem is the following indicated division:

$$\left(3 + \frac{1}{3}\right) \div \left(2 - \frac{3}{5}\right)$$

Rewriting the numerator and denominator as single fractions,

$$\left(\frac{9}{3} + \frac{1}{3}\right) \div \left(\frac{10}{5} - \frac{3}{5}\right) = \frac{10}{3} \div \frac{7}{5} = \frac{10}{3} \cdot \frac{5}{7} = \frac{50}{21}$$

Example 2 Express $\left(x + \frac{1}{y}\right) \div \left(x^2 - \frac{1}{y^2}\right)$ as a single fraction in lowest terms.

SOLUTION Write the divisor as a single fraction and the dividend as a single fraction:

$$x + \frac{1}{y} = \frac{xy + 1}{y}$$

$$x^2 - \frac{1}{y^2} = \frac{x^2y^2 - 1}{y^2}$$

continued

Now

$$\left(x + \frac{1}{y}\right) \div \left(x^2 - \frac{1}{y^2}\right) = \frac{xy + 1}{y} \div \frac{x^2y^2 - 1}{y^2}$$

$$\begin{aligned}\frac{xy + 1}{y} \div \frac{x^2y^2 - 1}{y^2} &= \frac{xy + 1}{y} \cdot \frac{y^2}{x^2y^2 - 1} \\ &= \frac{xy + 1}{y} \cdot \frac{y^2}{(xy + 1)(xy - 1)} \\ &= \frac{y \cdot y(xy + 1)}{y(xy - 1)(xy + 1)} \\ &= \frac{y}{xy - 1}\end{aligned}$$

Example 3 Simplify $\left(\frac{x}{x - y} - \frac{y}{x + y}\right) \div \left(\frac{x}{x + y} + \frac{y}{x - y}\right)$

SOLUTION First write the terms within parentheses as single fractions.

$$\frac{x}{x - y} - \frac{y}{x + y} = \frac{x(x + y) - y(x - y)}{(x - y)(x + y)} = \frac{x^2 + y^2}{x^2 - y^2}$$

and

$$\frac{x}{x + y} + \frac{y}{x - y} = \frac{x(x - y) + y(x + y)}{(x + y)(x - y)} = \frac{x^2 + y^2}{x^2 - y^2}$$

Now the problem becomes

$$\left(\frac{x^2 + y^2}{x^2 - y^2}\right) \div \left(\frac{x^2 + y^2}{x^2 - y^2}\right) = \frac{x^2 + y^2}{x^2 - y^2} \cdot \frac{x^2 - y^2}{x^2 + y^2} = 1$$

Example 4 Simplify $\dfrac{\frac{1}{x} + \frac{1}{2}}{\frac{1}{4}}$

SOLUTION Writing this as a division problem,

$$\begin{aligned}\frac{\frac{1}{x} + \frac{1}{2}}{\frac{1}{4}} &= \left(\frac{1}{x} + \frac{1}{2}\right) \div \frac{1}{4} \\ &= \left(\frac{2 + x}{2x}\right) \cdot \frac{4}{1} \\ &= \frac{(2 + x)2}{x} = \frac{2(x + 2)}{x}\end{aligned}$$

Another way to simplify a complex fraction involves the fundamental theorem of fractions. The main numerator and the main denominator of the complex fraction are each multiplied by the LCD of all fractions in the main numerator and main denominator. This method is shown in Example 5.

Example 5 Simplify $\dfrac{\dfrac{1}{x}+\dfrac{1}{2}}{\dfrac{1}{4}}$

SOLUTION (Alternate method)

1. The LCD of $\frac{1}{x}$ and $\frac{1}{2}$ and $\frac{1}{4}$ is $4x$.
2. Multiplying the main numerator and main denominator by the LCD,

$$\frac{4x\left(\frac{1}{x}+\frac{1}{2}\right)}{4x\left(\frac{1}{4}\right)} = \frac{4+2x}{x} = \frac{2(x+2)}{x}$$

Example 6 Simplify $\dfrac{1}{1+\dfrac{1}{1+\dfrac{1}{2}}}$

SOLUTION As a division problem this becomes

$$1 \div \left(1+\frac{1}{1+\frac{1}{2}}\right) = 1 \div \left(1+\frac{1}{\frac{3}{2}}\right)$$

$$= 1 \div \left(1+\frac{2}{3}\right)$$

$$= 1 \div \left(\frac{5}{3}\right)$$

$$= 1 \cdot \frac{3}{5} = \frac{3}{5}$$

Example 7 Use the alternate method to simplify $\dfrac{1}{x-\dfrac{x-1}{x-\dfrac{1}{x}}}$

continued

SOLUTION This complex fraction contains another complex fraction, which we simplify first.

$$\frac{x-1}{x-\frac{1}{x}} = \frac{x(x-1)}{x\left(x-\frac{1}{x}\right)} \qquad \left(\text{Fundamental theorem: } \frac{n}{d} = \frac{nk}{dk}\right)$$

$$= \frac{x(x-1)}{x^2-1}$$

$$= \frac{x}{x+1} \qquad \left(\text{Using } \frac{nk}{dk} = \frac{n}{d}\right)$$

Now the problem can be written as follows:

$$\frac{1}{x-\frac{\mathbf{x-1}}{\mathbf{x}-\frac{\mathbf{1}}{\mathbf{x}}}} = \frac{1}{x-\frac{\mathbf{x}}{\mathbf{x+1}}} \qquad \text{(Substituting the above result)}$$

$$= \frac{(x+1)1}{(x+1)\left(x-\frac{x}{x+1}\right)} \qquad \left(\text{Using } \frac{n}{d} = \frac{nk}{dk}\right)$$

$$= \frac{x+1}{x^2+x-x}$$

$$= \frac{x+1}{x^2}$$

Exercises 4.5

Simplify and write as single fractions in Exercises 1–30.

1. $\left(2+\frac{3}{4}\right) \div \left(3+\frac{7}{8}\right)$
2. $\left(2-\frac{3}{4}\right) \div \left(3-\frac{7}{8}\right)$
3. $\left(1+\frac{1}{3}\right) \div \left(1+\frac{1}{9}\right)$
4. $\left(3-\frac{5}{7}\right) \div \left(5+\frac{3}{14}\right)$
5. $\left(\frac{2}{3}+\frac{1}{4}\right) \div \left(\frac{5}{3}\right)$
6. $\frac{5}{3} \div \left(\frac{2}{3}+\frac{1}{4}\right)$
7. $\left(1+\frac{1}{y}\right) \div \left(1-\frac{1}{y^2}\right)$
8. $\left(1+\frac{3}{2t-3}\right) \div \left(\frac{1}{2t-3}\right)$
9. $\left(1-\frac{x^2}{9}\right) \div \left(1+\frac{x}{3}\right)$
10. $(a-b) \div \left(\frac{a}{b}-\frac{b}{a}\right)$
11. $\left(\frac{t}{2}+\frac{t}{3}\right) \div \left(t^2-\frac{t}{2}\cdot\frac{t}{3}\right)$
12. $\left(6+\frac{7}{x}-\frac{3}{x^2}\right) \div \left(9+\frac{3}{x}-\frac{2}{x^2}\right)$

13. $\left(\frac{x}{y} + \frac{y}{5}\right) \div \frac{xy}{10}$

14. $\left(\frac{x}{a} - \frac{x}{b}\right) \div \frac{x}{ab}$

15. $\left(a + \frac{3}{a}\right) \div \left(a^2 - \frac{9}{a^2}\right)$

16. $\left(3 - \frac{1}{y}\right) \div \left(9 - \frac{1}{y^2}\right)$

17. $\left(x + 3 - \frac{10}{x}\right) \div \left(x - \frac{25}{x}\right)$

18. $\left(y + 2 - \frac{8}{y}\right) \div \left(y - \frac{16}{y}\right)$

19. $\left(y - \frac{9}{y}\right) \div \left(y - 7 + \frac{12}{y}\right)$

20. $\left(\frac{1}{3} + \frac{1}{n}\right) \div \left(1 - \frac{1}{3n}\right)$

21. $\dfrac{1}{\frac{1}{x} + \frac{1}{y}}$

22. $\dfrac{1}{\frac{2}{a} + \frac{3}{b}}$

23. $\left(1 - \frac{1}{x + 1}\right) \div \left(1 + \frac{1}{x - 1}\right)$

24. $\left(1 + \frac{5}{x - 5}\right) \div \left(1 - \frac{5}{x + 5}\right)$

25. $\left(\frac{1}{a} - \frac{2}{a^2} - \frac{3}{a^3}\right) \div \left(1 - \frac{9}{a^2}\right)$

26. $\left(3 - \dfrac{2}{\frac{1}{5} + \frac{1}{3}}\right) \div \left(5 - \dfrac{2}{\frac{1}{5} + \frac{1}{3}}\right)$

27. $\left(2 + \dfrac{1}{\frac{1}{4} + \frac{1}{3}}\right) \div \left(4 - \dfrac{1}{\frac{1}{4} + \frac{1}{3}}\right)$

28. $\dfrac{1}{2 - \dfrac{1}{1 + \frac{1}{2}}}$

29. $\dfrac{1}{1 - \dfrac{1}{1 - \frac{1}{x}}}$

30. $\dfrac{1}{y - \dfrac{1}{y + \frac{1}{y}}}$

4.6 Changing Quotients into Sums

Division of a Polynomial by a Monomial

Since the division of a number by 2 means the multiplication of the number by $\frac{1}{2}$, the reciprocal of 2,

$$\frac{6x^2 - 4x + 3}{2} = \frac{1}{2}(6x^2 - 4x + 3)$$
$$= \frac{1}{2}(6x^2) - \frac{1}{2}(4x) + \frac{1}{2}(3)$$
$$= 3x^2 - 2x + \frac{3}{2}$$

In other words, the distributive axiom applies, and **each term** of the numerator must be divided by the divisor.

Example 1 Rewrite $\frac{9x^3 - 12x^2 + 6x + 2}{3x}$ as a sum.

SOLUTION Since each term of the numerator must be divided by $3x$,

$$\frac{9x^3 - 12x^2 + 6x + 2}{3x} = \frac{9x^3}{3x} - \frac{12x^2}{3x} + \frac{6x}{3x} + \frac{2}{3x}$$
$$= 3x^2 - 4x + 2 + \frac{2}{3x}$$

Division of a Polynomial by a Polynomial: Long Division

It is sometimes desirable to replace a single fraction by a sum or difference of two or more rational expressions. There are several applications (for example, problems in calculus) that require that a quotient of two polynomials be changed into the sum of fractions in which the numerator of each has a degree smaller than the degree of the denominator.

The method used to replace a quotient of two polynomials by a sum or difference is similar to the long-division algorithm used in arithmetic. (An algorithm is a method or pattern of doing a calculation.) For example, the division of 865 by 23 is done as follows:

```
     37
23)865
   69
   175
   161
    14
```

Thus

$$\frac{865}{23} = 37 + \frac{14}{23} = 37\frac{14}{23}$$

CHECK: $23(37) + 14 = 851 + 14 = 865.$

In arithmetic, the division process is stopped when the remainder is less than the divisor.

In algebra, the division process is stopped when the degree of the remainder polynomial is less than the degree of the divisor polynomial.

Example 2 Write $\dfrac{x^3 - 5x^2 + 4x + 7}{x + 2}$ as a sum of rational expressions with the degree of the polynomial in the numerator of any fraction smaller than the degree of the polynomial in its denominator.

SOLUTION

1. Divide x^3 by x to obtain the partial quotient, x^2.

$$\begin{array}{r|l} & x^2 \\ \hline x + 2 & x^3 - 5x^2 + 4x + 7 \end{array}$$

2. Subtract the product of x^2 and $x + 2$ from the dividend:

$$\begin{array}{r} x^3 + 2x^2 \qquad\qquad \\ \hline -7x^2 + 4x + 7 \end{array}$$

3. Repeat the process: Divide $-7x^2$ by x to obtain $-7x$.

$$\begin{array}{r|l} & x^2 - 7x + 18 \\ \hline x + 2 & x^3 - 5x^2 + 4x + 7 \end{array}$$

4. Subtract the product of $-7x$ and $x + 2$ from the remainder polynomial, $-7x^2 + 4x + 7$:

$$\begin{array}{r} x^3 + 2x^2 \qquad\qquad \\ \hline -7x^2 + 4x + 7 \\ -7x^2 - 14x \qquad \\ \hline 18x + 7 \end{array}$$

5. Repeat the process: Divide $18x$ by x to get 18.

$$\begin{array}{r} 18x + 36 \\ \hline -29 \end{array}$$

Subtract $18(x + 2)$.

6. Since the degree of the remainder, -29, is less than the degree of the divisor, $x + 2$, the process is stopped at this point, and the quotient is expressed as the following sum:

$$\frac{x^3 - 5x^2 + 4x + 7}{x + 2} = x^2 - 7x + 18 + \frac{-29}{x + 2}$$

Example 3 Check Example 2.

SOLUTION A division problem can be checked by multiplication.

$$(x + 2)\left(\frac{x^3 - 5x^2 + 4x + 7}{x + 2}\right) = x^3 - 5x^2 + 4x + 7$$

and

$$\begin{aligned}
&(x + 2)\left(x^2 - 7x + 18 + \frac{-29}{x + 2}\right) \\
&= (x + 2)(x^2 - 7x + 18) + (x + 2)\left(\frac{-29}{x + 2}\right) \\
&= x^3 - 5x^2 + 4x + 36 + (-29) \\
&= x^3 - 5x^2 + 4x + 7
\end{aligned}$$

Example 4 Divide and check: $\dfrac{5x^2 - 8x - 21}{x - 3}$

SOLUTION

$$\begin{array}{r}
5x + 7 \\
x - 3\overline{)5x^2 - 8x - 21} \\
\underline{5x^2 - 15x} \\
7x - 21 \\
\underline{7x - 21} \\
0
\end{array}$$

$\dfrac{5x^2}{x} = 5x$ **and** $5x(x - 3) =$ $5x^2 - 15x$

Subtracting $7x - 21$

$\dfrac{7x}{x} = 7$ **and** $7(x - 3) =$ $7x - 21$

Thus $\dfrac{5x^2 - 8x - 21}{x - 3} = 5x + 7$

CHECK (By factoring)

$$\frac{5x^2 - 8x - 21}{x - 3} = \frac{(5x + 7)(x - 3)}{(x - 3)} = 5x + 7$$

For some cases such as the one shown in Example 4, when the divisor is an exact factor of the dividend, it is much easier to use the method of reducing a fraction presented earlier than to use the long-division algorithm.

For the polynomial $x^3 + 1$ note that there is no x^2 term and no x term. However, when the long-division algorithm is used, space must be provided for these terms. They occur in the division process. This is shown in Example 5.

In general, space must be left for all terms of lower degree than the degree of the polynomial.

Example 5 Divide $\dfrac{x^3 + 1}{x + 1}$ by using the long-division algorithm.

SOLUTION The polynomial $x^3 + 1$ can also be considered as

$$x^3 + 0 \cdot x^2 + 0 \cdot x + 1$$

so places must be provided for the missing x^2 and x terms:

$$\begin{array}{r|llll}
 & x^2 & -x & +1 & \\
\hline
x + 1 & x^3 & & & +1 \\
 & x^3 + x^2 & & & \\
\hline
 & & -x^2 & & +1 \\
 & & -x^2 & -x & \\
\hline
 & & & +x & +1 \\
 & & & x & +1 \\
\hline
\end{array}$$

Thus $\dfrac{x^3 + 1}{x + 1} = x^2 - x + 1$

CHECK

$$\begin{aligned}
&(x + 1)(x^2 - x + 1) \\
&= (x + 1)(x^2 - x) + (x + 1)(1) \\
&= x^3 + x^2 - x^2 - x + x + 1 = x^3 + 1
\end{aligned}$$

In Example 5, the remainder is zero. Thus $x + 1$ is a factor of $x^3 + 1$.

It is important when using the long-division algorithm that both dividend and divisor be arranged in the same order—that is, the terms of both must be in descending order or the terms of both must be in ascending order.

Example 6 Divide and check $\dfrac{7y + y^4 - 7 + y^3}{3 - y + y^2}$

SOLUTION First arrange the polynomials in descending powers of y:

$$\frac{7y + y^4 - 7 + y^3}{3 - y + y^2} = \frac{y^4 + y^3 + 7y - 7}{y^2 - y + 3}$$

$$\begin{array}{r l}
 & y^2 + 2y \quad - 1 \\
y^2 - y + 3 \overline{)} & y^4 + y^3 \qquad\quad + 7y - 7 \\
 & \underline{y^4 - y^3 + 3y^2} \\
 & \qquad 2y^3 - 3y^2 + 7y - 7 \\
 & \qquad \underline{2y^3 - 2y^2 + 6y} \\
 & \qquad\qquad - y^2 + y - 7 \\
 & \qquad\qquad \underline{- y^2 + y - 3} \\
 & \qquad\qquad\qquad\quad - 4
\end{array}$$

Therefore

$$\frac{y^4 + y^3 + 7y - 7}{y^2 - y + 3} = y^2 + 2y - 1 + \frac{-4}{y^2 - y + 3}$$

QUICK CHECK Letting $y = 2$,

$$\frac{y^4 + y^3 + 7y - 7}{y^2 - y + 3} = \frac{16 + 8 + 14 - 7}{4 - 2 + 3} = \frac{31}{5}$$

$$y^2 + 2y - 1 + \frac{-4}{y^2 - y + 3} = 4 + 4 - 1 + \frac{-4}{4 - 2 + 3} = 7 - \frac{4}{5} = \frac{31}{5}$$

Exercises 4.6

Rewrite each quotient in Exercises 1–10 as a sum.

1. $\dfrac{x^3 + 3x^2 - 2x + 4}{x}$
2. $\dfrac{4x^3 - 2x^2 + 5x - 2}{x}$
3. $\dfrac{6y^2 - 4y + 3}{2y}$
4. $\dfrac{15y^3 + 9y^2 - 3y - 1}{3y}$
5. $\dfrac{6x^3 - 4x^2 + 5x + 2}{x^2}$
6. $\dfrac{12x^4 - 9x^2 + 4x - 5}{x^2}$
7. $\dfrac{x^4 - 5x^3 + 3x^2 + 2}{x^3}$
8. $\dfrac{x^4 - 5x^2 + 3}{x^4}$
9. $\dfrac{9x^3y^2 + 6x^2y - 2xy}{3xy}$
10. $\dfrac{x^3y - 3x^2y^2 + 6xy - 2}{xy}$

In Exercises 11–30, divide by using the long-division algorithm. Check by multiplication.

11. $\dfrac{2x^3 + 5x^2 - 3x + 7}{x + 4}$

12. $\dfrac{5x^4 + x^3 - 4x^2 + 2x - 3}{x + 2}$

13. $\dfrac{2x^3 - 13x^2 + 13x + 10}{x - 5}$

14. $\dfrac{6x^3 - 25x^2 + 3x + 4}{2x - 1}$

15. $\dfrac{4x^3 + 10x^2 - 16x - 14}{2x + 7}$

16. $\dfrac{4x^4 - 3x^2 + 7x - 1}{x - 1}$

17. $\dfrac{2x^3 - 3x^2 + 5}{x - 1}$

18. $\dfrac{9x^3 - 36x^2 - x + 4}{3x + 1}$

19. $\dfrac{x^3 + 4x^2 + 5x + 12}{x + 4}$

20. $\dfrac{x^3 + 3x^2 + 2x + 7}{x + 3}$

21. $\dfrac{8 + 5y^3 - 20y - 2y^2}{5y - 2}$

22. $\dfrac{4y + 12 - 7y^2 + y^3}{y - 6}$

23. $\dfrac{a^3 - 729}{a - 9}$

24. $\dfrac{a^3 + 64}{a + 4}$

25. $\dfrac{8t^3 + 1}{2t + 1}$

26. $\dfrac{y^3 - 125}{y - 5}$

27. $\dfrac{5x^4 - 30x^2 + 2x - 1}{x^2 - 6}$

28. $\dfrac{2x^4 - 14x^2 - 5}{x^2 - 7}$

29. $\dfrac{x^6 + x^4 + x^2 + 1}{x^2 + 1}$

30. $\dfrac{3x^4 + x^3 + 2x - 12}{x^2 + 2}$

In Exercises 31–42, divide and check as indicated.

31. $\dfrac{x^4 - 1}{x - 1}$
Check using $x = 2$.

32. $\dfrac{x^4 - 1}{x + 1}$
Check using $x = 2$.

33. $\dfrac{x^2 + 4}{x + 2}$
Check using $x = 5$.

34. $\dfrac{x^2 + 9}{x - 3}$
Check using $x = 4$.

35. $\dfrac{2x^4 + 4x - 6x^3 - 5}{x - 3}$
Check using $x = 2$.

36. $\dfrac{t^2 - 10t + 8 + t^3}{t + 4}$
Check using $t = 2$.

37. $\dfrac{x^4 + 4}{x^2 + 2x + 2}$
Check using $x = 2$.

38. $\dfrac{y^4 + 64}{y^2 - 4y + 8}$
Check using $y = 2$.

39. $\dfrac{y^4 - y^2 - 6y - 9}{y^2 - y - 3}$
Check using $y = 2$.

40. $\dfrac{y^4 - y^2 + 10y - 25}{y^2 - y + 5}$
Check using $y = 2$.

41. $\dfrac{t + t^3 - 4 - t^2}{4 + t^2}$
Check using $t = 2$.

42. $\dfrac{5 + t^4 - t - t^3}{t^2 - 5}$
Check using $t = 3$.

4.7 Fractional Equations

A **fractional equation** is an equation whose terms are rational expressions.

For example, $\frac{1}{x} = 3$ and $\frac{8}{x+2} - \frac{13}{x-3} = 15$ are fractional equations.

A fractional equation is not defined for a value of a variable that causes a denominator to become zero.

For example, $\frac{1}{x} = 3$ is not defined for $x = 0$, since $\frac{1}{0}$ is not defined.

Similarly,

$$\frac{8}{x+2} - \frac{13}{x-3} = 15$$

is not defined for $x = -2$ or $x = 3$.

Note that

$$\frac{8}{-2+2} = \frac{8}{0}$$

which is undefined. Also, for $x = 3$,

$$\frac{13}{x-3} = \frac{13}{3-3} = \frac{13}{0}$$

which is undefined.

We can change a fractional equation into a polynomial equation by multiplying each term of the equation by the LCD of all the fractions in the equation. However, we only get an equivalent equation (one that has the same solution set) when the LCD is not zero.

This means the variable may not equal any value that causes the LCD to equal zero.

A solution of LCD = 0 is called a **restricted value.**

A restricted value **cannot** be a solution of the fractional equation.

The Equivalence Theorem for Multiplication

If A, B, and C are any real numbers, then

$$A = B \text{ if and only if } AC = BC \text{ and } C \neq 0$$

This theorem is used to solve a fractional equation. Care must be taken to exclude any value of a variable for which $C = 0$. A value for which $C = 0$ is called a restricted value of the variable.

Example 1 State the restricted values of y and solve:

$$\frac{2}{y-5}+\frac{1}{y+5}=\frac{11}{y^2-25}$$

SOLUTION

1. Factor all denominators to find the LCD.

$$\frac{2}{y-5}+\frac{1}{y+5}=\frac{11}{(y-5)(y+5)}$$

2. Find the LCD.

$$\text{LCD}=(y-5)(y+5)$$

3. Solve LCD = 0 to find restricted values.

If $(y-5)(y+5)=0$, then
$y-5=0$ or $y+5=0$
$y=5$ and $y=-5$ are the restricted values.
$y \neq 5$ and $y \neq -5$

4. Multiply each side of the equation by the LCD.

$$\left(\frac{2}{y-5}+\frac{1}{y+5}\right)(y-5)(y+5)=\frac{11(y-5)(y+5)}{(y-5)(y+5)}$$

5. Solve the polynomial equation.

$$\begin{aligned} 2(y+5)+(y-5)&=11\\ 3y+5&=11\\ 3y&=6\\ y&=2 \end{aligned}$$

6. Reject any restricted value.

Since the common solution to $y \neq 5$, $y \neq -5$, and $y = 2$ is 2, $\{2\}$ is the solution set of the original equation.

7. Check in the original equation.

CHECK

Left side: $\frac{2}{y-5}+\frac{1}{y+5}=\frac{2}{2-5}+\frac{1}{2+5}=\frac{-2}{3}+\frac{1}{7}=\frac{-11}{21}$

Right side: $\frac{11}{y^2-25}=\frac{11}{(2)^2-25}=\frac{11}{4-25}=\frac{-11}{21}$

Example 2 State the restrictions and solve:

$$\frac{x+1}{x+2}-\frac{x+1}{x-3}=\frac{5}{x^2-x-6}$$

continued

SOLUTION

1–2. The LCD is $(x+2)(x-3)$.

3. Since $x+2 \neq 0$ and $x-3 \neq 0$, $x \neq -2$ and $x \neq 3$,

4. $$\left(\frac{x+1}{x+2}\right)(x+2)(x-3) - \left(\frac{x+1}{x-3}\right)(x+2)(x-3) = \frac{5}{(x+2)(x-3)}(x+2)(x-3)$$

5. $$\begin{aligned} (x+1)(x-3) - (x+1)(x+2) &= 5 \\ (x^2-2x-3) - (x^2+3x+2) &= 5 \\ -5x-5 &= 5 \\ -5x &= 10 \\ x &= -2 \end{aligned}$$

6. Thus $x \neq -2$ and $x \neq 3$ and $x = -2$. Since there is no common solution, the solution set is the empty set, $\varnothing$.

7. To check, we rework the problem very slowly and carefully.

Example 3 State the restrictions and solve:

$$\frac{x}{(x-4)^2} + \frac{2}{x-4} = \frac{3x-8}{(x-4)^2}$$

SOLUTION

1–2. LCD $= (x-4)^2$

3. $x \neq 4$, because $x - 4 \neq 0$.

4. $$(x-4)^2\frac{x}{(x-4)^2} + (x-4)^2\frac{2}{x-4} = \frac{3x-8}{(x-4)^2}(x-4)^2$$

5. $$\begin{aligned} x + 2(x-4) &= 3x-8 \\ 3x-8 &= 3x-8 \\ 3x &= 3x \\ x &= x \end{aligned}$$

6. Since $x = x$ is true for all real numbers, the solution set is the set of all real numbers except 4. For the solution set, we write R, $x \neq 4$.

7. Any number except 4 can be used to check. Let $x = 5$.

CHECK

Left side: $$\frac{5}{(5-4)^2} + \frac{2}{5-4} = 5 + 2 = 7$$

Right side: $$\frac{3(5)-8}{(5-4)^2} = \frac{15-8}{1} = 7$$

Example 4 shows another way to write the solution of a fractional equation. Each term of the equation is written as a fraction with the LCD as the denominator. This method often reduces the amount of writing necessary. It also helps to avoid errors.

Example 4 State the restrictions and solve.

$$\frac{x-1}{x-3}-\frac{2x}{x-2}=\frac{4}{x^2-5x+6}$$

SOLUTION

1. Factor denominators to find the LCD.

$$x^2-5x+6=(x-3)(x-2)$$

2. Find the LCD.

$$\text{LCD}=(x-3)(x-2)$$

3. Find restrictions.

$x \neq 3$ and $x \neq 2$, because $x-3 \neq 0$ and $x-2 \neq 0$.

4. Build up each term so the denominator is the LCD.

$$\frac{(x-1)\mathbf{(x-2)}}{(x-3)\mathbf{(x-2)}}-\frac{2x\mathbf{(x-3)}}{(x-2)\mathbf{(x-3)}}=\frac{4}{(x-2)(x-3)}$$

5. Multiply each side by the LCD.

$$x^2-3x+2-(2x^2-6x)=4$$

6. Solve the equation.

$$\begin{aligned} x^2-3x+2-2x^2+6x&=4\\ -x^2+3x-2&=0\\ x^2-3x+2&=0\\ (x-1)(x-2)&=0\\ x-1=0 \quad \text{or} \quad x-2&=0\\ x=1 \quad \text{or} \quad x&=2 \end{aligned}$$

7. Reject restricted values.

Since $x \neq 2$, the only possible solution is 1.
The solution set is $\{1\}$.

CHECK

Left side: $\frac{1-1}{1-3}-\frac{2(1)}{1-2}=0+2=2$

Right side: $\frac{4}{1-5+6}=\frac{4}{2}=2$

Example 5 State the restrictions and solve $\frac{a}{x}-\frac{b}{a}=\frac{1}{3x}$ for x where $a \neq 0$ and $b \neq 0$.

continued

SOLUTION

$$\text{LCD} = 3ax \quad \text{and} \quad x \neq 0$$

$$\frac{a}{x} - \frac{b}{a} = \frac{1}{3x}$$

$$3ax\left(\frac{a}{x}\right) - 3ax\left(\frac{b}{a}\right) = 3ax\left(\frac{1}{3x}\right)$$

$$3a^2 - 3bx = a$$

$$-3bx = a - 3a^2$$

$$x = \frac{a - 3a^2}{-3b} \quad \text{since } b \neq 0$$

$$x = \frac{a(3a - 1)}{3b}$$

Exercises 4.7

1–54. State the restrictions on the variable and solve. Check each solution.

1. $\frac{1}{2x} - \frac{2}{3x} = \frac{1}{24}$
2. $\frac{2}{5x} - \frac{1}{2x} = \frac{x + 2}{10x}$
3. $\frac{3}{x} - \frac{2}{3} = \frac{1}{3x}$
4. $\frac{1}{10} - \frac{1}{x} = \frac{1}{x} - \frac{1}{15}$
5. $\frac{x + 5}{x} = 0$
6. $\frac{x}{x - 6} = \frac{6}{x - 6}$
7. $\frac{5}{5 - x} = \frac{x}{5 - x}$
8. $\frac{x}{x - 6} = 0$
9. $\frac{x}{x - 5} = 0$
10. $\frac{x - 6}{x} = 0$
11. $\frac{2}{3x - 7} = \frac{5}{x + 2}$
12. $\frac{3}{2x - 1} = \frac{7}{3x + 1}$
13. $\frac{5}{x} = 20$
14. $\frac{3}{x} = 15$
15. $\frac{1}{x - 3} + \frac{1}{x - 7} = 0$
16. $\frac{1}{x - 4} + \frac{1}{x + 4} = 0$
17. $\frac{1}{x} + \frac{1}{x - 1} = \frac{2}{x + 1}$
18. $\frac{2}{x + 3} - \frac{1}{x - 3} = \frac{1}{x}$
19. $\frac{x + 2}{x + 5} = \frac{4}{5}$
20. $\frac{x - 2}{x - 3} = \frac{3}{5}$
21. $1 + \frac{6}{x - 4} = \frac{x + 2}{x - 4}$
22. $1 - \frac{x}{x + 3} = \frac{3}{x + 3}$

23. $\frac{5}{t}+\frac{1}{2t}-\frac{2}{3t}=29$

24. $\frac{2}{t}-\frac{1}{2t}+\frac{4}{5t}=23$

25. $\frac{2}{x+1}+\frac{1}{3x+3}=\frac{1}{6}$

26. $\frac{4t}{9t+18}+\frac{t+2}{3t-6}=\frac{7}{9}$

27. $\frac{6}{y}-\frac{1}{y-2}=\frac{3}{y^2-2y}$

28. $\frac{3}{y}-\frac{5}{y-1}=\frac{15}{y^2-y}$

29. $\frac{1}{t+4}+\frac{1}{t-4}=\frac{t+12}{t^2-16}$

30. $\frac{t}{t+2}+\frac{3}{t-2}=\frac{t^2+10}{t^2-4}$

31. $\frac{4}{3x+3}-\frac{4}{5x+5}=\frac{1}{15}$

32. $\frac{6}{2x-1}-\frac{5}{4x-2}=\frac{1}{2}$

33. $\frac{2}{x^2-3x-10}=\frac{1}{x^2-5x-14}$

34. $\frac{1}{3t^2-14t-5}=\frac{1}{2t^2-11t+5}$

35. $\frac{3}{x+4}+\frac{2}{x-1}=\frac{15}{x^2+3x-4}$

36. $\frac{5}{x-3}+\frac{3}{x+5}=\frac{10x}{x^2+2x-15}$

37. $\frac{x+1}{x-5}-\frac{x+3}{x+2}=\frac{2}{x^2-3x-10}$

38. $\frac{x-6}{x+6}-\frac{x-2}{x+2}=\frac{8}{x^2+8x+12}$

39. $\frac{3}{y+3}+\frac{y}{9-y^2}=\frac{1}{y-3}$

40. $\frac{3}{y-1}+\frac{y}{1-y^2}=\frac{3}{y+1}$

41. $\frac{x+6}{x-5}-\frac{x-6}{x+5}=\frac{11}{x^2-25}$

42. $\frac{x+3}{x-2}-\frac{5}{x^2-4}=\frac{x-3}{x+2}$

43. $\frac{1}{x^2+5x}+\frac{1}{5x+25}=\frac{1}{5x}$

44. $\frac{1}{x^2-7x}+\frac{1}{x^2+7x}=\frac{2}{x^2-49}$

45. $\frac{1}{t^2-1}-\frac{3}{t^2+2t+1}=0$

46. $\frac{3}{t^2+8t+16}-\frac{1}{t^2-16}=0$

47. $\frac{1}{x^2-9}+\frac{1}{x^2+6x+9}=\frac{2}{x^2-6x+9}$

48. $\frac{3}{x^2-4}-\frac{2}{x^2+4x+4}=\frac{1}{x^2-4x+4}$

49. $\frac{2x-3}{5x-2}=\frac{2x+5}{5x+2}$

50. $\frac{x-3}{3x-1}=\frac{x+11}{3x+1}$

51. $1+\frac{2}{x-1}=\frac{2}{x(x-1)}$

52. $2+\frac{3}{x-2}-\frac{3}{x(x-2)}=0$

53. $\frac{x}{x+3}-\frac{2}{x+1}=0$

54. $\frac{x}{x-2}-\frac{30}{(x+4)(x-2)}=\frac{5}{x+4}$

55. Solve $\frac{a-2x}{b-x}=\frac{3}{2}$ for x.

56. Solve $\frac{x}{a} - \frac{a}{b} = \frac{b}{a}$ for x.

57. Solve $\frac{1}{x} - \frac{1}{a} = \frac{1}{b} - \frac{1}{x}$ for x.

58. Solve $\frac{x + a}{x + b} = \frac{a}{b}$ for x.

59. Solve $\frac{1}{y - a} + \frac{1}{y - b} = 0$ for y.

60. Solve $\frac{1}{y - a} + \frac{1}{y + a} = \frac{a}{y^2 - a^2}$ for y.

4.8 Ratio and Proportion

At the beginning of this chapter we stated that a fraction, such as $\frac{5}{6}$, is also called the **ratio** of 5 to 6, or the ratio $\frac{5}{6}$. An equation that states that two ratios are equal is called a **proportion.** Thus a proportion has the form

$$\frac{a}{b} = \frac{c}{d}$$

The statement $\frac{a}{b} = \frac{c}{d}$ is expressed sometimes by saying, "a, b, c, and d are in proportion."

The terms a, b, c, and d of the proportion $\frac{a}{b} = \frac{c}{d}$ also have special names. In the order in which they are stated above, a is the first term, b the second, c the third, and d the fourth. The first and fourth terms are called the **extremes** of a proportion, and the second and third terms are called the **means** of the proportion.

Theorem

In a proportion the product of the means is equal to the product of the extremes. In symbols, if $\frac{a}{b} = \frac{c}{d}$, then $bc = ad$.

Example 1 If 3, 4, and 12 are the first three terms, in that order, of a proportion, find the fourth term.

SOLUTION Let $d =$ the fourth term. Then

$$\frac{3}{4} = \frac{12}{d}$$
$$3d = 48$$
$$d = \frac{48}{3} = 16$$

Thus the fourth term is 16.

Example 2 A sum of \$350 is to be divided between two partners in the ratio $\frac{3}{4}$. How much does each partner receive?

SOLUTION
Let $x =$ the share of one partner.
Then $350 - x =$ the share of the other partner. Then

$$\frac{x}{350 - x} = \frac{3}{4}$$
$$4x = 3(350 - x)$$
$$7x = 3(350)$$
$$x = \$150$$
$$350 - x = \$200$$

Example 3 The ratio of women students to men students at a certain college is $\frac{7}{9}$. If there are 2135 women students, how many men students are there?

SOLUTION Let $x =$ the number of men students. Then

$$\frac{7}{9} = \frac{2135}{x}$$
$$7x = 2135 \cdot 9$$
$$x = 305 \cdot 9$$
$$x = 2745$$

If two numbers are in the ratio $\frac{a}{b}$, then the numbers may be represented as ax and bx ($x \neq 0$) by applying the fundamental theorem of fractions: $\frac{ax}{bx} = \frac{a}{b}$.

Example 4 In a recent local election, the ratio of the number of persons voting for a certain measure to those who voted against the measure was 5 to 3. If 1600 persons voted on the measure, how many persons voted for the measure?

SOLUTION Let $5x$ = the number of persons who voted for the measure. Let $3x$ = the number of persons who voted against it. Note that $\frac{5x}{3x} = \frac{5}{3}$

Then
$$\begin{aligned} 5x + 3x &= 1600 \\ 8x &= 1600 \\ x &= 200 \\ 5x &= 1000 \end{aligned}$$

Thus, 1000 persons voted for the measure.

Proportions can also be used to solve certain percentage problems. For a statement of the form $P\%$ of a is b

the corresponding equation is $\frac{P}{100} = \frac{b}{a}$

Example 5 The postal rate increased from 15¢ to 20¢. What percent of the old rate was the increase of 5¢?

SOLUTION Let $x\%$ = the percent. Then

$$\frac{x}{100} = \frac{5}{15} \quad \text{and} \quad \frac{x}{100} = \frac{1}{3}$$
$$3x = 100$$
$$x = \frac{100}{3} = 33\frac{1}{3}$$

The percent increase was $33\frac{1}{3}\%$.

Example 6 In a certain city the average monthly payment on a home increased by \$150 over the average monthly payment the year before. If this was a 25% increase, find the monthly payment the year before.

SOLUTION Let x = the monthly payment the year before.

$$\frac{150}{x} = \frac{25}{100}$$
$$\frac{150}{x} = \frac{1}{4}$$
$$x = 4(150) = 600 \text{ dollars}$$

Exercises 4.8

1. The ancient Greeks considered the most beautiful rectangle to be the one whose sides were in the Golden Ratio. Approximating the Golden Ratio by the ratio $\frac{5}{8}$, find the dimensions of the most beautiful rectangle with a perimeter of 182 centimeters.

2. By weight, the ratio of oxygen to hydrogen in pure distilled water is $\frac{8}{1}$. How many grams of each element are in 100 grams of water?

3. Pie crust is made by combining shortening and flour in the ratio $\frac{2}{3}$ by volume. If 2 cups of flour are required per pie, how much shortening is required to make 15 pies? To make one pie?

4. A son and grandson are to divide an inheritance of \$5000 in the ratio $\frac{5}{3}$. How much does each receive if the inheritance tax is \$500 and the lawyer receives \$2700?

5. The batting average of a baseball player is the ratio of the number of hits he makes to the number of times he comes to bat. If a certain player has a batting average of 0.325, how many hits has he made in 40 times at bat?

6. A sampling revealed that on a certain day, for every 1000 tin cans manufactured, 15 were defective. If 250,000 cans were made on that day, how many were defective?

7. The ratio of sodium to chlorine in common table salt is $\frac{35}{23}$. Find the amount of each element in the salt compound weighing 290 pounds.

8. A piece of wire 84 inches long is cut so that the parts are in the ratio $\frac{4}{3}$. How long is each part?

9. The ratio of the size of a medium egg to a large egg is 2 to 3. If medium eggs cost 82¢ per dozen, what should be the equivalent cost for a dozen large eggs?

10. If 5 pounds of oranges cost \$1.80, what is the cost of 12 pounds of oranges?

11. Two companies contribute a total of $26,400 to a certain charity. If the ratio of their contributions is $\frac{5}{6}$, find the amount that each company contributes.

12. A blueprint has the scale 1 inch = 4 feet. What are the dimensions of a rectangular living room that measures $4\frac{1}{2}$ by 5 inches on the blueprint?

13. If the ratio of kilometers to miles is 8 to 5, what speed in miles per hour is equivalent to 90 kilometers per hour?

14. Referring to Exercise 13, find the speed in kilometers per hour if the speed in miles per hour is 25.

15. If the exchange ratio of Mexican pesos to United States dollars is 40 to 1, what is the price in dollars of an article costing 720 pesos?

16. If the exchange rate of Dutch guilders to United States dollars is 5 to 2, what is the price in dollars of a dozen tulip bulbs costing 18 guilders?

17. A motorist used 15 gallons of gasoline for a trip of 270 miles. How many gallons of gasoline would this motorist need for a trip of 432 miles? (Assume the miles per gallon is the same for each trip.)

18. At a certain market, the cost of 4 pounds of beef was $7.80. Assuming the same unit price, find the cost of 3 pounds of this beef.

19. The hourly salary of an employee increased from $8 an hour to $10 an hour. What percent of the old hourly rate was the increase?

20. By a proposed indexing method, a married couple's state income tax would be reduced from $1800 to $1530. What percent of the $1800 is the decrease?

21. On a test of 50 questions, a student received a grade of 86%. How many questions did the student answer correctly?

22. A baseball team played 35 games. It won 60% of the games it played. How many games did the team win?

23. In a certain city, the monthly garbage collection fee was increased by $1.35. This was an increase of 22.5%. Find the old fee and the new fee.

24. In a certain city, the commercial electricity prices for 1980 decreased by $40.50 from the 1979 prices. This was a 17% decrease. Find the price for 1979 and the price for 1980.

25–26. Clark's rule is one method used to determine the amount of medicine to give a child. It states that the ratio of the child's dosage to the adult dosage is equal to the ratio of the child's weight in pounds to 150 (150 represents the weight of an average adult).

$$\frac{\text{Child's dosage}}{\text{Adult dosage}} = \frac{\text{Child's weight in pounds}}{150}$$

25. The adult dosage of tincture of digitalis is 15 minims. Find the dosage for a child who weighs 40 pounds.

26. The adult dosage of a certain drug is 5 grains. Find the dosage for a child who weighs 60 pounds.

27–28. When a *P*% solution is prepared from a pure drug, the following proportion is used:

$$\frac{\text{Number of grams of pure drug}}{\text{Milliliters to be prepared}} = \frac{P}{100}$$

27. How many grams of pure boric acid crystals should be used to make 200 milliliters of 4% boric acid solution?

28. How many milliliters of a 2% silver nitrate solution can be made from 1 gram of pure silver nitrate crystals?

29. For a certain type of concrete mix, the ratio of cement to sand to rock is 1:3:5. Find the number of cubic feet of each component in a 108-cubic-foot mix of this concrete.

30. The profit of an investment is to be divided among the three partners in a business in the ratio of 5:3:2. How much does each receive if the profit is $48,000?

4.9 Work Problems

Verbal problems that involve work often lead to fractional equations, as shown in the following examples.

A formula for a work problem is $w = tr$, where w is the amount of work done, t the time spent working, and r the amount of work done per unit time (the rate). If the work done is a complete job, then $w = 1$.

When several persons work together, it is assumed that the total work done is the sum of the amounts of work done by those persons working.

In summary, $w = tr$ and

$$w \text{ (total)} = w \text{ (worker A)} + w \text{ (worker B)}$$

If the work done is the complete job, then $w = 1$ and $r = \frac{1}{t}$.

Example 1 One machine requires 80 minutes to complete a certain job. Another machine completes the same job in 120 minutes. If the machines worked together, how long would it take them to complete the job?

SOLUTION Let x = the number of minutes the machines work together.

Working Together

Formula:	t	$\cdot$	r	$=$	w
First machine	x		$\frac{1}{80}$		$\frac{x}{80}$
Second machine	x		$\frac{1}{120}$		$\frac{x}{120}$

EQUATION

work done by first + work done by second = total work

$$\frac{x}{80} + \frac{x}{120} = 1$$

$$240\left(\frac{x}{80} + \frac{x}{120}\right) = 240$$

$$3x + 2x = 240$$

$$5x = 240$$

$$x = \frac{240}{5} = 48 \text{ minutes}$$

Example 2 Working alone, a painter takes twice as much time to paint a job as his father does. If the father works alone for 2 days and then completes the job by working with his son for 3 days, how long would it have taken the father to paint the job alone?

SOLUTION

Let $x =$ the time of the father working alone.
Then $2x =$ the time of the son working alone.

Working Together

Formula:	t	$\cdot$ r	$=$ w
Father	5	$\frac{1}{x}$	$\frac{5}{x}$
Son	3	$\frac{1}{2x}$	$\frac{3}{2x}$

EQUATION

father's work + son's work = total work

$$\frac{5}{x} + \frac{3}{2x} = 1$$

$$10 + 3 = 2x$$

$$2x = 13$$

$$x = \frac{13}{2} = 6\frac{1}{2} \text{ days}$$

Exercises 4.9

1. An experienced typist can type 3 times as fast as a new one. Working together, they complete a certain job in 9 hours. How long would it have taken each one alone?

2. A new machine can process checks 5 times as fast as an old one. With both machines operating, 1000 checks can be processed in $\frac{1}{2}$ hour. How long would it take each machine alone to process 1000 checks?

3. A mason can lay the same amount of brick in 6 days that his helper can in 12 days. If they worked together, how long would it take them to lay this amount of brick?

4. A roofer requires 10 hours to shingle a roof. His helper can do the same work in 15 hours. How long would it take them if they worked together?

5. Three machines can manufacture a certain article in 12 minutes, 15 minutes, and 20 minutes, respectively. If the three machines worked simultaneously, how long would it take them to manufacture 1000 articles?

6. One pipe can fill a tank in 18 minutes, a second in 30 minutes, and a third in 45 minutes. If all three pipes were opened at the same time, how long would it take to fill the tank?

7. One crew of men can sheetrock a certain building in 15 days. With the help of a second crew, the building can be sheetrocked in 6 days. How long would it take the second crew, working alone, to sheetrock the building?

8. A gardener takes 36 hours to do a certain landscaping job. Working with his partner, they can do the same job in 20 hours. How long would it take the partner to do the same job, working alone?

9. A man can mow his lawn in 40 minutes. His son can mow the lawn in 60 minutes. The man starts to mow the lawn; 10 minutes later his son joins him. How long does it take them to complete the job (assuming they have two lawnmowers)?

10. An old machine can do a certain job in 70 minutes. A new machine takes only 28 minutes to do the same job. One day, when both machines were working together on the job, the old machine broke down after working for 15 minutes. How long did it take the new machine to complete the job?

11. Two pipes can fill a tank in 15 and 20 hours, respectively. A third pipe can drain the tank in 12 hours. How long would it take to fill the tank if all three pipes were open?

12. Two inlet pipes can fill a swimming pool in 8 hours and 12 hours, respectively. One day the outlet pipe was left open, and it took 6 hours to fill the pool. How long does it take the outlet pipe to drain the filled pool when the inlet pipes are closed?

13. One number is 4 times another. The sum of their reciprocals is $\frac{3}{4}$. Find the numbers.

14. The sum of a number and its negative reciprocal is $\frac{3}{2}$. Find the number.

15. The focal length F of a concave mirror can be found by using the formula:

$$\frac{1}{F} = \frac{1}{a} + \frac{1}{b}$$

where a is the distance of the object from the mirror and b is the distance of the image from the mirror. A certain concave mirror has a focal length of 8 inches. At what distance from this mirror will an object be if the image distance is $\frac{1}{4}$ the object distance?

16. When two resistors, R_1 and R_2, are connected in parallel, the total resistance R can be found by using the formula

$$\frac{1}{R} = \frac{1}{R_1} + \frac{1}{R_2}$$

If two resistors connected in parallel are such that one has 3 times the resistance of the other, and if the total resistance is 15 ohms, find the number of ohms in each resistor.

Review Exercises

Simplify in Exercises 1–10.

1. $\dfrac{10x^3yz^2}{15xy^3z}$

2. $\dfrac{15a^2b}{4ax^2y} \cdot \dfrac{8a^2x^3y^2}{12b^2x}$

3. $\dfrac{a - b}{b - a}$

4. $\dfrac{9x^2 - 36y^2}{x^2 - 4xy + 4y^2}$

5. $\dfrac{2x^2 + 5x - 150}{2x^2 + 17x - 30}$

6. $\dfrac{x^2 - 2x + 1}{15x^2} \cdot \dfrac{20x^5}{1 - x^2}$

7. $\dfrac{x^3 + 5x^2 + 6x}{x^2 - x} \cdot \dfrac{x^2 - 3x + 2}{(x^2 - 4)(x + 3)}$

8. $\dfrac{5xy + 15y}{15y} \div \dfrac{x^2 + 4x + 3}{x^2 - 1}$

9. $\dfrac{a^2 + 3a}{a^2 - 7a + 12} \div \dfrac{a^2 - a - 12}{a^2 - 3a}$

10. $\left(\dfrac{x^2 - x}{x^2 + 2x - 3} \cdot \dfrac{x^2 + 2x + 1}{x^2 + 4x}\right) \div \dfrac{x^2 - 3x - 4}{x^2 - 16}$

In Exercises 11–20 write each as a single fraction reduced to lowest terms.

11. $\dfrac{5}{x + 3} - \dfrac{x}{x + 3}$

12. $\dfrac{x}{x + 3} + \dfrac{5x^2}{x^2 - 9}$

13. $\dfrac{x}{x^2 - 6x + 5} + \dfrac{3}{x - 1}$

15. $\dfrac{1}{x} + \dfrac{1}{2} + \dfrac{5}{x + 2}$

14. $\dfrac{2}{x - 1} - \dfrac{4}{x} + \dfrac{2}{x + 1}$

16. $\dfrac{3}{x^2 + 5x + 4} - \dfrac{2}{x^2 + 4x + 3}$

17. $\dfrac{1}{(x + 2)^2} - \dfrac{2}{x^2 - 4} + \dfrac{1}{(x - 2)^2}$

18. $\left(x + \dfrac{3}{y}\right) \div \left(x - \dfrac{3}{y}\right)$

19. $\left(\dfrac{1}{x} - \dfrac{1}{3}\right) \div \left(x - \dfrac{9}{x}\right)$

20. $\left(\dfrac{3x - 1}{3x} - \dfrac{3x}{3x + 1}\right) \div \left(\dfrac{5x - 2}{5x} - \dfrac{5x}{5x + 2}\right)$

In Exercises 21–25 divide by using the long-division algorithm. Check by multiplication.

21. $\dfrac{x^3 - 4x^2 + 3x - 1}{x - 3}$

23. $\dfrac{39 + 20x^2 + 15x^3}{5x + 10}$

22. $\dfrac{x^4 + x^2 - 2}{x^2 + 3}$

24. $\dfrac{a^3 - 1}{a - 1}$

25. $\dfrac{7x^3 - 3x + x^2 + 3x^3 - 5x - 10}{1 + 2x}$

Solve and check Exercises 26–35. State the restricted values of the variable.

26. $\dfrac{1}{x - 6} - \dfrac{3}{x + 6} = \dfrac{6}{x^2 - 36}$

28. $\dfrac{2}{x + 3} - \dfrac{2}{x - 3} = \dfrac{1}{3 - x}$

27. $\dfrac{x + 3}{x + 2} = \dfrac{3}{2}$

29. $\dfrac{2x - 3}{x + 1} = \dfrac{2x + 5}{x - 3}$

30. $\dfrac{1}{x^2 - 3x} - \dfrac{1}{x^2 + 5x} = \dfrac{2}{x^2 + 2x - 15}$

31. $\dfrac{x}{x + 4} + \dfrac{4}{x - 4} = \dfrac{x^2 + 16}{x^2 - 16}$

34. Solve $\dfrac{x - a}{x + b} = \dfrac{x + b}{x - a}$ for x.

32. $\dfrac{1}{x - 1} - \dfrac{x}{1 - x} = 1$

35. Solve $\dfrac{p - x}{x} = c + \dfrac{1}{x}$ for x.

33. $x + \dfrac{5x}{x - 3} = \dfrac{15}{x - 3}$

36. The numerator of a fraction is 3 less than the denominator. If 1 is added to the denominator and 2 is subtracted from the numerator, the value of the new fraction formed is $\dfrac{1}{4}$. Find the original fraction.

37. The ratio of the numerator of a certain fraction to the denominator is $\frac{3}{5}$. If 20 is subtracted from the numerator and 10 is added to the denominator, the resulting fraction is equal to $\frac{1}{2}$. Find the original fraction.

38. A swimming pool can be filled by a pipe in 14 hours and by a hose in 35 hours. How long does it take to fill the pool using both pipe and hose, assuming there is no loss in water pressure?

39. A mechanic can do a certain repair job in 4 hours. He and his assistant can complete the job in 3 hours. How long would it take the assistant working alone?

40. If the ratio of pounds of kilograms is 20 to 9, how many kilograms are equivalent to 120 pounds?

41. If the ratio of gallons to liters is 1 to 3.8, how many liters are there in 5 gallons?

42. If the ratio of centimeters to feet is 30.5 to 1, how tall in centimeters is a person who is 6 feet?

43. A piece of wire 65 centimeters long is to be cut into two pieces. The ratio of the lengths of the two pieces is to be 8 to 5. How long should each piece be?

5 Graphing and Linear Systems

Objectives

OBJECTIVES	TYPICAL PROBLEMS	LOCATIONS
Graph ordered pairs of real numbers.	1. Graph $A(6, 2)$ and $B(-1, 9)$. Join A to B with a straight line. 2. Graph $C(4, 8)$ and $D(-1, -7)$. Join C to D with a straight line. 3. Name the point of intersection of line AB and line CD.	*Section 5.1* *Examples 1–5* *Exercises 1–58*
Graph a linear equation.	4. Graph $3x - 4y = 12$ on a number plane.	*Section 5.2* *Examples 1–5* *Exercises 1–31*
Find the slope of a line, given two points on the line.	5. Find the slope of the line joining $(5, -3)$ and $(-2, 4)$.	*Section 5.3* *Examples 1, 2* *Exercises 1–10*
Find the slope and y-intercept of a line using the equation $y = mx + b$.	6. Write in the form $y = mx + b$ and state the slope and y-intercept: $5y - 4x = 20$	*Section 5.3* *Examples 3–6* *Exercises 11–50*
Find the equation of a line.	7. Write in standard form the equation of the line that passes through the points $(5, 2)$ and $(-3, 6)$.	*Section 5.4* *Examples 1, 2* *Exercises 1–36*
Solve a system of two linear equations in two variables by the graphic method.	8. Solve by the graphic method: $2x + 3y = 6$ $x - 2y = 10$	*Section 5.5* *Examples 1–7* *Exercises 1–24*
Solve a system of two linear equations in two variables by the substitution method.	9. Solve by the substitution method: $2x - y = 10$ $3x - 4y = 5$	*Section 5.6* *Examples 1–3* *Exercises 1–30*
Solve a system of two linear equations in two variables by the addition method.	10. Solve by the addition method: $2x + 5y = 30$ $4x - 3y = 8$	*Section 5.7* *Examples 1, 2* *Exercises 1–30*

Solve a verbal problem using two variables.	11. A certain city charged \$320 for the inspection of 5 single homes and 2 duplexes. It charged \$340 for the inspection of 4 single homes and 3 duplexes. Find the inspection fee for 1 single home and 1 duplex.	*Section 5.8* *Examples 1–3* *Exercises 1–18*
(Optional) Graphically solve a linear inequality in two variables.	12. Solve graphically: $\{(x, y) \mid 5y \geq 15 - 3x\}$	*Section 5.9* *Examples 1–6* *Exercises 1–26*

Answers to Typical Problems

1–3.

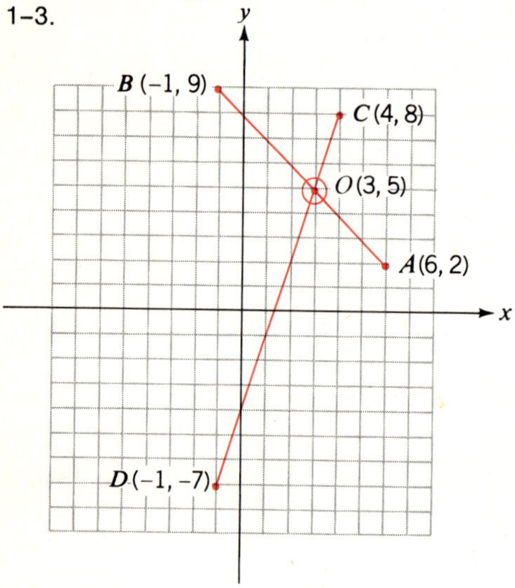

4.

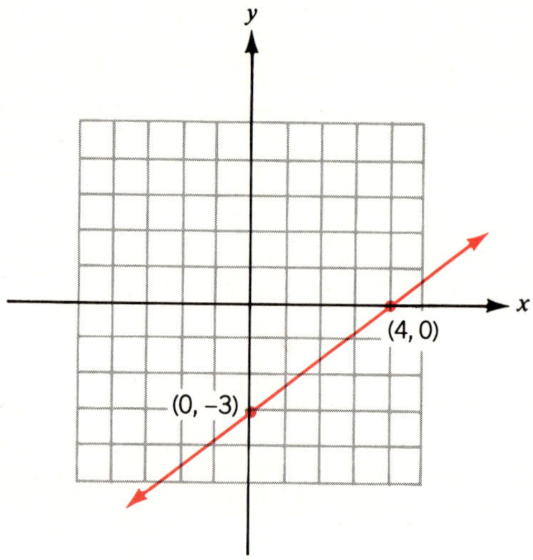

5. -1
6. $y = \frac{4}{5}x + 4$; $m = \frac{4}{5}$, $b = 4$
7. $x + 2y - 9 = 0$
8. $(6, -2)$
9. $(7, 4)$
10. $(5, 4)$
11. \$40, single home
    \$60, duplex

12.

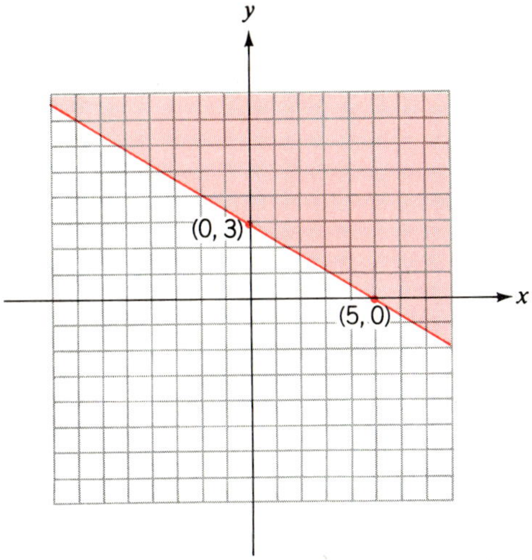

In geometry, a straight line is thought of as an infinite number of points, each of which is said to be on the line. In Chapter 1 we saw how the geometric line can be used to visualize the set of real numbers and their properties.

That discussion focused on linear equations in *one* variable. In this chapter the concept of linear equation will be extended to include the case of *two* variables.

The geometric plane is used to visualize equations in two variables and thereby to provide a better understanding of such equations.

In this chapter you will be introduced to a rectangular coordinate system, in which a one-to-one correspondence is established between the set of points on a plane and the set of ordered pairs of real numbers. You will then study linear equations in two variables and their graphs. Finally, you will be shown how to solve a system of linear equations in two variables and how to use these techniques for a variety of practical applications.

5.1 Rectangular Coordinates

Ordered Pairs and Solutions

An **ordered pair** is an expression having the form (**a, b**), where **a** is called the **first component** (or first member) of the ordered pair and **b** is called the **second component** (or second member) of the ordered pair.

The order in which the components of an ordered pair are written is important. For example, the ordered pair (3, 5) is *not* the same as the ordered pair (5, 3).

Definition

A **solution of an open equation in two variables** x and y is an ordered pair (a, b) such that the equation becomes true when x is replaced by a and y is replaced by b.

Example 1 Which of the following is a solution of $3x - y = 10$?
a. (4, 2) b. (2, 4)

SOLUTION

a. For (4, 2), $x = 4$ and $y = 2$. Then

$$3x - y = 3(4) - 2 = 12 - 2 = 10$$

Since $3x - y = 10$ is true for $x = 4$ and $y = 2$, (4, 2) is a solution.

b. For (2, 4), $x = 2$ and $y = 4$. Then

$$3x - y = 3(2) - 4 = 6 - 4 = 2$$

Since $3x - y = 10$ is false for $x = 2$ and $y = 4$, (2, 4) is not a solution.

Example 2 Find solutions of $y = 2x + 3$ for
a. $x = 5$ b. $x = 0$ c. $x = -5$ d. $y = 7$

SOLUTION

a. For $x = 5$, $y = 2(5) + 3 = 10 + 3 = 13$, and (5, 13) is a solution.
b. For $x = 0$, $y = 2(0) + 3 = 3$, and (0, 3) is a solution.
c. For $x = -5$, $y = 2(-5) + 3 = -10 + 3 = -7$, and $(-5, -7)$ is a solution.
d. For $y = 7$, $7 = 2x + 3$, $2x + 3 = 7$, $2x = 4$, $x = 2$. Therefore, (2, 7) is a solution.

In general, most equations in two variables have infinitely many solutions that are ordered pairs of real numbers. As an example of an exception, $x^2 + y^2 = -1$ has no solutions in the set of ordered pairs of real numbers, since x^2 and y^2 and $x^2 + y^2$ must be positive or zero.

Rectangular Coordinates

In Chapter 1 a one-to-one correspondence was established between the set of real numbers and the set of points on a line: Every point on the number line represents a unique real number, and for every real number there exists a corresponding unique point on the number line.

A similar one-to-one correspondence can be established between the set of ordered pairs of real numbers and the set of points on a plane.

A **number plane** is a plane whose points have been placed in one-to-one correspondence with the set of ordered pairs of real numbers.

A **rectangular coordinate system** is obtained by taking two perpendicular number lines, called the **axes,** intersecting at their origins.

The point of intersection of the axes is called the **origin** of the coordinate system.

It is customary that one axis is horizontal, called the ***x*-axis,** with its positive direction to the right, and the other axis is vertical, called the ***y*-axis,** with its positive direction upward.

Unless specified otherwise, the unit segment on the *x*-axis is selected to be the same length as the unit segment on the *y*-axis—that is, the same scale is used for both axes.

The axes separate the plane into four regions, called **quadrants,** that are numbered consecutively, starting with the upper right quadrant and proceeding counterclockwise (see Fig. 5.1).

To each ordered pair (a, b) of real numbers is associated a unique point P, the point of intersection of a vertical line through point a on the x-axis and a horizontal line through point b on the y-axis.

Conversely, to each point P in the plane is associated a unique ordered pair (a, b), where a is the coordinate of the point of intersection

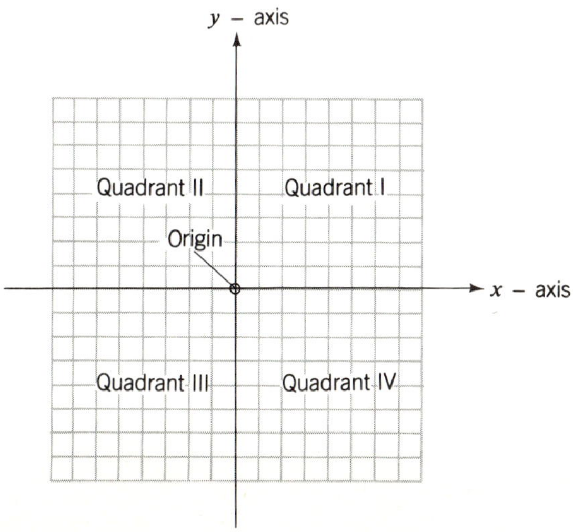

Figure 5.1

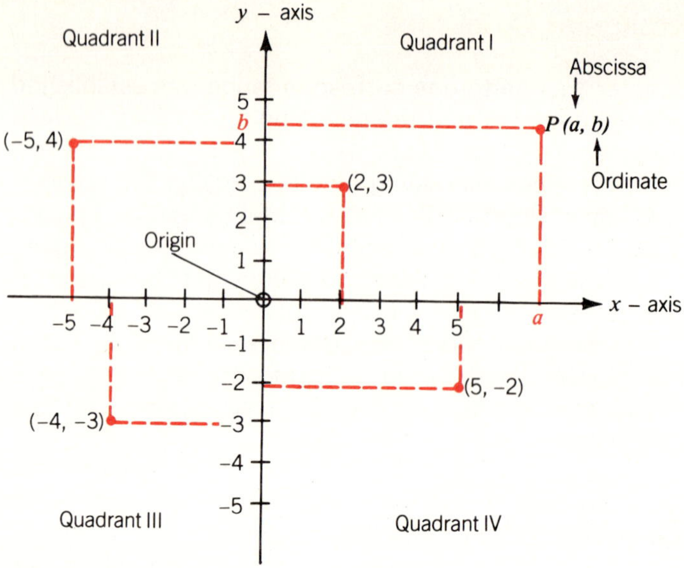

Figure 5.2

of the vertical line through *P* and the *x*-axis and *b* is the coordinate of the point of intersection of the horizontal line through *P* and the *y*-axis.

A point *a* on the *x*-axis is assigned the ordered pair (*a*, 0), and a point *b* on the *y*-axis is assigned the ordered pair (0, *b*). The origin is assigned the ordered pair (0, 0).

The numbers of the ordered pair (*a*, *b*) are called the **coordinates** of *P*, with the first component, *a*, called the *x*-coordinate, or **abscissa,** and the second component, *b*, called the *y*-coordinate, or **ordinate** (see Fig. 5.2). A point will be designated symbolically by *P*(*a*, *b*).

The association in which two perpendicular coordinate lines are used to establish a one-to-one correspondence between the set of points on a plane and the set of ordered pairs of real numbers is called a **rectangular** (or **cartesian**) **coordinate system.**

Example 3 Graph *P* (3, 4).

SOLUTION To locate the point *P,* whose coordinates are (3, 4), draw a vertical line through point 3 on the *x*-axis and a horizontal line through point 4 on the *y*-axis. The point *P* is the intersection of these two lines (see Fig. 5.3). This point is located in the first quadrant.

Example 4 Graph *Q* (−3, 2).

SOLUTION To locate the point *Q*, whose coordinates are (−3, 2), draw a vertical line through point −3 on the *x*-axis and a horizontal line through point 2 on the *y*-axis. The intersection of these two lines is point Q. This point is in the second quadrant (see Fig. 5.3).

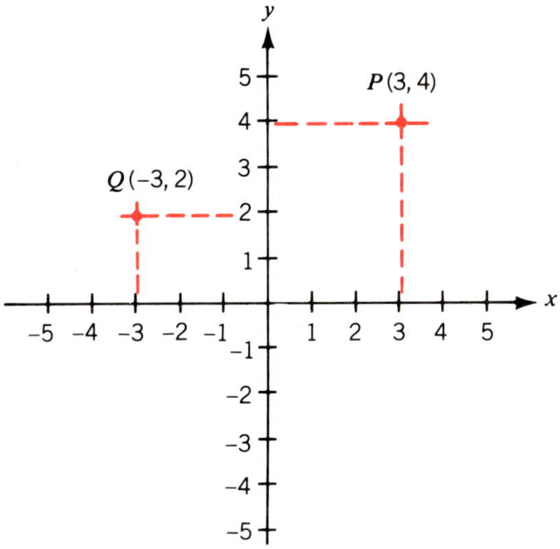

Figure 5.3

If vertical lines are drawn through every point that corresponds to an integer on the *x*-axis and horizontal lines are drawn through every point that corresponds to an integer on the *y*-axis, the intersections of these lines will represent a one-to-one correspondence between the points of intersection of these lines and the set of ordered pairs of integers. The lines form a grid, as shown in Fig. 5.4.

There are many more points on a number plane other than those corresponding to ordered pairs of integers, just as there are many more points on a number line other than those corresponding to integers.

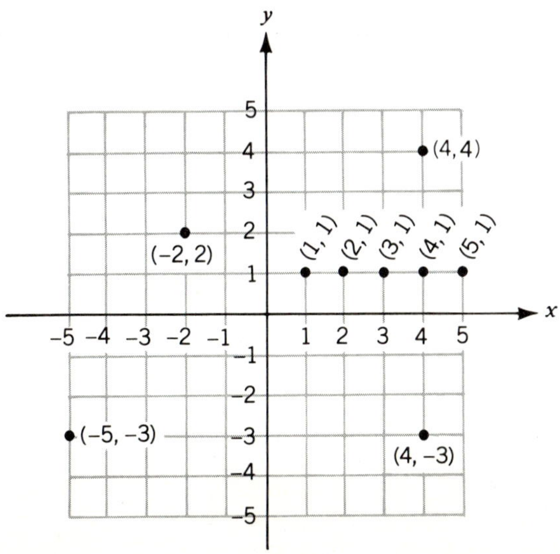

Figure 5.4

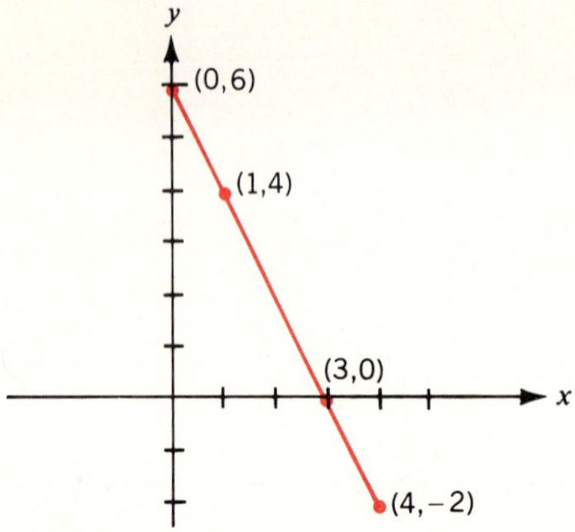

Figure 5.5

The set of points on a number plane is in one-to-one correspondence with the set of ordered pairs of real numbers.

Example 5 Graph the solutions of the equation $2x + y = 6$ for a. $x = 0$, b. $x = 1$, c. $y = 0$, d. $y = -2$. Join these points in the order a, b, c, d.

SOLUTION

a. For $x = 0$, $2(0) + y = 6$ and $y = 6$. The solution is (0, 6).
b. For $x = 1$, $2(1) + y = 6$ and $y = 4$. The solution is (1, 4).
c. For $y = 0$, $2x + 0 = 6$ and $x = 3$. The solution is (3, 0).
d. For $y = -2$, $2x + (-2) = 6$, $2x = 8$, and $x = 4$. The solution is (4, −2).

The graph is shown in Fig. 5.5.

Exercises 5.1

For Exercises 1–10, state the coordinates of each of the points in Fig. 5.6.

1. *A*
2. *B*
3. *C*
4. *D*
5. *E*
6. *F*
7. *G*
8. *H*
9. *K*
10. *L*

For Exercises 11–22, locate the points whose coordinates are given on one set of axes. Label each point with its coordinates.

11. (2, 3)
12. (3, 2)
13. (−1, −4)
14. (1, −3)
15. (−1, 3)
16. (0, 4)

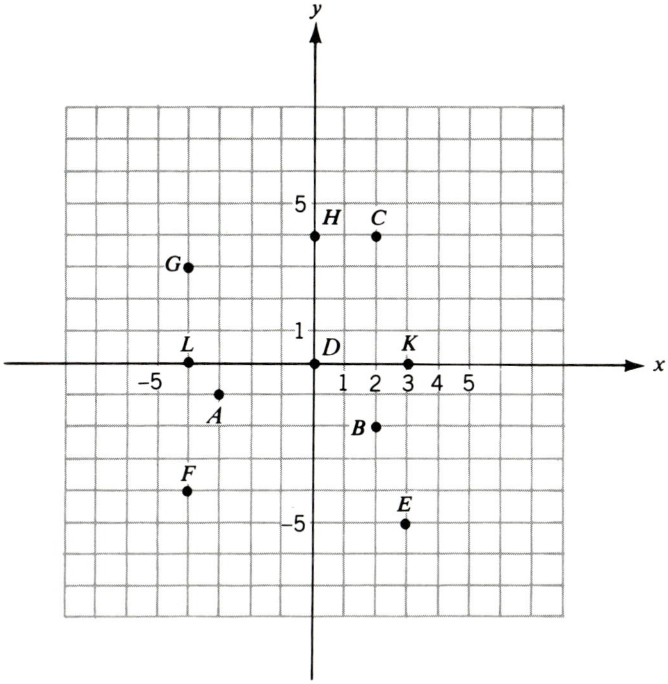

Figure 5.6

17. (4, 0)

18. (0, 0)

19. (3, −2)

20. (−2, 1)

21. (0, −2)

22. (−2, −4)

23. Locate the points (3, 2) and (−1, −2) and connect them with a straight line. Locate the points (2, −2) and (−1, 4) and connect them with a straight line. What are the coordinates of the point of intersection?

24. Locate the points (3, 5) and (6, −4) and connect them with a straight line. Locate the points (7, 1) and (1, −5) and connect them with a straight line. What are the coordinates of the point of intersection?

Which of Exercises 25–30 are solutions of $y = 2x + 3$?

25. (0, 3)

26. (1, 5)

27. (2, 6)

28. (3, 0)

29. (5, 13)

30. (−4, −1)

Which of Exercises 31–36 are solutions of $2x + y = 6$?

31. (3, 0)

32. (0, 3)

33. (−3, 0)

34. (−3, 12)

35. (4, −2)

36. (4, 1)

Which of Exercises 37–42 are solutions of $3x - y = 12$?

37. (0, 12)
38. (0, −12)
39. (4, 0)
40. (0, 4)
41. (−1, 15)
42. (−2, −18)

From the set {(3, 5), (5, 3), (0, 5), (3, 0), (5, −3)} select one or more solutions for each equation in Exercises 43–48.

43. $5x + 3y = 15$
44. $2y = 3x - 9$
45. $y = 17 - 4x$
46. $5x + 2y = 25$
47. $y = 5$
48. $x = 3$

From the set {(2, 6), (−2, 6), (6, −2), (−2, 0), (0, 6)} select one or more solutions for each equation in Exercises 49–54.

49. $3x + 2y = 6$
50. $y = 3x + 6$
51. $y = 10 - 2x$
52. $3x + y + 6 = 0$
53. $x = -2$
54. $y = 6$

For Exercises 55–58, copy and complete the table of values; then plot the points on a set of coordinate axes. If the points were connected, what geometric term would describe this graph?

55. $y = 3x - 6$

x	y
0	?
?	0
1	?
−2	?
?	−3

56. $3x + 5y = 15$

x	y
0	?
?	0
−5	?
?	−6
10	?

57. $x - y = 8$

x	y
0	?
?	0
−2	?
?	−4
3	?

58. $3x - 2y = 12$

x	y
0	?
?	0
2	?
−4	?
?	6

5.2 Graphs of Linear Equations

In the previous section we saw that a solution of an equation in two variables is an ordered pair of numbers for which the equation becomes a true statement.

Definition
The **solution set** of an equation in two variables is the set of all solutions of the equation.

For an equation having the form $y = mx + b$ there is exactly one real number y for each real number x. Consequently, an equation in the form $y = mx + b$ has infinitely many solutions.

A **linear function** is the set of ordered pairs (x, y) defined by a rule having the form $y = mx + b$—in other words, the solution set of $y = mx + b$.

The **graph of a linear function** is the graph of the set of ordered pairs belonging to the linear function.

The graph of a linear function is a straight line.

The **set-builder notation** is very useful for describing a set of ordered pairs. In set-builder notation, a linear function is defined as follows.

$$\{(x, y) \mid y = mx + b\}$$

This is read: "The set of all ordered pairs (x, y) such that $y = mx + b$." Note that the vertical bar is read "such that."

Example 1 Graph $\{(x, y) \mid y = 2x + 1\}$.

SOLUTION This means we are to graph all ordered pairs whose y-coordinate is obtained by adding 1 to twice the x-coordinate.

Solutions can be found by arbitrarily assigning values for x and calculating the corresponding values for y. It is convenient to make a table of values:

x	$2x + 1 =$	y	(x, y)
1	$2(1) + 1 =$	3	(1, 3)
2	$2(2) + 1 =$	5	(2, 5)
3	$2(3) + 1 =$	7	(3, 7)
0	$2(0) + 1 =$	1	(0, 1)
-2	$2(-2) + 1 =$	-3	$(-2, -3)$

Thus the ordered pairs (1, 3), (2, 5), (3, 7), (0, 1), and $(-2, -3)$ are solutions for the given equation. Figure 5.7 is the graph of these five points.

continued

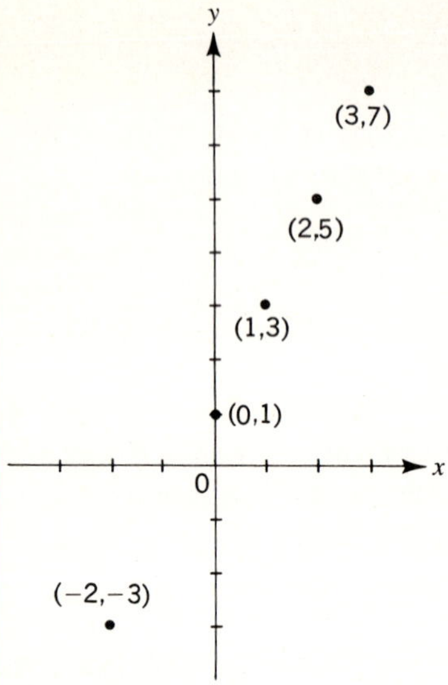

Figure 5.7

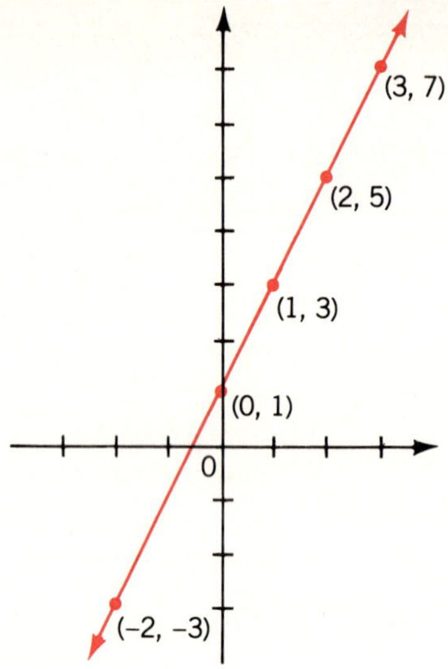

Figure 5.8

The plotted points in Fig. 5.8 seem to lie in a straight line, and in fact they do. There are infinitely many solutions to the equation $y = 2x + 1$, and only part of the graph representing the solutions can be shown. The arrows at each end of the line representing the solution set indicate that the line continues indefinitely in both directions.

Definition

Any equation in two variables x and y that can be expressed in the form

$$Ax + By + C = 0$$

where A and B are real numbers, not both zero, is called a **linear equation in two variables.**

If $B \neq 0$, then this equation can be solved for y.

$$Ax + By + C = 0$$

$$By = -Ax - C$$

$$y = -\frac{A}{B}x - \frac{C}{B}$$

$$y = mx + b \quad \text{where} \quad m = -\frac{A}{B}$$

$$\text{and} \quad b = -\frac{C}{B}$$

Thus if $B \neq 0$, then $Ax + By + C = 0$ defines a linear function. If $B = 0$, then $A \neq 0$ and the general equation becomes

$$Ax + C = 0$$

Solving for x,

$$x = -\frac{C}{A}$$

or $x = c$, a real number.

Definition
The **graph of a linear equation** is the graph of its solution set, and for any linear equation in two variables, this graph is a straight line.

To verify that the graph of a linear equation in two variables actually is a straight line, two things must be shown:

1. The set of points whose coordinates satisfy a given linear equation $Ax + By + C = 0$ (A and B real numbers, not both zero) all lie on a straight line.
2. The coordinates (x, y) of any point lying on a straight line satisfy a linear equation in x and y.

Both these statements can be proved, but for now they will be taken for granted, and it will be assumed that in a cartesian coordinate system the graph of a linear equation in two variables is a line and every line in the plane is the graph of a linear equation in two variables.

Since two points uniquely determine a line, it is necessary to find only two solutions (ordered pairs) that satisfy a given linear equation, plot the points corresponding to the coordinates, and draw a line through these points. This line represents the graph of the given equation. It is a good idea, however, to find a third solution as a check.

Example 2 Graph $2x + y = 6$.

SOLUTION Make a table of solutions. Any value may be selected for x, and then the corresponding y value is computed, using the equation.

Similarly, any value may be selected for y and its x value computed.

It is convenient to select $x = 0$ for one point and $y = 0$ for another point, because the resulting calculations are relatively easy. These resulting points are called the ***intercept points.***

Selecting $x = 0$, $y = 0$, and $x = 5$, we make the following table:

x	y	$2x + y = 6$	(x, y)
0		$2(0) + y = 6$, $y = 6$	$(0, 6)$
	0	$2x + 0 = 6$, $x = 3$	$(3, 0)$
5		$2(5) + y = 6$, $y = -4$	$(5, -4)$

The points are plotted, and a straight line is drawn through them (see Fig. 5.9).

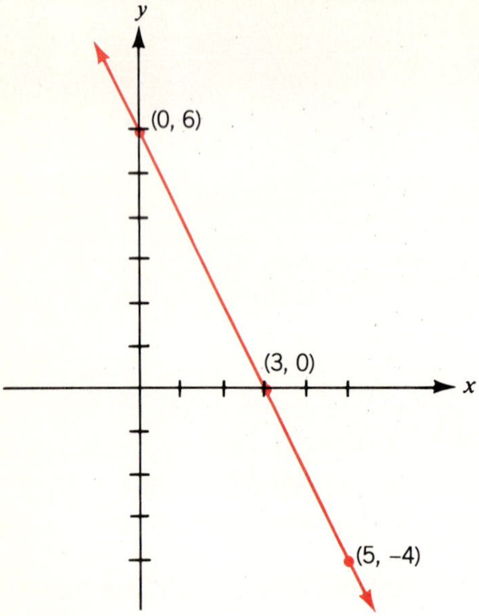

Figure 5.9

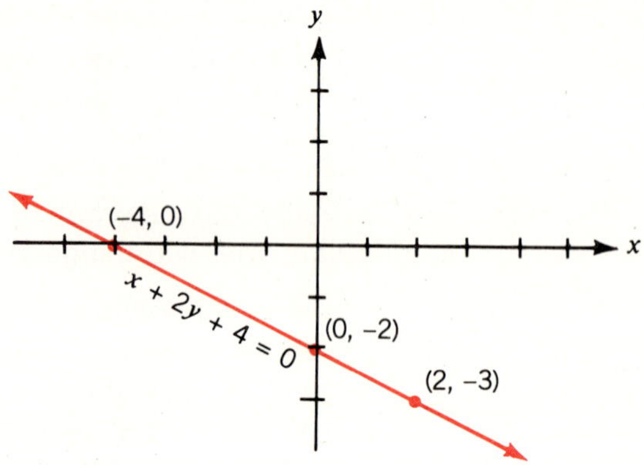

Figure 5.10

Definition: Intercepts

If a point having coordinates $(a, 0)$ is on a graph, then a is called an ***x*-intercept** of the graph. If a point having coordinates $(0, b)$ is on a graph, then b is called a ***y*-intercept** of the graph.

Example 3 Graph $\{(x, y) | x + 2y + 4 = 0\}$.

SOLUTION

Selecting $x = 0$, $y = 0$, and $y = -3$, we make the following table:

x	y	$x + 2y + 4 = 0$	(x, y)
0		$0 + 2y + 4 = 0,\ 2y = -4,\ y = -2$	$(0, -2)$
	0	$x + 2(0) + 4 = 0,\ x + 4 = 0,\ x = -4$	$(-4, 0)$
	-3	$x + 2(-3) + 4 = 0,\ x - 6 + 4 = 0,\ x - 2 = 0,\ x = 2$	$(2, -3)$

The plotted points and the line through them are shown in Fig. 5.10.

Example 4 Graph $\{(x, y) | y = 3\}$.

SOLUTION All ordered pairs whose second component is 3 are solutions of $y = 3$: for example, $(2, 3)$, $(0, 3)$, $(-3, 3)$. The graph is a horizontal line, as shown in Fig. 5.11.

Example 5 Graph $\{(x, y) | x = 2\}$.

SOLUTION All ordered pairs whose first component is 2 are solutions of $x = 2$: for example, $(2, 4)$, $(2, 0)$, $(2, -3)$. The graph is a vertical line, as shown in Fig. 5.12.

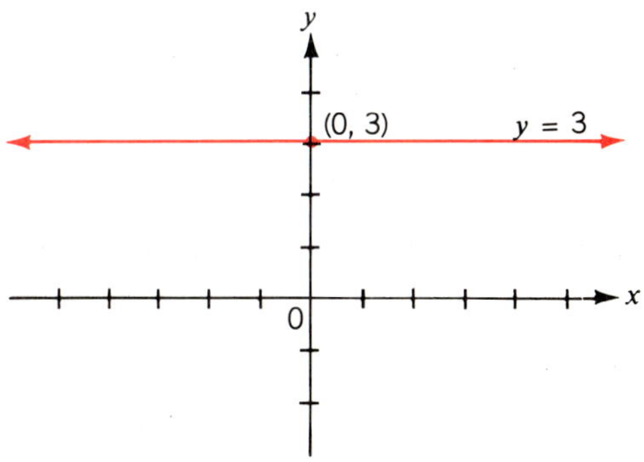

Figure 5.11

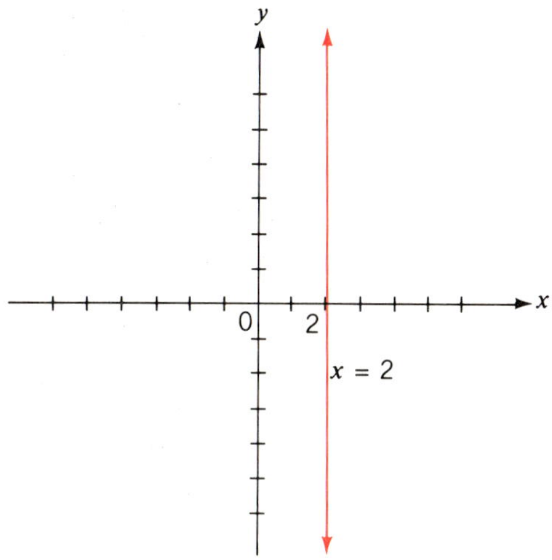

Figure 5.12

Exercises 5.2

In Exercises 1–6, determine the x-intercept and the y-intercept and use these values to graph each equation.

1. $y = 2x - 6$
2. $y = 3x + 6$
3. $x + y = 4$
4. $x - y = 4$
5. $3x - 2y = 6$
6. $2x - 3y = 12$

Graph each of Exercises 7–24 on a number plane.

7. $y = 2x + 6$
8. $y = 3x - 9$
9. $y = 4 - x$
10. $y = 8 - 4x$
11. $x + 2y = 0$
12. $3x - 4y = 0$
13. $x - y = 5$
14. $x + 2y = 8$
15. $2x + 3y = 6$
16. $5x - 2y = 10$
17. $x = -3$
18. $y + 1 = 0$
19. $x - 5y = 0$
20. $5x - 2y = 0$
21. $x - 3y = 9$
22. $x + 4y = 12$
23. $3x + 5y = 15$
24. $4x - 5y = 10$

In Exercises 25 and 26, draw the graphs of each of the following equations on one set of coordinate axes. What characteristic do these graphs have in common? How do these graphs differ from one another?

25. a. $y = x + 2$
    b. $y = x - 1$
    c. $y = x$
    d. $y = x + 4$
26. a. $y = 2x + 1$
    b. $y = 3x + 1$
    c. $y = -5x + 1$
    d. $x + y - 1 = 0$

27. a. For what value of k will the point $(k, 3)$ be on the graph of $y = 2x - 7$?
    b. For what value of k will the point $(2, k)$ be on the graph of $y = 5x - 1$?

28. The number of hours H that a growing child should sleep may be related to the age A of the child in years by the formula

$$H = 17 - \frac{A}{2}$$

Represent this relation graphically by selecting the horizontal axis as the A-axis. Graph from $A = 0$ to $A = 18$.

29. The following table shows the changes in temperature in a town on a certain day from 9 A.M. to 3 P.M.

Hour	9	10	11	12	1	2	3
Temperature	62°	66°	71°	74°	70°	69°	67°

   a. Draw a graph to show these changes in temperature. Join the points consecutively by straight lines.
   b. Mark the points on the graph that indicate the approximate time when the temperature was 68°.
   c. Determine from the graph for how long the temperature was above 68°.

30. The formula $A = 100(1 + 0.06t)$ expresses the amount of money A (in dollars) to which a sum of \$100 accumulates if invested at 6% simple interest for t years.
   a. Graph this equation, finding points for $t = 2, 4, 6, 8$, and 10. Join the points by a straight line.
   b. Find out from the graph how much money has accumulated at the end of 5 years.
   c. Find out from the graph how long it would take the sum to reach \$142.

31. According to the straight-line depreciation method, a car that costs \$3000 new and that has a probable scrap value of \$500 at the

end of 10 years will have a book value of B dollars at the end of n years, where $B = 3000 - 250n$.

a. Graph this equation for $0 \le n \le 10$.
b. From the graph determine the book value at the end of $5\frac{1}{2}$ years.
c. When will the book value be $1875?
d. On the same set of axes, graph $A = 250n$, the amount in the depreciation fund.
e. When is the amount in the depreciation fund equal to the book value?

5.3 Slope and the Linear Equation

Slope

If a roof rises 1 foot for every 4 feet of horizontal distance, the steepness or pitch of the roof is $\frac{1}{4}$ (Fig. 5.13). The ratio $\frac{1}{4}$ represents the **slope** of the roof, and it is the ratio of the rise to the run. It is often important to evaluate the slope of a line, and the computation is similar to the roof problem.

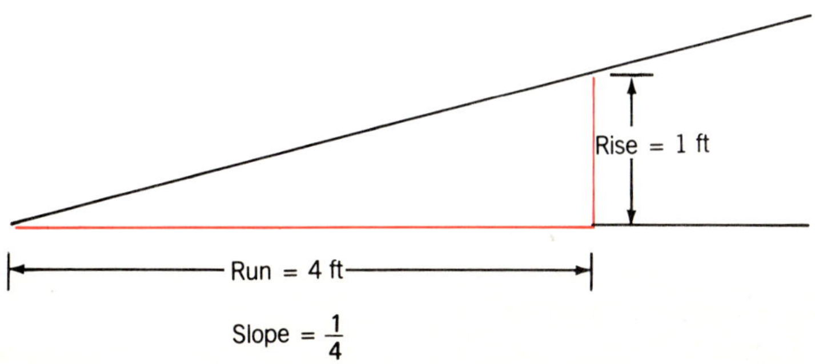

Figure 5.13

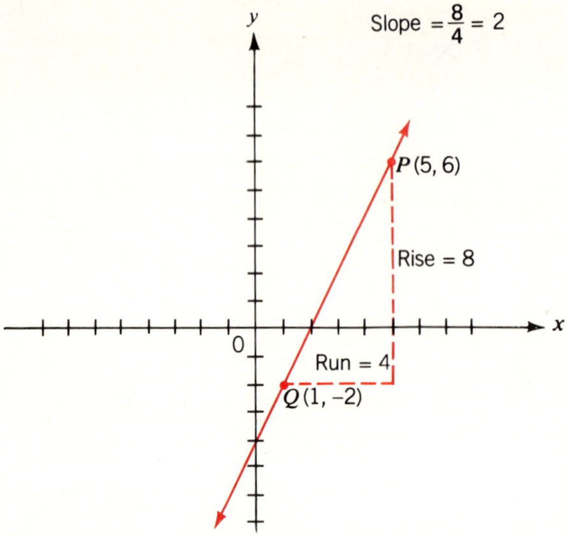

Figure 5.14

Consider the graph of the equation $2x - y = 4$ (Fig. 5.14). Choose any two points on the graph—for example, point $P(5, 6)$ and point $Q(1, -2)$.

$$\text{Slope} = \frac{\text{rise}}{\text{run}} = \frac{\text{change in ordinate } (y)}{\text{change in abscissa } (x)}$$
$$= \frac{6 - (-2)}{5 - 1} = \frac{8}{4} = \frac{2}{1}$$

Therefore, the slope is $\frac{2}{1}$, or 2. The slope of a line is the same for all points on the line, not for just the two specific points selected.

A definition for the slope of a general line can be conveniently expressed by using subscripts to indicate two points on the line. For example, P_1 (read "P sub one") and P_2 (read "P sub two") name two points. The numeral 1 at the lower right of P_1 and the numeral 2 at the lower right of P_2 are subscripts used to indicate that P_1 is the first point and P_2 is the second point.

Similarly, the coordinates of P_1 can be expressed as (x_1, y_1) and those of P_2 as (x_2, y_2).

Definition

If P_1 (x_1, y_1) and P_2 (x_2, y_2) are any two points on a line and if $x_1 \neq x_2$, then the **slope** m of the line joining P_1 and P_2 is given by

$$m = \frac{y_2 - y_1}{x_2 - x_1}$$

See Fig. 5.15.

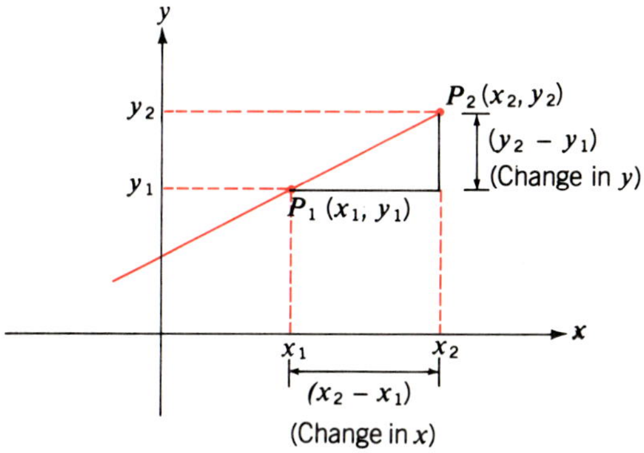

Figure 5.15

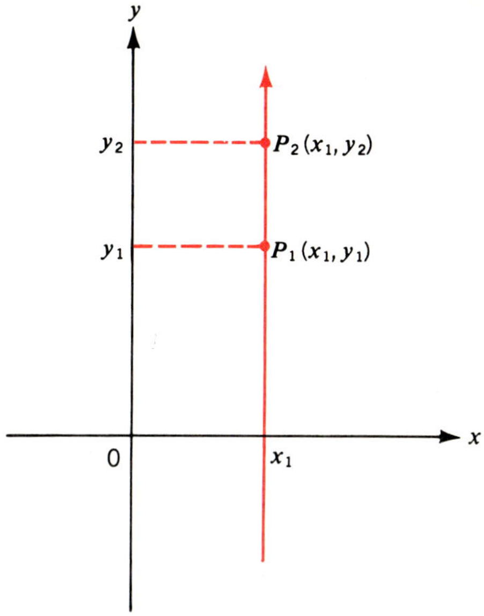

Figure 5.16

If a line is vertical (parallel to the y-axis), then

$$x_1 = x_2 \quad \text{and} \quad x_2 - x_1 = 0.$$

Since the denominator of the slope ratio is zero, the **slope of a vertical line is undefined** (see Fig. 5.16).

If a line is horizontal (parallel to the x-axis), then

$$y_1 = y_2 \quad \text{and} \quad y_2 - y_1 = 0.$$

In this case, the numerator of the slope ratio is zero, and **the slope of a horizontal line is 0** (see Fig. 5.17).

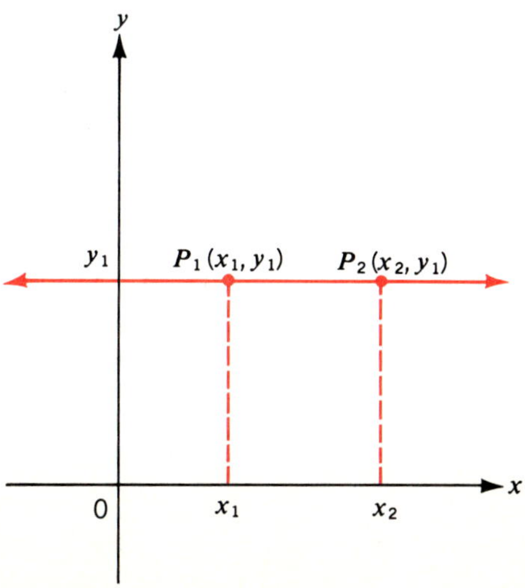

Figure 5.17

Example 1 Find the slope of the line passing through the points whose coordinates are (3, 2) and (5, 1), respectively.

SOLUTION Using

$$m = \frac{y_2 - y_1}{x_2 - x_1} \quad \text{with} \quad (x_2, y_2) = (5, 1)$$
$$\text{and} \quad (x_1, y_1) = (3, 2)$$
$$m = \frac{1 - 2}{5 - 3} = -\frac{1}{2}$$

ALTERNATE SOLUTION Using

$$m = \frac{y_2 - y_1}{x_2 - x_1} \quad \text{with} \quad (x_2, y_2) = (3, 2)$$
$$\text{and} \quad (x_1, y_1) = (5, 1)$$
$$m = \frac{2 - 1}{3 - 5} = -\frac{1}{2}$$

Note in Example 1 that the value of the slope does not depend on which point is called the first point and which the second.

Note also that the slope is negative, which means that the line "falls to the right" (see Fig. 5.18).

Example 2 Find the slope of the line passing through (2, 1) and (3, 4).

SOLUTION

$$m = \frac{y_2 - y_1}{x_2 - x_1} = \frac{4 - 1}{3 - 2} = 3$$

In this case the slope is positive, and the line "rises to the right" (see Fig. 5.19).

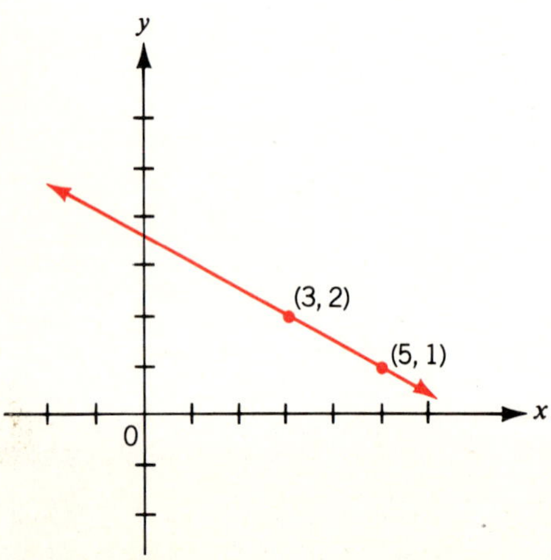

Figure 5.18

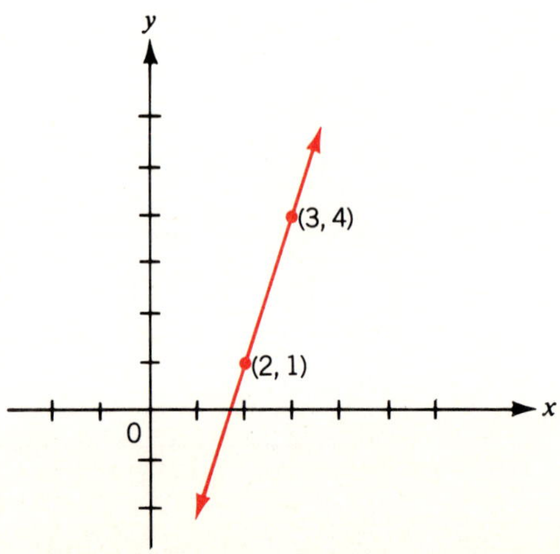

Figure 5.19

The Origin of Coordinates

The concept of a coordinate system probably originated with the ancient Egyptian surveyors. The hieroglyphic symbol used to designate the districts into which Egypt was divided was a grid symbol.

Records indicate that the Greeks used the ideas of longitude and latitude to locate points in the sky and on the earth. The Romans, who were noted for their surveying techniques, arranged the streets of their cities on a rectangular coordinate system.

The Arab and Persian mathematicians were the first to use geometric figures for algebraic problems. Examples are found in the works of the Arab al-Khowarizmi (ca. 825) and the Persian Omar Khayyam (ca. 1100). This usage is again found in the writings of Fibonacci (1220), Pacioli (1494), and Cardan (1545).

René Descartes (1596–1650) is credited with the invention of analytic geometry, since he used a rectangular coordinate system to establish a relationship between equations and curves. Pierre de Fermat, another great French mathematician, formulated coordinate geometry at the same time and made a considerable contribution in this field. The modern terms *coordinates*, *abscissa*, and *ordinate* were contributed by the German mathematician Gottfried Wilhelm Leibniz in 1692.

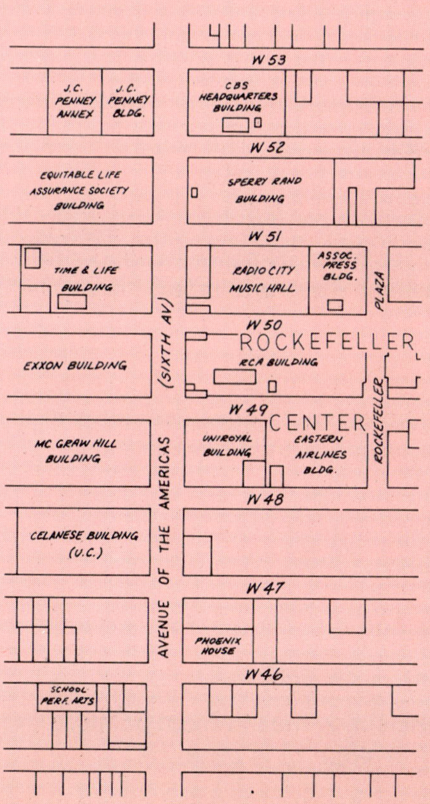

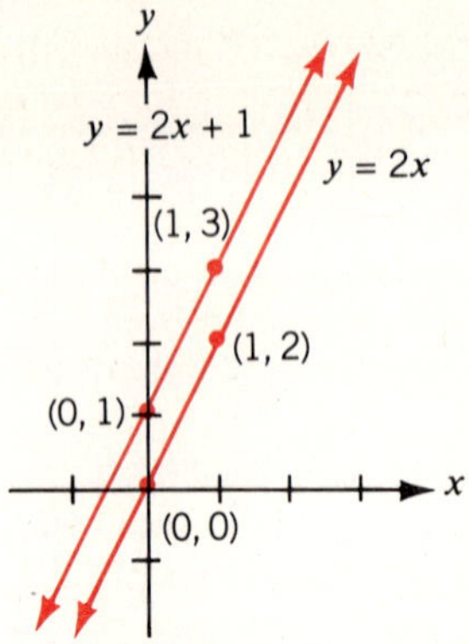

Figure 5.20

Although the slope of any given nonvertical line is uniquely determined, a given slope does not determine a unique line. Consider the equations

$$y = 2x + 1 \tag{1}$$

and

$$y = 2x \tag{2}$$

Select two points on the graph of equation (1) and two points on the graph of equation (2). The coordinates (0, 1) and (1, 3) satisfy equation (1). (See Fig. 5.20.) The slope of line (1) is $\frac{3-1}{1-0} = 2$. The coordinates (0, 0) and (1, 2) satisfy equation (2). The slope of line (2) is $\frac{2-0}{1-0} = 2$. Both lines have the same slope, but from Fig. 5.20 it is clear that the lines are distinct. They are in fact parallel, which leads to the following theorem.

Theorem

Two distinct nonvertical lines in the same plane are parallel if and only if they have the same slope.

The Slope-Intercept Form of a Linear Equation

Let us find the slope of the line represented by the equation

$$2x + 3y - 12 = 0$$

Selecting the intercept points, (0, 4) and (6, 0), as points on the line, then the slope m is given by

$$m = \frac{4-0}{0-6} = \frac{-2}{3}$$

Now let us solve $2x + 3y - 12 = 0$ for y.

$$y = \frac{-2}{3}x + 4$$

Note that the slope, $\frac{-2}{3}$, is the coefficient of x. Note also that the constant term 4 is the y-intercept.

In general, if a line is represented by the equation

$$y = mx + b$$

then **m, the coefficient of x, is the slope of the line and b is the y-intercept.**

$$y = \underset{\substack{\uparrow \\ \text{slope}}}{mx} + \underset{\substack{\uparrow \\ y\text{-intercept}}}{b}$$

Definitions

$y = mx + b$ is called the **slope-intercept form** of a linear equation.

$Ax + By + C = 0$ is called the **standard form** of a linear equation.

Example 3 Find the slope and the y-intercept of the line whose equation is $5x - 4y = 20$.

SOLUTION Solving the equation for y,

$$\begin{aligned} 5x - 4y &= 20 \\ 5x &= 4y + 20 \\ 4y + 20 &= 5x \\ 4y &= 5x - 20 \\ y &= \frac{5}{4}x - 5 \end{aligned}$$

Comparing with $y = mx + b$, the slope of the line is $m = \frac{5}{4}$, and the y-intercept is $b = -5$.

Example 4 Check the value of the slope found in Example 3 by selecting any two points on the line and using the slope formula.

SOLUTION

Selecting $y = 5$ and using $5x - 4y = 20$,

$$\begin{aligned} 5x - 4(5) &= 20 \\ 5x &= 40 \text{ and } x = 8 \end{aligned}$$

Let $P_1 = (8, 5)$.

continued

Selecting $y = -10$ and using $5x - 4y = 20$,

$$5x - 4(-10) = 20$$
$$5x = -20 \text{ and } x = -4$$

Let $P_2 = (-4, -10)$.

$$m = \frac{y_2 - y_1}{x_2 - x_1} = \frac{-10 - 5}{-4 - 8} = \frac{-15}{-12} = \frac{5}{4}$$

Example 5 Which of the following lines is parallel to $y = 3x - 5$?

a. $3x - y + 6 = 0$ b. $x + 3y - 5 = 0$

SOLUTION

1. The slope of $y = 3x - 5$ is 3, the coefficient of x, since the equation is in the slope-intercept form.
2. Solving each equation for y to obtain the slope-intercept form,

a. $3x - y + 6 = 0$

$$3x + 6 = y$$
$$y = 3x + 6 \text{ and slope } m = 3$$

Since $y = 3x - 5$ and $3x - y + 6 = 0$ have the same slope, these lines are parallel.

b. $x + 3y - 5 = 0$

$$3y = -x + 5$$
$$y = -\frac{1}{3}x + \frac{5}{3} \text{ and } m = -\frac{1}{3}$$

Since $-\frac{1}{3} \neq 3$, the lines $y = 3x - 5$ and $x + 3y - 5 = 0$ are not parallel.

Example 6 Find the value of k so that the line through $(1, k)$ and $(5, 4k)$ has the property stated:

a. Slope -6 b. Slope 0 c. Parallel to $3x - y = 2$

SOLUTION

a. $m = \dfrac{4k - k}{5 - 1} = \dfrac{3k}{4}$ and $m = -6$

$$\frac{3k}{4} = -6,\ 3k = -24,\ k = -8$$

b. $\dfrac{3k}{4} = 0,\ k = 0$

c. Solving $3x - y = 2$ for y,

$$3x = y + 2$$
$$y = 3x - 2$$

and the slope of this line is 3.

Since parallel lines have the same slope,

$$\frac{3k}{4} = 3,\ 3k = 12,\ k = 4$$

Exercises 5.3

For Exercises 1–10, find the slope of the line segment joining the two given points.

1. (4, 2) and (2, 1)
2. (0, 1) and (−2, −2)
3. (0, 4) and (3, 0)
4. (−3, 2) and (2, −3)
5. (−1, −2) and (−3, −4)
6. (3, 2) and (1, 6)
7. (3, 2) and (4, 1)
8. (1, 0) and (−4, −3)
9. (−2, −1) and (2, 1)
10. (−5, −4) and (−2, −3)

Write each of the equations in Exercises 11–30 in the slope-intercept form, $y = mx + b$, if possible, and determine the slope and the y-intercept of the graph of the equation. Check the value of the slope by selecting any two points on the line and by using the slope formula.

11. $2x + y = 5$
12. $3x + 2y = 12$
13. $2x - y = 8$
14. $\frac{1}{2}(x + y + 10) = 0$
15. $5x - 2y = 10$
16. $2x - 3y = 9$
17. $5(x - y) + 3 = 0$
18. $2(x - y) - 3(x + y) = 5$
19. $3y - 15 = 0$
20. $3x - 9 = 0$
21. $y - 2x + 4 = 0$
22. $2y - 4x = 5$
23. $3x + 2y = 4$
24. $4x + 16 = 2y$
25. $3(x + 1) = 2y$
26. $x - (y - 1) + 5 = 0$
27. $2x + 8 = 0$
28. $4y + 8 = 0$
29. $y = x$
30. $x + y = 0$

For each pair of lines in Exercises 31–42, state whether the lines are parallel. If they are parallel, give the slope.

31. $2x + 2y = 1$
    $3x + 3y = 1$
32. $x = 2y - 6$
    $y = 2x - 6$
33. $2x = 3y$
    $3x = 2y$
34. $4x = 2y + 1$
    $2x - y = 4$
35. $5x + 3y = 15$
    $3x + 5y = 30$
36. $x + y = 0$
    $x - y = 0$
37. $x = y - 2$
    $y = x - 2$
38. $5x - 2y = 3$
    $10x - 4y = 5$
39. $2(x + 3y) = 3(x + 2y - 1)$
    $5x + 10 = 0$
40. $3x - 4y = 12$
    $4x - 3y = 12$
41. $x = 3y + 6$
    $2x - 6y = 3$
42. $(x + y) - (x - y) = 6$
    $2y - 8 = 0$

In Exercises 43–48, find the value of k so that the line through (2, k) and (5, 3k) has the property stated.

43. Slope 4

44. Slope $\frac{1}{2}$

45. Slope 0

46. Slope -2

47. Parallel to $y = x + 5$

48. Parallel to $2x + y = 4$

49. As an example of a slope, a highway grade expressed as a percentage means the number of feet the road changes in elevation for 100 feet measured horizontally.
    a. A certain highway has a $2\frac{1}{2}\%$ grade. How many feet does it rise in a 1-mile stretch (horizontal distance)? (*Note:* 1 mile = 5280 feet.)
    b. How many feet does a $-3\frac{1}{4}\%$ grade highway drop for a $\frac{1}{2}$-mile horizontal stretch?

50. A certain county specification requires that an inclined water pipe must have a slope greater than or equal to $\frac{1}{4}$. Which of the water pipes whose rises and runs are given below meets this specification?
    a. Rise = 20 feet, run = 64 feet
    b. Rise = 125 feet, run = 500 feet
    c. Rise = 60 feet, run = 250 feet

5.4 Determining the Equation of a Line

In the preceding discussion, the emphasis was on graphing the line whose equation was given. Now consider the problem of finding the equation of a line whose geometric conditions are given.

In order to accomplish this task, one of the following conditions must be given:

1. One point on the line and the slope of the line or
2. Two points on the line

If the coordinates of a point on a line are given and if the slope of the line is known, then the equation of the line can be found by using the definition of the slope. Since (x, y) is to be a general point on the line, then

$$m = \frac{y - y_1}{x - x_1}$$

where (x_1, y_1) and m are the given point and slope, respectively.

Multiplying both sides of this equation by $x - x_1$ yields the point-slope form of a line, $y - y_1 = m(x - x_1)$.

Definition

The **point-slope form** of a line with slope m passing through the point (x_1, y_1) is

$$y - y_1 = m(x - x_1)$$

Example 1 Find an equation of a line whose slope is -2 and that passes through the point (3, 4).

SOLUTION $m = -2$ and $(x_1, y_1) = (3, 4)$

Using the point-slope form of a line,

$$\begin{aligned} y - y_1 &= m(x - x_1) \\ y - 4 &= -2(x - 3) \\ y - 4 &= -2x + 6 \\ y &= -2x + 10 && \text{(Slope-intercept form)} \\ 2x + y - 10 &= 0 && \text{(Standard form)} \end{aligned}$$

It is conventional to express the answer in standard form—that is, the linear equation is written in the form $Ax + By + C = 0$, where $A \geq 0$.

Example 2 Find an equation of a line passing through the points (3, 4) and (2, 1).

SOLUTION

1. Find the slope:

$$m = \frac{y_2 - y_1}{x_2 - x_1} = \frac{4 - 1}{3 - 2} = 3$$

2. Selecting (x_1, y_1) as (2, 1), and using the point-slope form,

$$\begin{aligned} y - y_1 &= m(x - x_1) \\ y - 1 &= 3(x - 2) \\ y - 1 &= 3x - 6 \\ y &= 3x - 5 && \text{(Slope-intercept form)} \\ 3x - y - 5 &= 0 && \text{(Standard form)} \end{aligned}$$

ALTERNATE SOLUTION AND CHECK Selecting (x_1, y_1) as (3, 4) and (x_2, y_2) as (2, 1), the slope,

$$m = \frac{y_2 - y_1}{x_2 - x_1} = \frac{1 - 4}{2 - 3} = \frac{-3}{-1} = 3$$

Using the point-slope form,

$$\begin{aligned} y - y_1 &= m(x - x_1) \\ y - 4 &= 3(x - 3) \\ y - 4 &= 3x - 9 \\ y &= 3x - 5 && \text{(Slope-intercept form)} \\ 3x - y - 5 &= 0 && \text{(Standard form)} \end{aligned}$$

Note in the preceding example that when two points on the line are given, it makes no difference which is called the first and which is called the second in finding the equation of the line. Since two possibilities are available, one choice can serve as a check for the other choice.

Exercises 5.4

In Exercises 1–10, write an equation of the line with the given slope and passing through the point whose coordinates are given.

1. $m = -2$; (2, 4)
2. $m = 3$; (1, 5)
3. $m = 6$; (−1, 3)
4. $m = -1$; (2, −1)
5. $m = \frac{1}{2}$; (4, −2)
6. $m = \frac{1}{3}$; (−6, 1)
7. $m = -\frac{3}{4}$; (−3, −4)
8. $m = -\frac{2}{5}$; (−5, −12)
9. $m = 0$; (2, 7)
10. $m = 0$; (5, 0)

In Exercises 11–20, find the slope, if it exists, and write an equation of the line that contains the points whose coordinates are given.

11. (2, 1) and (4, 2)
12. (2, 3) and (−2, −3)
13. (−2, −3) and (4, −1)
14. (−3, −4) and (−4, −3)
15. (5, 2) and (−3, 0)
16. (−2, 5) and (0, −3)
17. (2, 3) and (−5, 3)
18. (4, 6) and $\left(4, -\frac{1}{2}\right)$
19. (−2, 5) and (−2, −2)
20. (6, −7) and (−5, −7)

For Exercises 21–30, find an equation of the form $Ax + By + C = 0$ for the line satisfying the stated conditions.

21. Parallel to $3x + 2y = 6$ and passing through the point (1, 2)
22. Slope $-\frac{1}{2}$, y-intercept 3
23. Slope $-\frac{1}{2}$, passing through (3, 0)
24. Slope $\frac{3}{4}$, passing through the origin
25. Parallel to the x-axis, passing through (1, 2)
26. Parallel to the y-axis, passing through (1, 2)
27. Parallel to $2x - y + 3 = 0$ and passing through the point (−2, −3)
28. $m = \frac{4}{3}$, y-intercept −2
29. $m = -\frac{4}{3}$, passing through (−2, 0)
30. Passing through the origin and the point (−3, 5)

31. Do the points A: $(1, -2)$, B: $(3, 0)$, and C: $(0, 3)$ lie on a straight line? Why?

32. Do the points P: $(0, -1)$, Q: $(2, 0)$, and R: $(4, 1)$ lie on a straight line? Why?

33. For what value of k will the points $(8, 2)$, $(5, 3)$, $(2, k)$ lie on a straight line?

34. For what value of k will the points $(5, 0)$, $(-2, 4)$, $(k, -4)$ lie on a straight line?

35. The results of a study made by a physiologist for the purpose of trying to predict adult height from height as a child are shown in the following table, where C = the heights of 2-year-old children in inches and A = adult heights in inches.

C	30	31	33	34	34	35	37	38
A	60	61	65	67	66	70	73	74

a. Graph the ordered pairs, using the horizontal axis as the C-axis.
b. Draw a straight line that comes closest to passing through all the points plotted. (This is called the *line of best fit.*)
c. Write the equation for this line.
d. Using the equation in (c), predict the adult height of a 2-year-old child whose height is
   1. 32 inches 2. 36 inches 3. 39 inches

36. The "bromine number" test is a test used in chemistry to determine the number of double bonds of carbon in a compound. Measurements made at the end of 5 minutes, 10 minutes, and 15 minutes for the chemical reaction yielded the results shown in the following table:

t **Time in Minutes**	***B*** **Bromine Number**
5	35
10	50
15	65

a. Graph the ordered pairs (t, B).
b. Join the ordered pairs with a straight line and extend this line so it intersects the vertical B-axis.
c. From the graph, find the value of B for $t = 0$. (This indicates the number of double bonds of carbon in the compound.)
d. Check the result in (c) by writing the equation of the line in the graph and evaluating B for $t = 0$.

5.5 Systems of Linear Equations: Graphic Method

In geometry one learns that two distinct straight lines on a plane either intersect in exactly one point or are parallel. The algebraic counterpart of a set of two straight lines is a set, or system, of two linear equations in two variables. Whether the two linear equations have a common solution corresponds to whether their graphs have a common point of intersection or are parallel lines. *Intersection* is a key concept here, so first the intersection of sets will be explained.

Definition

The **intersection of two sets** A and B is the set consisting of all elements that are in *both* A and B. In symbols,

$A \cap B$ (read "A intersection B")

designates the intersection of A and B.

Example 1 If $A = \{1, 2, 3, 4\}$ and $B = \{3, 4, 5\}$, list the elements in $A \cap B$.

SOLUTION $A \cap B = \{3, 4\}$

Example 2 List the elements in $A \cap B$ if

$$A = \{(2, 5), (3, 8), (4, 7)\} \quad \text{and} \quad B = \{(3, 8), (-2, 5), (5, 2)\}$$

SOLUTION $A \cap B = \{(3, 8)\}$

Example 3 List the elements in $A \cap B$ if

$$A = \{(1, -1), (2, -2), (3, -3)\} \quad \text{and}$$
$$B = \{(-1, 1), (-2, 2), (-3, 3)\}$$

SOLUTION $A \cap B = \varnothing$, the empty set, and sets A and B are **disjoint.**

Example 4 List the elements in $A \cap B$ if

$$A = \{(x, y) | y = x\} \quad \text{and} \quad B = \{(x, y) | y = x + 1\}$$

SOLUTION If $y = x$ and $y = x + 1$, then

$$x = x + 1$$

and

$$-x + x = -x + x + 1$$
$$0 = 1$$

continued

Since this is impossible, there are no values of x such that $x = x + 1$.

Therefore, the sets do not have an element in common and $A \cap B = \varnothing$, the empty set.

Example 5 List the elements in $A \cap B$ if

$$A = \{(x, y) | y = x + 1\} \quad \text{and} \quad B = \{(x, y) | y = 9 - x\}$$

SOLUTION If $y = x + 1$ and $y = 9 - x$, then

$$x + 1 = 9 - x$$
$$2x = 8 \quad \text{and} \quad x = 4$$

For $y = x + 1$ and $x = 4$, $y = 5$ and $(4, 5)$ is a solution of $y = x + 1$.

For $y = 9 - x$ and $x = 4$, $y = 9 - 4 = 5$ and $(4, 5)$ is also a solution of $y = 9 - x$.

Therefore, $A \cap B = \{(4, 5)\}$.

It has been seen that the solution set of a linear equation in two variables is an infinite set of ordered pairs. It was also stated that the graph of a linear equation in two variables is a straight line.

When the graphs of two linear equations in two variables are drawn on the same set of axes, then there are three possibilities:

1. The lines intersect in exactly one point.
2. The lines are parallel and there is no point in common.
3. The lines coincide and have all their points in common.

These three possibilities are illustrated by the graphs in Fig. 5.21.

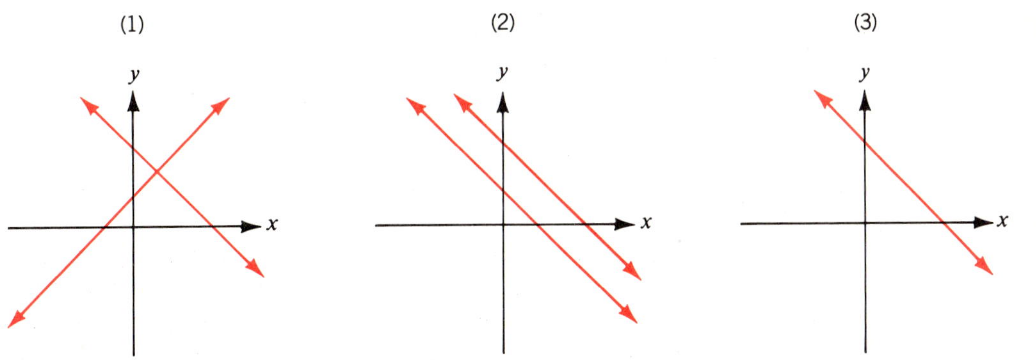

(1) $A = \{(x, y) \mid x + y = 5\}$, $B = \{(x, y) \mid y = x + 1\}$, $A \cap B = \{(2, 3)\}$ — Intersecting lines

(2) $A = \{(x, y) \mid x + y = 5\}$, $B = \{(x, y) \mid x + y = 2\}$, $A \cap B = \phi$ — Parallel lines

(3) $A = \{(x, y) \mid x + y = 5\}$, $B = \{(x, y) \mid 3x + 3y = 15\}$, $A \cap B = \{(x, y) \mid x + y = 5\}$ — Coincident lines

Figure 5.21

Definition
A **system of equations** is a set of equations in two or more variables.

A system of equations is also referred to as a set of **simultaneous equations.**

Definition
The **solution set of a system of equations** in two variables is the set of all ordered pairs that are common solutions to all the equations in the system.

Using the language of set theory, the solution set of a system of two equations in two variables is the intersection of the solution set of one of the equations with the solution set of the other. In symbols, if $A = \{(x, y) | ax + by = c\}$ and $B = \{(x, y) | dx + ey = f\}$, then $A \cap B$ is the solution of the system.

By examining the graphs in Fig. 5.21, the conclusions shown in Table 5.1 can be made.

It has been stated that two nonvertical lines are parallel if and only if their slopes are equal. It was also stated that the slope and *y*-intercept of a line completely determine the line. Thus two nonvertical lines coincide if and only if their slopes and their *y*-intercepts are equal. Two vertical

TABLE 5.1

The System	Geometric Meaning	Algebraic Meaning
1. $x + y = 5$ $y = x + 1$ $A \cap B = \{(2, 3)\}$	The lines have exactly one point in common. The lines intersect. The lines have different slopes.	The equations have exactly one solution in common.
2. $x + y = 5$ $x + y = 2$ $A \cap B = \varnothing$	The lines have no point in common. The lines are parallel. The lines have the same slope and different *y*-intercepts.	The equations have no common solution—that is, the intersection of their solution sets is the empty set, $\varnothing$.
3. $x + y = 5$ $3x + 3y = 15$ $A \cap B$ $= \{(x, y) \| x + y = 5\}$	The lines have all their points in common. The lines coincide. The lines have the same slope and the same *y*-intercept.	The equations have infinitely many solutions in common: the solution set of either equation.

lines either are parallel or coincide if their x-intercepts are equal. Vertical lines have the form $x = a$, so they can be identified immediately.

Thus it may be concluded that **two nonvertical lines intersect in exactly one point if and only if their slopes are unequal.**

Example 6 Determine whether the lines represented by each of the following systems of equations intersect, are parallel, or are coincident:

a. $3x - 5y = 9$
$3x + 2y = 6$

b. $2x - 2y = 7$
$3x - 3y = 5$

c. $5x + y = 4$
$x + 0.2y = 0.8$

SOLUTION Solve each equation for y to find the slope of the line. If $y = mx + b$, then m = the slope and b = the y-intercept.

a.
$$3x - 5y = 9$$
$$-5y = 9 - 3x$$
$$y = \frac{3}{5}x - \frac{9}{5}$$

Thus the slope is $\frac{3}{5}$.

$$3x + 2y = 6$$
$$2y = 6 - 3x$$
$$y = -\frac{3}{2}x + 3$$

Thus the slope is $-\frac{3}{2}$.

Since the slopes are not equal, the lines intersect in exactly one point.

b.
$$2x - 2y = 7$$
$$-2y = -2x + 7$$
$$y = x - \frac{7}{2}$$

Thus the slope is 1, and the y-intercept is $-\frac{7}{2}$.

$$3x - 3y = 5$$
$$-3y = -3x + 5$$
$$y = x - \frac{5}{3}$$

Thus the slope is 1, and the y-intercept is $-\frac{5}{3}$.

Since the slopes are equal and the y-intercepts are not equal, the lines are parallel.

c.
$$5x + y = 4$$
$$y = -5x + 4$$

Thus the slope is -5, and the y-intercept is 4.

$$x + 0.2y = 0.8$$
$$0.2y = -x + 0.8$$
$$y = -\frac{1}{0.2}x + 4$$
$$y = -5x + 4$$

Thus the slope is -5, and the y-intercept is 4.

Since the slopes are equal and the y-intercepts are equal, the lines coincide.

Two linear equations in two variables whose graphs intersect are solved by the **graphic method** with the following procedure:

1. Graph both equations on the same set of axes.
2. Determine the point of intersection. (It may be necessary to extend the lines to a considerable length to obtain the point of intersection.) The ordered pair that names the point of intersection is the solution set of the system.
3. Check the solution in both equations. It is necessary to check the solution because the point of intersection cannot always be read accurately from the graph.

Example 7 Using the graphic method, solve the system

$$2x + 3y - 5 = 0$$
$$x - 2y + 8 = 0$$

SOLUTION

$2x + 3y - 5 = 0$

x	y
1	1
4	−1
7	−3

$x - 2y + 8 = 0$

x	y
−8	0
−6	1
−4	2

Solution set: $\{(-2, 3)\}$ See Fig. 5.22.

CHECK

$$2x + 3y - 5 = 2(-2) + 3(3) - 5$$
$$= -4 + 9 - 5 = 0$$
$$x - 2y + 8 = -2 - 2(3) + 8$$
$$= -2 - 6 + 8 = 0$$

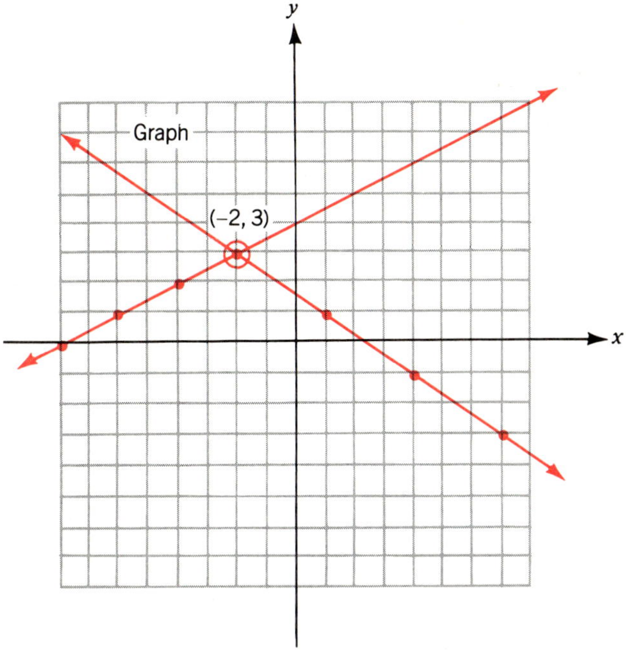

Figure 5.22

Exercises 5.5

For each pair of lines represented by the equations in Exercises 1–12,

a. Find the slope of each line.
b. Find the y-intercept of each line.
c. State whether the lines intersect, are parallel, or coincide.
d. State whether each system of equations has exactly one solution, no solution, or a line of solutions.

1. $3x + 3y = 4$
   $3x - 3y = 4$
2. $3x + 3y = 4$
   $2x + 2y = 5$
3. $3x + 3y = 4$
   $6x + 6y = 8$
4. $x + 3 = 0$
   $y - 4 = 0$
5. $x + 3 = y$
   $3x - y = 2x$
6. $3x - 4y = 12$
   $4y - 3x + 12 = 0$
7. $3x - 2y = 3$
   $5x + 2y = 5$
8. $3x - 2y = 3$
   $3x - 2 = 3y$
9. $2x - 1 = y$
   $3y + 3 = 6x$
10. $3x - 6y = 3$
    $2x - 4y = 2$
11. $x - 2y = 5$
    $2y - x = 5$
12. $x = y - 6$
    $y = x - 6$

René Descartes: Soldier, Philosopher, Mathematician

René Descartes (1596–1650) was born to a French aristocratic family, from whom he inherited enough money to spend his life studying and traveling. Because of delicate health, when he was in school at the Jesuit college of La Flèche he was allowed to stay in bed in the morning as long as he liked, a habit he retained throughout his life. Descartes said that he formulated his greatest ideas in mathematics and philosophy during these periods.

When he completed his studies at age 16, he was disenchanted with the philosophy he had learned. He wanted first-hand knowledge of the real world, so he went to Paris, where for a time he enjoyed the busy social life. When he tired of this, he became a soldier. For several years he alternated army life, which included fighting in several battles, with a life of travel and meditation. When he was 32, he ended his career as a soldier and spent the next 20 years in Holland, traveling from one small village to another, wanting only peace and quiet for his thoughts and research.

By 1634 Descartes had completed his work *Le Monde,* a philosophical and scientific account of the creation of the universe. He was about to publish it when he learned that in 1633 the Inquisition had forced Galileo to swear on his knees that the earth did not revolve about the sun. Not only had Descartes accepted Galileo's ideas, he had used them in his work. Intimidated and deeply hurt, Descartes stopped publication and vowed he would never publish anything.

Fortunately, however, his friends made him change his mind, and in 1637 his masterpiece was printed. This treatise, called the *Discours de la Méthode* but commonly referred to as *The Method,* contained 3 appendices, the third being the famous "La Géométrie." Almost all of the algebraic symbols that Descartes introduced there are the ones we use today. More than this, "La Géométrie" contains the introduction to modern analytic geometry.

The publication caused a sensation. His name became known throughout Europe. Kings and queens wanted him at their courts. He routinely rejected such offers, preferring only a quiet, peaceful life, but one ruler, Queen Christine of Sweden, was not to be denied. Descartes eventually accepted her offer, but life in Sweden was his downfall. Descartes was accustomed to rising at 11 A.M., but in Sweden the daily instruction began at 5 A.M. in cold, unheated quarters. Four months after his arrival he contracted pneumonia and died.

Solve each system in Exercises 13–22 using the graphic method.

13. $x + y = 7$
    $x - y = 1$

14. $4x - 3y = 0$
    $2x - 3y = 6$

15. $2x + y = 3$
    $x - 2y = 4$

16. $x + 3y = 9$
    $2x + y = 8$

17. $x - 4y = 4$
    $x + 2y = 10$

18. $x + 2y = 4$
    $2x - y = 3$

19. $x + y = 0$
    $2x + 3y = 6$

20. $3x + y - 3 = 0$
    $x + y = 5$

21. $3x - 2y = 6$
    $x - 2y = 10$

22. $5x + 2y = 10$
    $x + 2y + 6 = 0$

23. A dietitian wants to combine two foods so that the combination will supply 120 units of vitamins and 80 units of minerals. Food A supplies 2 units of vitamins and 2 units of minerals per ounce. Food B supplies 3 units of vitamins and 1 unit of minerals per ounce. If x = the number of ounces of Food A needed and y = the number of ounces of Food B needed, then $2x + 3y = 120$ and $2x + y = 80$. Solve this system graphically and state how many ounces of each food should be used.

24. A machine costing \$4800 will have a scrap value of \$800 at the end of 10 years. Using the straight-line method of depreciation, the book value y at the end of the xth year is given by the equation $y = 4800 - 400x$. To offset the depreciation, a contribution is placed in a fund each year. The amount of money y in the fund at the end of the xth year is given by the equation $y = 400x$. Solve this system of equations graphically, thereby finding the year when the amount in the depreciation fund is equal to the book value of the machine.

5.6 Systems of Linear Equations: Substitution Method

The graphic method gives us a valuable visual representation of the solution of a system of equations. However, it has the disadvantage that the solution cannot always be obtained accurately. Thus it is desirable to investigate other methods of solution.

It should be obvious that the solution of the system

$$x = 2$$
$$y = 3$$

can be recognized immediately. The solution is (2, 3), since this is the only ordered pair that simultaneously satisfies both equations. This special case suggests the possibility of transforming a system whose solution is not obvious into the simpler equivalent form

$$x = p$$
$$y = q$$

Definition

Two systems of equations are equivalent if and only if the systems have the same solution set.

The substitution method is one method for solving a system of two linear equations in two variables. This method is also used to solve systems of equations that are not linear. The procedure is as follows:

1. Solve one of the equations for one variable as a function of the other.
2. Substitute the expression that is obtained into the other equation.
3. Solve the resulting equation in one variable.
4. Substitute the solution obtained in step 3 into the equation obtained in step 1 and simplify.
5. Check the solution obtained in each of the original equations.

Example 1 Using the substitution method, solve the system

$$2x + y = 8$$
$$8x - 5y = 5$$

SOLUTION

1. Solve $2x + y = 8$ for y:

$$y = 8 - 2x$$

continued

2. Substitute $8 - 2x$ for y in $8x - 5y = 5$:

$$8x - 5(8 - 2x) = 5$$

3. Solve this equation for x:

$$\begin{aligned} 8x - 40 + 10x &= 5 \\ 18x - 40 &= 5 \\ 18x &= 45 \\ x &= \frac{45}{18} = \frac{5}{2} \end{aligned}$$

4. Substitute $\frac{5}{2}$ for x in $y = 8 - 2x$:

$$y = 8 - 2\left(\frac{5}{2}\right) = 8 - 5 = 3$$

5. Check the solution $\left(\frac{5}{2}, 3\right)$ in both equations:

$$2x + y = 2\left(\frac{5}{2}\right) + 3 = 8$$

$$8x - 5y = 8\left(\frac{5}{2}\right) - 5(3) = 20 - 15 = 5$$

The substitution axiom of the equal relation and the operational theorems of equivalence previously used in the solution of a linear equation in one variable guarantee that the system

$$\begin{aligned} x &= \frac{5}{2} \\ y &= 3 \end{aligned}$$

is equivalent to

$$\begin{aligned} 2x + y &= 8 \\ 8x - 5y &= 5 \end{aligned}$$

Sometimes it is more convenient to solve for one variable rather than the other. As a general rule, if the coefficient of a variable is 1 in one of the equations, then this variable and this equation are selected for the first step of the solution process.

Example 2 Using the substitution method, solve the system

$$4x - 3y = 26$$
$$x - 5y = 15$$

SOLUTION In the second equation, the coefficient of x is 1, and this equation is first solved for x:

1. $x - 5y = 15$ and $x = 5y + 15$
2. $4x - 3y = 26$
$$4(5y + 15) - 3y = 26$$
$$20y + 60 - 3y = 26$$
$$17y = -34$$
$$y = -2$$
3. $x = 5y + 15$
$$x = 5(-2) + 15 = 5$$

The solution is $(5, -2)$.

CHECK For $4x - 3y = 26$, $4(5) - 3(-2) = 20 + 6 = 26$.
For $x - 5y = 15$, $5 - 5(-2) = 5 + 10 = 15$.

For some equations it may be more convenient to solve for a *multiple* of x or of y instead of x or y. This idea is demonstrated in the following example.

Example 3 Using the substitution method, solve the system

$$5x - 2y = 8$$
$$5x - 3y = 2$$

SOLUTION Note that $5x$ occurs in each equation, so either equation can be solved for $5x$, and the resulting expression in y can be substituted in the other equation. Solving $5x - 3y = 2$ for $5x$, then $5x = 3y + 2$. Substituting in $5x - 2y = 8$,

$$(3y + 2) - 2y = 8$$
$$y + 2 = 8$$
$$y = 6$$

Since

$$5x = 3y + 2,$$
$$5x = 3(6) + 2$$
$$5x = 20 \quad \text{and} \quad x = 4$$

The solution is $(4, 6)$.

CHECK For $5x - 2y = 8$, $5(4) - 2(6) = 20 - 12 = 8$.
For $5x - 3y = 2$, $5(4) - 3(6) = 20 - 18 = 2$.

Exercises 5.6

Solve each system in Exercises 1–30 using the substitution method and check each solution set.

1. $2x + 3y = 16$
   $y = 2x$
2. $3x + 2y = 36$
   $y = 3x$
3. $5x - 3y = 5$
   $y = 2x - 3$
4. $5x + 2y = 25$
   $y = 3x - 4$
5. $3x - 5y = 14$
   $4x + y = 11$
6. $3x + y = 2$
   $5x + 2y = 6$
7. $x + 2y = 8$
   $4x + 3y = 7$
8. $3x + 2y + 1 = 0$
   $x - 3y = 18$
9. $2x - y = 8$
   $3x - 2y = 10$
10. $5x - 3y = 7$
    $3x - y = 9$
11. $3x + 4y = 4$
    $x - 2y = 8$
12. $x + 4y + 10 = 0$
    $2x - 3y = 13$
13. $x = 2y - 3$
    $y = x + 5$
14. $x = 4 - y$
    $y = 2x + 10$
15. $3x = 5y + 15$
    $3y = x - 1$
16. $3x - 4y = 12$
    $3x = y - 6$
17. $8y - 5x = 1$
    $2y - x = 1$
18. $5y - 3x = 8$
    $2y - x = 3$
19. $x + 3y = 0$
    $y = 4x$
20. $x = 3y$
    $2x + 5y = 0$
21. $y = 3x - 1$
    $6x - 2y = 2$
22. $2x - y = 4$
    $y = 2x + 3$
23. $2x - 2y = 1$
    $y = x + 5$
24. $x = 6 - y$
    $3x + 3y = 18$
25. $r - t + 3 = 0$
    $2r + 2t = 8$
26. $5r - 2t = 1$
    $5r = t + 7$
27. $3r + 6t = 12$
    $3r = 4 + 2t$
28. $2r + 6t = 10$
    $3t = 2 - 4r$
29. $3r - 2s = 1$
    $3r = 4s - 7$
30. $2r - 3s = 17$
    $2r + 8s = 6$

5.7 Systems of Linear Equations: Addition Method

Another method for solving a system of two linear equations in two variables is the addition method. This method is sometimes more convenient to use than the substitution method. Consider the system

$$\begin{aligned} 7x + 3y - 13 &= 0 \\ 2x + 3y + 2 &= 0 \end{aligned}$$

If it is assumed that the ordered pair (p, q) is a solution of both equations, then $7p + 3q - 13 = 0$ and $2p + 3q + 2 = 0$. Moreover, by selecting a and b to be two real numbers that are not both zero, then it follows that (p, q) is also a solution of

$$a(7x + 3y - 13) + b(2x + 3y + 2) = 0$$

since

$$a(7p + 3q - 13) + b(2p + 3q + 2) = a(0) + b(0) = 0 + 0 = 0$$

The objective is to obtain an equivalent system having the form

$$\begin{aligned} x &= p \\ y &= q \end{aligned}$$

Either equation of this system can be obtained by choosing a and b in such a manner that the coefficients of one of the variables become additive inverses.

Choosing $a = 1$ and $b = -1$ and writing the system as

$$\begin{aligned} 7x + 3y &= 13 \\ 2x + 3y &= -2 \end{aligned}$$

then

$$\left.\begin{aligned} 1(7x + 3y) &= 1(13) \\ -1(2x + 3y) &= -1(-2) \end{aligned}\right\} \rightarrow \begin{aligned} 7x + 3y &= 13 \\ -2x - 3y &= 2 \\ \hline 5x &= 15 \\ x &= 3 \end{aligned}$$

Adding,

If $a = 2$ and $b = -7$, then the coefficients of x become additive inverses, and

$$\left.\begin{aligned} 2(7x + 3y) &= 2(13) \\ -7(2x + 3y) &= -7(-2) \end{aligned}\right\} \rightarrow \begin{aligned} 14x + 6y &= 26 \\ -14x - 21y &= 14 \\ \hline -15y &= 40 \\ y &= \frac{-8}{3} \end{aligned}$$

Adding,

Thus the system

$$7x + 3y - 13 = 0$$
$$2x + 3y + 2 = 0$$

is equivalent to

$$x = 3$$
$$y = \frac{-8}{3}$$

and the solution is $\left(3, \frac{-8}{3}\right)$.

CHECK

For $7x + 3y - 13 = 0$, $7(3) + 3\left(\frac{-8}{3}\right) - 13 = 21 - 8 - 13 = 0$.

For $2x + 3y + 2 = 0$, $2(3) + 3\left(\frac{-8}{3}\right) + 2 = 6 - 8 + 2 = 0$.

Example 1 Using the addition method, solve the system

$$2x - 3y = 18$$
$$7x - 2y = -5$$

SOLUTION

1. Multiply each equation so the y-terms become opposites.
2. Add and solve.
3. Multiply each equation so the x-terms become opposites.
4. Add and solve.

1. $\left.\begin{array}{r} 2(2x - 3y = 18) \\ -3(7x - 2y = -5) \end{array}\right\} \rightarrow \begin{array}{r} 4x - 6y = 36 \\ -21x + 6y = 15 \\ \hline -17x \qquad = 51 \end{array}$

2. $x = -3$

3. $\left.\begin{array}{r} 7(2x - 3y = 18) \\ -2(7x - 2y = -5) \end{array}\right\} \rightarrow \begin{array}{r} 14x - 21y = 126 \\ -14x + 4y = 10 \\ \hline -17y = 136 \end{array}$

4. $y = -8$

The solution is $(-3, -8)$.

CHECK

$$\begin{aligned} 2x - 3y &= 2(-3) - 3(-8) \\ &= -6 + 24 = 18 \text{ and } 18 = 18 \\ 7x - 2y &= 7(-3) - 2(-8) \\ &= -21 + 16 = -5 \text{ and } -5 = -5 \end{aligned}$$

Example 2 Using the addition method, solve the system

$$4x - y = 3$$
$$5x + 2y = 20$$

SOLUTION

$$\left.\begin{array}{r} 2(4x - y = 3) \\ 1(5x + 2y = 20) \end{array}\right\} \rightarrow \begin{array}{rl} 8x - 2y &= 6 \\ 5x + 2y &= 20 \\ \hline 13x \qquad &= 26 \\ x &= 2 \end{array}$$

Although this addition process can be used to find the value for y, it is also possible to replace x by 2 in either of the two equations and then solve this equation for y. Selecting the first,

$$\begin{aligned} \text{for } x = 2 \text{ and } 4x - y &= 3 \\ 4(2) - y &= 3 \\ 8 - y &= 3 \\ -y &= 3 - 8 = -5 \\ y &= 5 \end{aligned}$$

The solution is (2, 5).

CHECK For $x = 2$ and $y = 5$,

$$4x - y = 4(2) - 5 = 8 - 5 = 3$$
$$5x + 2y = 5(2) + 2(5) = 10 + 10 = 20$$

Exercises 5.7

Solve each system in Exercises 1–30 using the addition method and check each solution set.

1. $x + y = 7$
   $x - y = 3$
2. $x - y = 4$
   $x + y = 10$
3. $2x - 3y = 12$
   $4x + 3y = 6$
4. $3x + 2y = -4$
   $5x - 2y = 20$
5. $3x + y = 2$
   $2x + y = 3$
6. $x + 2y = 11$
   $x + 3y = 18$
7. $3x + 4y = 20$
   $3x + 2y = 16$
8. $2x - 3y = 3$
   $4x - 3y = 15$
9. $2x - y = 3$
   $3x + 2y = 8$
10. $5x + y = 8$
    $7x - 2y = 1$
11. $2x + 3y = 12$
    $3x - 2y = 5$
12. $4x - 3y = 10$
    $3x + 4y = 20$
13. $2x - y - 9 = 0$
    $3x - 4y - 1 = 0$
14. $4x - 3y - 2 = 0$
    $2x - y - 6 = 0$

15. $4x + 3y = 2$
    $3x + 2y = 4$

16. $2x + 3y = 5$
    $3x + 5y = 6$

17. $5x + 4y = 9$
    $4x + 5y = 18$

18. $3x + 2y = 5$
    $2x + 3y = 10$

19. $3x - 8y = 4$
    $2x - 5y = 3$

20. $7x - 3y = 5$
    $5x - 2y = 5$

21. $x + 3y = 4$
    $2x + 6y = 6$

22. $2x - 3y = 2$
    $6x - 9y = 6$

23. $x - 3y = 4$
    $2x - 6y = 8$

24. $2x + 3y = 2$
    $6x + 9y = 9$

25. $3r - 5s - 7 = 0$
    $2r + 4s + 10 = 0$

26. $2r - 3s + 8 = 0$
    $3s - r + 2 = 0$

27. $3r - 4t + 5 = 0$
    $3t - 2r = 6$

28. $5r + 3t = 10$
    $2t + 3r = 5$

29. $5a = 2b + 2$
    $3b = 8a - 5$

30. $3a = 2b - 2$
    $4b = 5a + 10$

31–38. Solve for x and y.

31. $x + y = 3a$
    $2x - y = 3b$

32. $2x + y = 5a$
    $x - 2y = 5b$

33. $3x - 2y = a + 5b$
    $2x + 3y = 5a - b$

34. $3x + 4y = a + b$
    $2x + 3y = a - b$

35. $ax + by = 5$
    $2ax - by = 1$

36. $ax - by = 4$
    $ax + 2by = 7$

37. $ax - by = a^2 + b^2$
    $bx + ay = a^2 + b^2$

38. $ax - by = a^2 - b^2$
    $bx + ay = 2ab$

5.8 Verbal Problems

Verbal problems that involve two unknown numbers often can be solved more conveniently by using two variables to form two equations.

Example 1 Find two numbers whose sum is 135 and whose difference is 61.

SOLUTION Let x = one number and y = the other number.

Then
$$\begin{aligned} x + y &= 135 \\ x - y &= 61 \\ \hline 2x &= 196 \\ x &= 98 \end{aligned}$$
(Solving by using the addition method)

Now if $x = 98$ and $x + y = 135$, then $98 + y = 135$ (by using the substitution axiom).

Thus $y = 135 - 98 = 37$, and the two numbers are 98 and 37.

CHECK

$$x + y = 98 + 37 = 135$$
$$x - y = 98 - 37 = 61$$

Example 2 A jet plane traveling with the wind flies 2325 miles in 3 hours. Against the wind, it takes the jet 4 hours to go 2900 miles. Find the speed of the plane in still air and the speed of the wind.

SOLUTION

Let x = the speed in still air of the plane and
y = the speed of the wind.

Formula	r ·	t =	d
With the wind	$x + y$	3	2325
Against the wind	$x - y$	4	2900

EQUATIONS

$$\left.\begin{aligned} 3(x + y) &= 2325 \\ 4(x - y) &= 2900 \end{aligned}\right\} \rightarrow \begin{aligned} x + y &= 775 \\ x - y &= 725 \\ \hline 2x &= 1500 \\ x &= 750 \\ 750 + y &= 775 \\ y &= 25 \end{aligned}$$

Thus the speed in still air of the jet plane is 750 mph, and the speed of the wind is 25 mph.

Example 3 A restaurant bought 3 tablecloths and 5 dozen napkins for \$54. A month later it bought 5 tablecloths and 2 dozen napkins for \$52. Find the cost of 1 tablecloth and the cost of 1 dozen napkins if these were the same for both purchases.

SOLUTION

Let $x =$ the cost of 1 tablecloth.
Let $y =$ the cost of 1 dozen napkins.
Then $3x + 5y = 54$ and $5x + 2y = 52$.
Using the addition method,

$$\left.\begin{aligned} -2(3x + 5y) &= -2(54) \\ 5(5x + 2y) &= 5(52) \end{aligned}\right\} \rightarrow \begin{aligned} -6x - 10y &= -108 \\ 25x + 10y &= 260 \\ \hline 19x \qquad &= 152 \\ x &= 8 \end{aligned}$$

Replacing x by 8 in $5x + 2y = 52$,

$$\begin{aligned} 5(8) + 2y &= 52 \\ 2y &= 12 \quad \text{and} \quad y = 6 \end{aligned}$$

Therefore, 1 tablecloth cost \$8, and 1 dozen napkins cost \$6.

Exercises 5.8

Solve Exercises 1–18 by using ***two*** *variables.*

1. The sum of two numbers is 9, and their difference is 19. Find the two numbers.
2. The sum of two numbers is 20, and their difference is 6. Find the numbers.
3. Twice the larger of two numbers is 3 less than 3 times the smaller. The sum of the two numbers is 10 more than 4 times the difference when the smaller is subtracted from the larger. Find the numbers.
4. When the larger of two numbers is increased by twice the smaller, the sum is 12. When twice the larger number is decreased by the smaller, the difference is 9. Find the numbers.
5. Three loaves of bread and 2 pounds of butter cost \$6. Five loaves of bread and 3 pounds of butter cost \$9.50. Find the price of 1 loaf of bread and the price of 1 pound of butter if these were the same for both purchases.
6. In a certain boys' store, 4 sweaters and 3 shirts cost \$63, whereas 3 sweaters and 4 shirts cost \$56. If sweaters are one fixed price and shirts are another fixed price, find the price of 1 sweater and 1 shirt.

7. A bill for 5 quarts of milk and 2 dozen eggs was $4.30. Another bill for 3 quarts of milk and 3 dozen eggs was $4.20. Find the price of 1 quart of milk and the price of 1 dozen eggs if these were the same for both purchases.

8. A sale rack in a department store had the following sign: "All coats on this rack one low price." A second rack advertised all slacks, regardless of original price, for a different single low price. If 4 coats and 5 pairs of slacks could be bought for $138.30, whereas 2 coats and 8 pairs of slacks could be bought for $115.90, find the sale price of 1 coat and the sale price of 1 pair of slacks.

9. A man can walk uphill at 2 mph and downhill at 4 mph on a road no part of which is level. If he walked 24 miles in 9 hours, how much of the road was uphill?

10. During part of its trip, a jet plane traveling 600 mph ran into a storm that reduced its speed by 80 mph. If the total trip of 2380 miles was made in 4 hours, how far did the plane travel during the storm?

11. A crew rows 30 miles downstream in 4 hours. Returning, they cover the same distance in 6 hours and 40 minutes. Find the rate of the current and the rate of the boat in still water.

12. A boat travels 120 miles up a river in 6 hours. Its return trip downstream takes 4 hours. Find the speed of the current and the speed of the boat in still water.

13. A plane with a headwind flies 300 miles from city A to city B in 2 hours. On the same day and under the same weather conditions, a plane with the same speed in still air flies from city B to city A in 1 hour and 30 minutes. Find the speed of the wind.

14. A plane with a tailwind completes a 2600-mile trip in 5 hours. On the same day and under the same weather conditions, a second plane flies in the *opposite* direction for 5 hours but completes only 2400 miles. Find the rate of the wind and the still air rate of the planes if both planes fly at the same still air rate.

15. Twenty pounds of ore A combined with 30 pounds of ore B produce 26.5 pounds of silver. Ten pounds of ore A combined with 20 pounds of ore B produce 15.5 pounds of silver. Find the percentage of silver in each ore.

16. An alloy containing 74 grams of gold is the combination of two other alloys containing gold: 40 grams of alloy A and 70 grams of alloy B. If 50 grams of alloy A and 90 grams of alloy B are combined, the resulting alloy contains 94 grams of gold. Find the percentage of gold in alloy A and alloy B.

17. A manufacturer must use two different machines for the production of two products. The number of hours required on each machine to produce the two products and the maximum number of hours each machine can be run per week are given in the following table.

Machine	Product A	Product B	Maximum Hours
I	3	2	48
II	1	4	46

How many items of each product will be manufactured each week if the machines are used at their maximum capacity?

18. A mail order company charges a fixed fee for shipping and handling the first 10 ounces of merchandise plus an additional fee for each ounce over 10 ounces. If the shipping and handling charge for 30 ounces of merchandise is $2.49 and the charge for 42 ounces is $3.45, find the fixed fee and the additional fee.

5.9 Linear Inequalities in Two Variables (Optional)

Equations and inequalities in one variable can be associated with points on a line, and the solution sets can be shown as line graphs.

Equations in two variables can be associated with points in a plane, using horizontal and vertical axes as references. Inequalities in two variables can also be graphed in a plane.

The graph of the linear equation $y = ax + b$ divides the plane into three mutually exclusive regions, or three disjoint sets of points:

1. $y = ax + b$
2. $y > ax + b$
3. $y < ax + b$

In Fig. 5.23 the region above the line $y = ax + b$ represents all points satisfying the condition $\{(x, y) | y > ax + b\}$, whereas the region below the line $y = ax + b$ represents all points satisfying the condition $\{(x, y) | y < ax + b\}$. The line is the graph of the equation $y = ax + b$.

To graph an inequality of form $y < ax + b$ or $y > ax + b$, first graph the equality, use a broken line to indicate that the graph of the equality is not part of the solution set, and then shade the desired region.

Example 1 Graph the relation $\{(x, y) | y < 2x + 3\}$.

SOLUTION First graph the equation $y = 2x + 3$ and use a broken line to indicate that the line is *not* to be included in the solution set.

continued

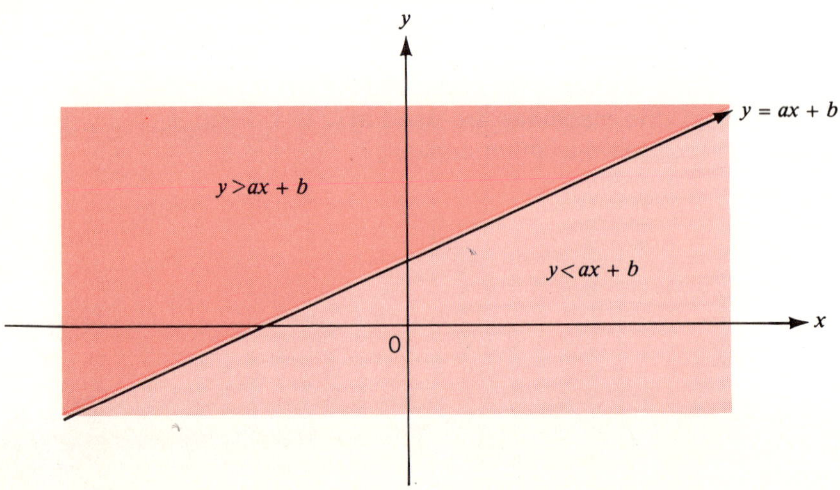

Figure 5.23

To graph $y = 2x + 3$, use a table of values such as

x	0	1	2
y	3	5	7

to obtain the ordered pairs (0, 3), (1, 5), (2, 7). Plot these points and draw a dashed line through them.

Since $y < 2x + 3$, shade the region *below* the broken line. See Fig. 5.24.

CHECK

1. Select any point in the shaded region.
2. Test whether the ordered pair satisfies the given inequality.
   a. Let $(x, y) = (1, 2)$.
   b. Is $y < 2x + 3$ for $x = 1$ and $y = 2$?

$$2 < 2(1) + 3$$
$$2 < 5, \quad \text{true}$$

Example 2 Graph $3x - 2y > x + y - 1$.

SOLUTION Solve the inequality for y:

$$\begin{aligned} 3x - 2y &> x + y - 1 \\ -3y &> -2x - 1 \\ 3y &< 2x + 1 \\ y &< \frac{2}{3}x + \frac{1}{3} \end{aligned}$$

continued

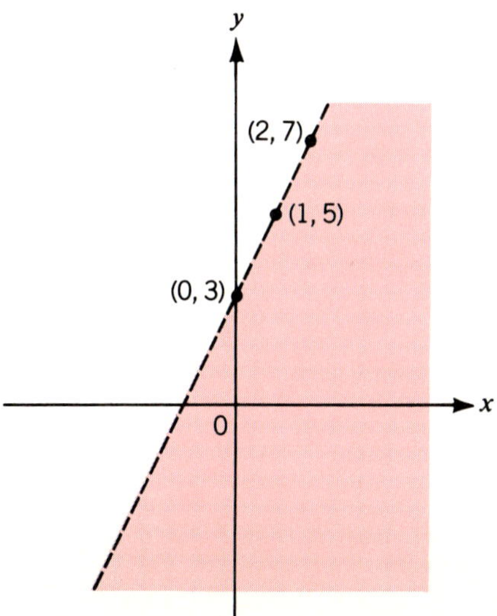

Figure 5.24

Graph the equality $y = \frac{2}{3}x + \frac{1}{3}$, using a broken line (see Fig. 5.25). All points *below* the broken line are in the solution set

$$\left\{(x, y) \middle| y < \frac{2}{3}x + \frac{1}{3}\right\}$$

Therefore, the shaded region is the graphic solution. (*Note:* The inequality does not define a function, but it does define a relation.)

Example 3 Find the graphical solution for the inequality

$$y \geq -2x + 4$$

SOLUTION Graph the equation $y = -2x + 4$. Use a solid line, since the line is part of the solution set. Shade the region *above* the line for the points

$$\{(x, y) | y > -2x + 4\}$$

The line and the shaded region in Fig. 5.26 represent the solution set

$$\{(x, y) | y \geq -2x + 4\}$$

Example 4 Graph the relation $\{(x, y) | x > 3\}$.

SOLUTION The graph of the equation $x = 3$ is a vertical line through the point (3, 0). All points (x, y) such that $x > 3$ lie to the *right* of this line (Fig. 5.27).

Example 5 Graph the relation $\{(x, y) | y > 2\}$.

SOLUTION The graph of the constant function $y = 2$ is a horizontal line through the point (0, 2). Since $y > 2$, the shaded region in Fig. 5.28 includes all points *above* this horizontal line.

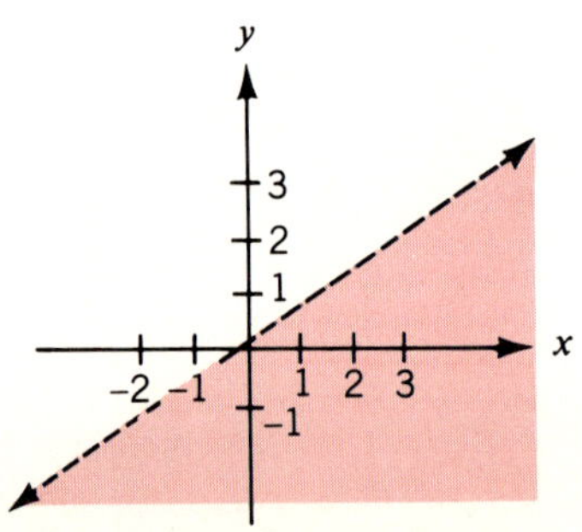

Figure 5.25

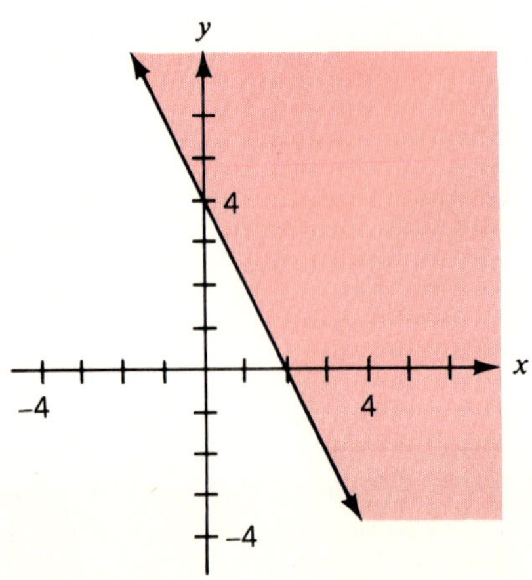

Figure 5.26

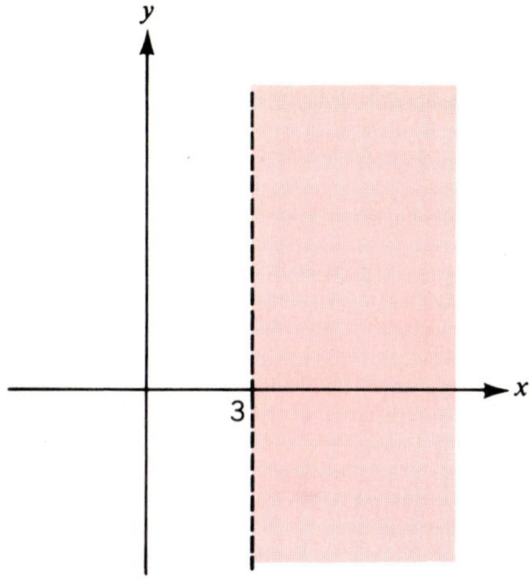

Figure 5.27

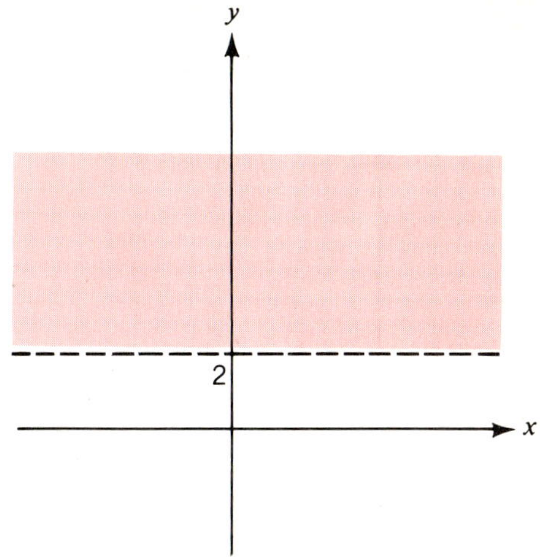

Figure 5.28

Example 6 Find the graphic solution for the inequality

$$\{(x, y) \mid -2 \leq x < 3\}.$$

SOLUTION First graph the vertical lines $x = -2$ and $x = 3$. Use a solid line to indicate the inclusion of $x = -2$ and a broken line to show that $x \neq 3$. Then shade the region *between* these lines, since $-2 \leq x < 3$ for *all* values of y. See Fig. 5.29.

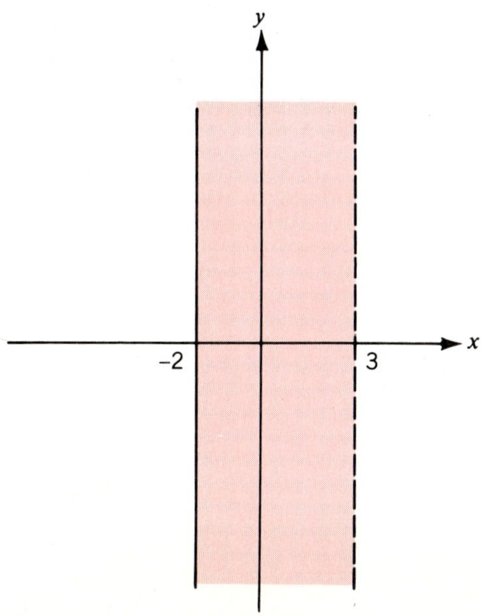

Figure 5.29

Exercises 5.9

Show a graphic solution for each of the sets in Exercises 1–26.

1. $\{(x, y) | x \geq 2\}$
2. $\{(x, y) | x < -1\}$
3. $\{(x, y) | y > -1\}$
4. $\{(x, y) | y \leq 4\}$
5. $\{(x, y) | x > 4\}$
6. $\{(x, y) | x \leq -2\}$
7. $\{(x, y) | y \geq 4\}$
8. $\{(x, y) | y < 3\}$
9. $\{(x, y) | 1 \leq x \leq 3\}$
10. $\{(x, y) | 2 < y \leq 5\}$
11. $\{(x, y) | x < 0\}$
12. $\{(x, y) | y \geq 0\}$
13. $\{(x, y) | -2 < x \leq 1\}$
14. $\{(x, y) | 1 \leq x < 3\}$
15. $\{(x, y) | y < 3x + 2\}$
16. $\{(x, y) | y > x - 5\}$
17. $\{(x, y) | y \geq 2x - 3\}$
18. $\{(x, y) | y > x\}$
19. $\{(x, y) | x + y \leq 2\}$
20. $\{(x, y) | y \geq x + 2\}$
21. $\{(x, y) | y \leq x\}$
22. $\{(x, y) | x + 2y < 2\}$
23. $\{(x, y) | 4 \geq 2y - x\}$
24. $\{(x, y) | 3x \leq y + 2\}$
25. $\{(x, y) | x - y < 2\}$
26. $\{(x, y) | y \leq x + 2\}$

Review Exercises

1. Define the following terms.

a. Coordinate axes
b. Ordinate
c. Abscissa
d. Origin
e. Quadrants
f. Slope of a line

2. Which of the following points lie on the graph of $x + 2y - 5 = 0$?

a. (1, 4)
b. (7, −1)
c. (0, 0)
d. (−4, −5)
e. (5, 0)

3. Graph each of the following and find the slope and y-intercept from the graph.

a. $3x + 2y = 12$
b. $\frac{x}{2} + \frac{y}{5} = 1$
c. $\frac{1}{2}(x - 4) = \frac{1}{4}(y + 2)$
d. $x - 3y - 6 = 0$

4. Check the answers for Exercise 3 by changing each equation into an equivalent equation of the form $y = mx + b$.

5. Find an equation of the line through the given points with slope m.

a. $(0, 3)$, $m = \frac{1}{2}$

b. $(1, 5)$, $m = -3$

c. $(-3, -1)$, $m = 0$

d. $(-3, 1)$, $m = -\frac{1}{3}$

e. $(2, -1)$, $m = \frac{1}{4}$

f. $(5, 2)$, slope undefined

6. Find an equation of the line through the given points.

a. $(-3, 0)$ and $(6, 2)$

b. $(-3, 6)$ and $(3, -8)$

c. $(-2, -3)$ and $(-4, -5)$

d. $(7, 5)$ and $(-1, 5)$

7. Which of the following lines are parallel to $2x - 5y = 10$?

a. $5x - 2y = 10$

b. $5y = 2x - 1$

c. $4x - 10y = 5$

d. $5y = 1 - 2x$

8. Find an equation of the line that is parallel to the line whose equation is $2x - 3y = 5$ and that passes through the point $(-1, 4)$.

9. For each of the following determine an equation of the line satisfying the stated conditions:

a. Passing through $(3, -2)$ and $(4, 1)$

b. Parallel to the x-axis with y-intercept 2

c. Passing through the origin with slope $-\frac{3}{5}$

d. Passing through $(5, -7)$ with slope undefined

10. Match column B with column A.

COLUMN A	COLUMN B
a. $x + y = 50$, $y + 5x = 90$	1. Equations with no solution
b. $7x - 5y = 0$, $8x + 3y = 0$	2. Equations whose graphs are coincident lines
c. $2x + 6y = 5$, $3x + 9y = 1$	3. A system of equations that can be used to find the number of nickels and cents in a collection of 50 coins worth 90 cents
d. $y = 2 - x$, $x = 2 - y$	4. A system equivalent to $x = 15$, $y = 75$
e. $x + y = 90$, $y = 5x$	5. Equations whose linear graphs intersect at the origin

11. State the coordinates of the point of intersection of the graphs of the equations $y = 3x - 9$ and $x + 2y + 4 = 0$.

12. Solve by the substitution method and check by the graphic method.

a. $2x - 3y = 6$
$x - 2y = 2$

b. $3x + y = 9$
$2x - 3y = 6$

13. Solve by the addition method and check.

a. $3x + 5y = 3$
   $2x + 3y = 1$

b. $2x - 3y = 16$
   $4x + 3y = 5$

14. One week a plumber worked 20 hours, and his helper worked 12 hours on a job. They sent the contractor of the job a bill of $520 for their combined wages. The next week the plumber worked 15 hours, and his helper worked 10 hours. This time the bill for their combined wages was $400. Find the hourly wage of each.

15. Cruising at half speed against a wind, a traffic helicopter flies 36 miles in 40 minutes. Flying with the same wind at full speed, the helicopter flies 93 miles in 30 minutes. Find the speed of the wind.

16. A rush order was received in an office. On the first day the regular typist worked 10 hours, an extra typist 5 hours, and together they completed two thirds of the order. The next day both typists worked 4 hours, and they finished the order. How long would it have taken the regular typist to do the work alone?

17. The admission prices at a movie theater were such that one person paid $24.50 for 4 adults and 6 children, whereas another person paid $17.50 for 3 adults and 4 children. What was the admission price for an adult and what was the admission price for a child?

18. A man received a total income of $1440 from $5000 invested in bonds and $4000 invested in stocks. The same year his brother received a total income of $780 from $3000 invested in the same bonds and $2000 invested in the same stocks. Find the interest rate they received from each investment.

19. In competitive business, y, the price for each unit of a commodity, depends on x, the number of units demanded by consumers. The equation relating the price and the number of units demanded is called the *demand law*. A certain demand law is represented by $5x + 8y = 60$.

a. Graph this equation for $x \geq 0$ and $y \geq 0$.

b. Find the highest price that will be paid for this commodity—that is, the value of the y-intercept (where $x = 0$).

c. Find the greatest amount that will be demanded—that is, the value of the x-intercept (where $y = 0$).

20. Referring to Exercise 19, y, the price for each unit of a commodity, also depends on x, the number of units that can be supplied. The equation that relates price and number of units that can be supplied is called the *supply law*. A supply law corresponding to the demand law in Exercise 19 is $8y = 6x + 16$.

a. Graph this equation on the same set of axes you used for Exercise 19. Again, use $x \geq 0$ and $y \geq 0$.

b. Find the lowest price at which the supplier will sell the commodity—that is, the value of the y-intercept.

c. Find the point of intersection of the demand curve and the supply curve. (This point corresponds to *market equilibrium;* the values of x and y are called *equilibrium quantity* and *equilibrium price,* respectively. In this case the quantity demanded equals the quantity supplied.)

21. Graph each of the following.

a. $\{(x, y) \mid -2 < x \leq 1\}$

b. $\{(x, y) \mid -2 \leq y < 1\}$

c. $\{(x, y) \mid x + y \geq 2\}$

d. $\{(x, y) \mid 2x - 5y > 10\}$

6 Quadratic Equations

Objectives

OBJECTIVES	TYPICAL PROBLEMS	LOCATIONS
Simplify a square root radical by using a table.	1–4. Simplify. 1. $(\sqrt{6})^2$ 2. $\sqrt{(-6)^2}$ 3. $(-\sqrt{6})^2$ 4. $\sqrt{\frac{49}{64}}$	*Section 6.1* *Examples 1, 2* *Exercises 1–26*
Approximate a square root radical.	5. Approximate to the nearest hundredth. $10 - \sqrt{30}$	*Section 6.1* *Example 3* *Exercises 27–40*
Use the theorem of Pythagoras to find a leg (*a* or *b*) or the hypotenuse *c* of a right triangle.	6. Find c if $a = 7$ and $b = 24$. 7. Find b if $a = 6$ and $c = 10$.	*Section 6.1* *Examples 4, 5* *Exercises 41–56*
Simplify a square root radical.	8–9. Simplify. 8. $\sqrt{500}$ 9. $\frac{30}{\sqrt{12}}$	*Section 6.2* *Examples 1–9* *Exercises 1–80*
Simplify square root radical expressions.	10–12. Simplify. 10. $\sqrt{50} - \sqrt{8}$ 11. $(5 - \sqrt{7})^2$ 12. $\frac{10 + \sqrt{12}}{4}$	*Section 6.3* *Examples 1–3, 5* *Exercises 1–40, 59–66*
Evaluate expressions for a radical value of the variable.	13–14. Evaluate. 13. $x^2 - 6x$ for $x = 3 + \sqrt{5}$ 14. $\sqrt{b^2 - 4ac}$ for $a = 2, b = -2, c = -3$	*Section 6.3* *Examples 4, 6, 7* *Exercises 41–58, 67–76*
Solve a quadratic equation by factoring.	15–16. Solve by factoring. 15. $5x^2 = 35x$ 16. $3x^2 = 14x + 5$	*Section 6.4* *Examples 1–9* *Exercises 1–44*
Solve a quadratic equation by completing the square.	17. Solve by completing the square. $x^2 - 12x + 16 = 0$	*Section 6.5* *Examples 1–3* *Exercises 1–30*
Solve a quadratic equation by using the quadratic formula.	18. Solve by using the quadratic formula. $3x^2 + 2x - 3 = 0$	*Section 6.6* *Examples 1–5* *Exercises 1–50*

Solve verbal problems by using a quadratic equation.	19–20. Solve. 19. The distance, D, in feet, needed for a car to come to a stop safely is related to the speed, r, of the car by $20D = r^2 + 22r$ Find r for $D = 70$ feet. 20. A house lot has the shape of a right triangle. The longest side of the lot is 150 feet. The shortest side of the lot is 30 feet less than the third side. Find the lengths of the sides of the lot.	*Section 6.7* *Examples 1–5* *Exercises 1–30*

Answers to Typical Problems

1. 6
2. 6
3. 6
4. $\frac{7}{8}$
5. 4.52
6. 25
7. 8
8. $10\sqrt{5}$
9. $5\sqrt{3}$
10. $3\sqrt{2}$
11. $32 - 10\sqrt{7}$
12. $\frac{5 + \sqrt{3}}{2}$
13. -4
14. $2\sqrt{7}$
15. $\{0, 7\}$
16. $\left\{\frac{-1}{3}, 5\right\}$
17. $\{6 \pm 2\sqrt{5}\}$
18. $\left\{\frac{-1 \pm \sqrt{10}}{3}\right\}$
19. 28 mph
20. 90 feet, 120 feet, 150 feet

Any algebraic equation that can be written in the form $ax^2 + bx + c = 0$, where $a \neq 0$, is called a **quadratic equation in one variable.** A quadratic equation in one variable is an equation that contains a square of the variable.

The following are examples of quadratic equations:

$$x^2 = 9 \quad , \quad 4x^2 - 25 = 0 \quad , \quad x^2 + 2x - 24 = 0 \quad ,$$
$$3x^2 + 6x = 0 \quad , \quad 5x^2 - x - 6 = 0$$

In Chapter 3 we saw that a polynomial equation can readily be solved if the linear factors of the polynomial can be found easily. However, not all polynomials with integral coefficients can be factored over the set of integers; $x^2 - 2$ and $x^2 + x - 1$ are examples. Finding the factors in these cases is more difficult. It is desirable, then, to develop other techniques for solving equations.

Earlier we noted that if $x^2 = 2$ and if x is positive, then $x = \sqrt{2}$. Similarly, if $x^2 = 5$ and if x is positive, then $x = \sqrt{5}$.

Since a quadratic polynomial involves terms of the form ax^2, it is reasonable to conclude that the solutions of some quadratic equations will involve radicals such as $\sqrt{2}$ and $\sqrt{5}$. Indeed, this is the case. Therefore, to develop methods for solving quadratic equations, it is first necessary to study radicals. Then we study solving a quadratic equation by factoring, by completing the square, and by using the quadratic formula. Verbal applications are shown at the end.

6.1 Square Roots

Symbolic Representation of Square Roots

Definition
The number a is a **square root** of b if and only if $a^2 = b$.

This definition implies that the operations of squaring and extracting square roots are inverses of each other, just as addition and subtraction are inverse operations and multiplication and division are inverse operations. For example,

$$3 \text{ is a square root of } 9 \text{ since } 3^2 = 9$$
$$-3 \text{ is a square root of } 9 \text{ since } (-3)^2 = 9$$
$$\frac{5}{8} \text{ is a square root of } \frac{25}{64} \text{ since } \left(\frac{5}{8}\right)^2 = \frac{25}{64}$$
$$\frac{-5}{8} \text{ is a square root of } \frac{25}{64} \text{ since } \left(\frac{-5}{8}\right)^2 = \frac{25}{64}$$
$$0 \text{ is a square root of } 0 \text{ since } 0^2 = 0$$

Examination of these examples and similar ones reveals the following properties:

1. If x is a real number, then $x^2 \geq 0$. (In other words, the square of a real number is never negative.)
2. Every positive real number has two square roots, a positive real number and its negative. Thus if a is a positive real number ($a > 0$), then the square roots of a^2 are a and $-a$.

Now it is useful to denote a square root by the symbol $\sqrt{x}$. However, the symbolic expression $\sqrt{x}$ must represent exactly one number. Suppose, for example, that $\sqrt{9} = 3$ and $\sqrt{9} = -3$. Then it would follow that $3 = -3$, and this is impossible. Therefore, $\sqrt{9}$ and, in general, $\sqrt{x}$ must represent exactly one number in order to avoid contradictions. Mathematicians agree that $\sqrt{9}$ shall be used to designate $+3$, the positive square root of 9. The negative square root, -3, is indicated by $-\sqrt{9}$.

In general, the following definition is made.

Definition of $\sqrt{x}$

If $x > 0$, then $\sqrt{x}$ is the unique **positive** real number such that $(\sqrt{x})^2 = x$.

The expression $\sqrt{x}$ is called a **radical,** and x is called the **radicand.**

Note: If $x < 0$, then $\sqrt{x}$ is not a real number.

The following useful theorem aids in the solution of problems that involve square roots.

Theorem

$\sqrt{x^2} = x$ if and only if $x \geq 0$

The positive real number $\sqrt{x}$ is called the **principal square root** of x.

Irrational Square Roots

Although every positive real number x has two real square roots, namely, $\sqrt{x}$ and $-\sqrt{x}$, it does not follow that every positive rational number has two rational roots.

Rational numbers such as 49, 81, and $\frac{9}{16}$ are called **perfect squares** because their square roots are also rational. The number 2, on the other hand, is not a perfect square; its square roots, $\sqrt{2}$ and $-\sqrt{2}$, are not rational but irrational.

Q, the set of rational numbers, was defined as the set of quotients of integers $\frac{p}{q}$ where $q \neq 0$. It can be shown that $\sqrt{2}$ cannot be expressed as the quotient of two integers, and thus $\sqrt{2}$ is irrational.

The terminating decimals and the nonterminating repeating decimals represent the rational numbers. Therefore, it is convenient to think of the irrational numbers as the set of nonterminating, nonrepeating decimals and of the real numbers as the union of these two sets—that is, the set of all decimals.

The table of squares and square roots inside the front book cover is useful in providing a first approximation of the square root of a number.

The square root values in the table are not exact but are approximations to the nearest thousandth.

Example 1 Simplify:

a. $(\sqrt{9})^2$ b. $(\sqrt{5})^2$ c. $\sqrt{(-5)^2}$

SOLUTION

a. $(\sqrt{9})^2 = 9$ by definition

b. $(\sqrt{5})^2 = 5$ by definition

c. $\sqrt{(-5)^2} = \sqrt{25} = 5$

Example 2 Simplify $\sqrt{64} - \sqrt{(-4)^2}$.

SOLUTION

$$\begin{aligned}\sqrt{64} - \sqrt{(-4)^2} &= \sqrt{64} - \sqrt{16} \\ &= 8 - 4 = 4\end{aligned}$$

Example 3 Approximate $\sqrt{12} - \sqrt{20}$ to the nearest hundredth by using the table inside the book cover or by using a pocket calculator.

SOLUTION From the table,

$$\begin{aligned}\sqrt{12} &= 3.464 \\ \sqrt{20} &= 4.472 \\ \sqrt{12} - \sqrt{20} &= 3.464 - 4.472 \\ &= -1.008 \\ &= -1.01 \text{ to the nearest hundredth}\end{aligned}$$

The Theorem of Pythagoras

The theorem of Pythagoras is a theorem that has many practical applications. We introduce it here to develop familiarity with it and to provide practice in working with square roots. Later in the chapter we shall meet other problems that require its use.

The Theorem of Pythagoras
The square of the length of the hypotenuse of a right triangle is equal to the sum of the squares of the lengths of its legs.

The **hypotenuse** of a right triangle, which is opposite the right angle, is the longest side. The **legs** of a right triangle are the other two sides—that is, the sides that form the right angle.

Thus if c represents the length of the hypotenuse in Fig. 6.1 and if a and b represent the lengths of the legs, then

$$c^2 = a^2 + b^2$$

Therefore, a length of $\sqrt{2}$ can be represented by the diagonal of a square (see Fig. 6.2).

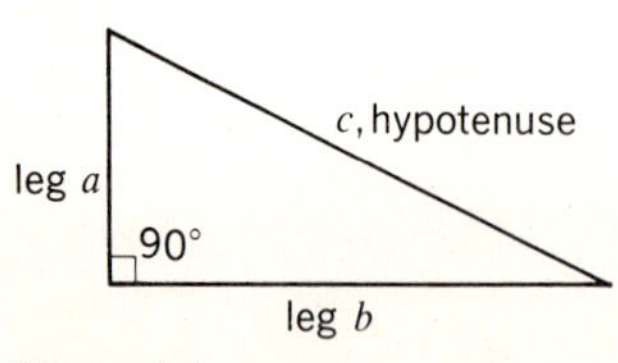

Figure 6.1

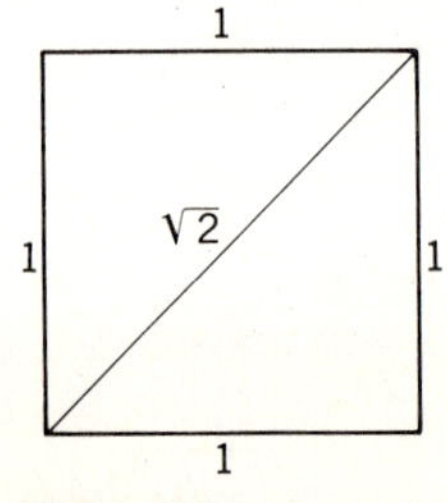

Figure 6.2

Since

$$c^2 = 1^2 + 1^2$$
$$c^2 = 2$$

Thus

$$c = \sqrt{2}$$

Example 4 Find the hypotenuse c of a right triangle if the lengths of its legs a and b are such that $a = 12$ and $b = 35$.

SOLUTION Using the theorem of Pythagoras,

$$c^2 = a^2 + b^2$$
$$c^2 = (12)^2 + (35)^2$$
$$c^2 = 144 + 1225 = 1369$$
$$c = \sqrt{1369} = 37 \quad \text{(See the table.)}$$

Example 5 If a, b, and c are the sides of a right triangle whose hypotenuse is c, find b if $a = 16$ and $c = 18$.

SOLUTION Using

$$a^2 + b^2 = c^2$$
$$(16)^2 + b^2 = (18)^2$$
$$b^2 = (18)^2 - (16)^2 = (18 + 16)(18 - 16)$$
$$b^2 = 68$$
$$b = 8.25 \text{ to the nearest hundredth}$$

Exercises 6.1

Simplify in Exercises 1–26. If necessary, use the table of squares and square roots inside the front cover.

1. $\sqrt{16}$
2. $\sqrt{169}$
3. $\sqrt{(-3)^2}$
4. $\sqrt{(-1)^2}$
5. $-\sqrt{4}$
6. $-\sqrt{(-2)^2}$
7. $\sqrt{400}$
8. $\sqrt{144}$
9. $(\sqrt{3})^2$
10. $(\sqrt{16})^2$
11. $(-\sqrt{4})^2$
12. $(-\sqrt{5})^2$
13. $-(\sqrt{4})^2$
14. $-(\sqrt{5})^2$
15. $\sqrt{5^2} - \sqrt{4^2}$
16. $\sqrt{(25)^2} - \sqrt{7^2}$
17. $\sqrt{(-5)^2} + \sqrt{(12)^2}$
18. $\sqrt{(-8)^2} + \sqrt{(-15)^2}$
19. $\sqrt{(-5)^2 + (12)^2}$
20. $\sqrt{(-8)^2 + (-15)^2}$
21. $\sqrt{\dfrac{25}{64}}$
22. $\sqrt{\dfrac{9}{16}}$

23. $\sqrt{0.01}$

24. $\sqrt{0.0001}$

25. $\dfrac{\sqrt{64 + 36}}{\sqrt{64} + \sqrt{36}}$

26. $\dfrac{4\sqrt{225 - 144}}{4(\sqrt{225} - \sqrt{144})}$

In Exercises 27–40, use the table of squares and square roots inside the front cover or a pocket calculator. Approximate to the nearest hundredth.

27. $\sqrt{45}$

28. $\sqrt{10}$

29. $\sqrt{86}$

30. $\sqrt{3}$

31. $2\sqrt{5}$

32. $3\sqrt{7}$

33. $5 - \sqrt{3}$

34. $3 - \sqrt{5}$

35. $5\sqrt{8}$

36. $\sqrt{2}\sqrt{6}$

37. $2 + \sqrt{2}$

38. $\sqrt{2} + \sqrt{6}$

39. $\sqrt{5} - \sqrt{3}$

40. $\sqrt{7} + \sqrt{11}$

If a and b represent the lengths of the legs of a right triangle and c represents the length of the hypotenuse of that triangle, determine the length of the missing side or sides in Exercises 41–56.

41. $a = 12, b = 5$

42. $b = 5, c = 13$

43. $b = 5, c = 15$

44. $a = 12, b = 35$

45. $a = 1, b = 1$

46. $a = 2, c = 5$

47. $a = 15, c = 17$

48. $a = 2, c = \sqrt{30}$

49. $a = \sqrt{5}, b = \sqrt{3}$

50. $b = \sqrt{10}, c = \sqrt{30}$

51. $a = b, c = 10$

52. $a = b, c = \sqrt{10}$

53. $c = 2a, b = \sqrt{3}$

54. $c = 2b, a = 6$

55. $a = 2b, c = 10$

56. $b = 2a, c = \sqrt{10}$

6.2 Radicals: Simplification, Products and Quotients

In order to solve quadratic equations, you need to know how to add, subtract, multiply, and divide expressions that contain radicals. In this section we will learn how to simplify a square root radical and how to multiply and divide square root radicals.

Theorem 1

The product of two square root radicals is the square root of the product of their radicands. In symbols, $\sqrt{r}\sqrt{s} = \sqrt{rs}$ for $r \geq 0$ and $s \geq 0$.

Example 1 Simplify $\sqrt{3}\sqrt{12}$.

SOLUTION

$$\sqrt{3}\sqrt{12} = \sqrt{3 \cdot 12} = \sqrt{36} = 6$$

Example 2 Simplify $\sqrt{2}\sqrt{5}$.

SOLUTION

$$\sqrt{2}\sqrt{5} = \sqrt{10}$$

A **radical** of the form $\sqrt{M}$, where M is a monomial with an integer for its coefficient, is said to be **simplified** if no perfect squares are factors of the **radicand** M.

The following theorem is useful for simplifying square root radicals.

Theorem 2

If x and y are any nonnegative real numbers,

$$\sqrt{x^2y} = \sqrt{x^2}\sqrt{y} = x\sqrt{y}$$

Example 3 Simplify $\sqrt{75}$.

SOLUTION The basic idea is to factor the radicand to find the perfect square factors:

$$75 = 3 \cdot 5^2$$
$$\sqrt{75} = \sqrt{5^2 \cdot 3} = \sqrt{5^2}\sqrt{3} = 5\sqrt{3}$$

Pythagoras

Although there are many stories about the life of Pythagoras (ca. 580–501 B.C.), little is known for certain. He was probably born on the island of Samos, and he probably traveled to Egypt, and Babylonia. He is known to have settled in Crotona on the Italian coast, where he founded a brotherhood composed of some 300 wealthy young aristocrats. This group, known as the Pythagoreans, became the prototype of all the secret societies of Europe and America. Their motto, "Number rules the universe," expressed the combination of mathematics and mysticism in which they believed. Shakespeare refers to the Pythagorean belief in immortality and transmigration of the soul in *The Merchant of Venice:*

Thou almost mak'st me waver in my faith,
To hold opinion with Pythagoras,
That souls of animals infuse themselves,
Into the trunks of men.

The name of Pythagoras is most famous in connection with the relationship of the squares of the sides of a right triangle. Although Pythagoras did not discover this relationship (it was already known to the Babylonians), he may have offered the first proof of this statement.

Example 4 Simplify $\sqrt{6}\sqrt{30}$.

SOLUTION

$$\begin{aligned}\sqrt{6}\sqrt{30} = \sqrt{6(30)} &= \sqrt{(2\cdot 3)(2\cdot 3\cdot 5)}\\ &= \sqrt{6^2(5)}\\ &= \sqrt{6^2}\sqrt{5}\\ &= 6\sqrt{5}\end{aligned}$$

Theorem 3

The quotient of two square root radicals is the square root of their radicands.

In symbols, for $r \geq 0$ and $s > 0$,

$$\frac{\sqrt{r}}{\sqrt{s}} = \sqrt{\frac{r}{s}}$$

Example 5 Simplify $\dfrac{\sqrt{18}}{\sqrt{50}}$.

SOLUTION

$$\frac{\sqrt{18}}{\sqrt{50}} = \sqrt{\frac{18}{50}} = \sqrt{\frac{9}{25}} = \frac{3}{5}$$

Example 6 Simplify $\sqrt{\dfrac{15}{4a^2}}$ for $a > 0$.

SOLUTION

$$\sqrt{\frac{15}{4a^2}} = \frac{\sqrt{15}}{\sqrt{4a^2}} = \frac{\sqrt{15}}{2a}$$

A radical expression is not in simplified form if there is a radical in a denominator or if a fraction occurs in a radicand.

The following theorem is useful for simplifying radical expressions that involve quotients.

Theorem 4

If r is any real number and s and t are positive real numbers,

$$\frac{r}{\sqrt{s^2t}} = \frac{r\sqrt{t}}{st}$$

Note that $\dfrac{r}{\sqrt{s^2t}} = \dfrac{r}{s\sqrt{t}}$ (By Theorem 2)

$$= \frac{r\sqrt{t}}{s\sqrt{t}\sqrt{t}}$$ (Fundamental theorem of fractions)

$$= \frac{r\sqrt{t}}{st}$$

Example 7 Simplify $\frac{1}{\sqrt{5}}$.

SOLUTION

$$\frac{1}{\sqrt{5}} = \frac{\sqrt{5}}{\sqrt{5}\sqrt{5}} = \frac{\sqrt{5}}{5}$$ (Multiplying numerator and denominator by $\sqrt{5}$)

Example 8 Simplify $\sqrt{\frac{1}{6}}$.

SOLUTION

$$\sqrt{\frac{1}{6}} = \frac{\sqrt{1}}{\sqrt{6}} = \frac{1\sqrt{6}}{\sqrt{6}\sqrt{6}} = \frac{\sqrt{6}}{6}$$ (Multiplying numerator and denominator by $\sqrt{6}$)

Example 9 Simplify $\frac{15}{\sqrt{50}}$.

SOLUTION

$$\frac{15}{\sqrt{50}} = \frac{15}{\sqrt{25(2)}}$$

$$= \frac{15}{\sqrt{25}\sqrt{2}}$$ (Simplifying the radical)

$$= \frac{15\sqrt{2}}{5\sqrt{2}\sqrt{2}}$$ (Removing the radical from the denominator)

$$= \frac{3\sqrt{2}}{2}$$ (Simplifying the fraction)

Exercises 6.2

Simplify the radicals in Exercises 1–40.

1. $\sqrt{2}\sqrt{50}$
2. $\sqrt{18}\sqrt{2}$
3. $\sqrt{5}\sqrt{20}$
4. $\sqrt{48}\sqrt{3}$
5. $\sqrt{15}\sqrt{5}$
6. $\sqrt{100}\sqrt{20}$
7. $\sqrt{5}\sqrt{25}$
8. $\sqrt{3}\sqrt{12}$
9. $\sqrt{125}$
10. $\sqrt{27}$
11. $\sqrt{32}$
12. $\sqrt{24}$
13. $\sqrt{8}$
14. $\sqrt{12}$
15. $\sqrt{40}$
16. $\sqrt{54}$
17. $\sqrt{162}$
18. $\sqrt{28}$
19. $\sqrt{98}$
20. $\sqrt{45}\sqrt{5}$

21. $\sqrt{150}$
22. $\sqrt{405}$
23. $\sqrt{35}\sqrt{14}$
24. $\sqrt{3}\sqrt{60}$
25. $\sqrt{6}\sqrt{30}$
26. $\sqrt{8}\sqrt{30}$
27. $\sqrt{108}$
28. $\sqrt{363}$
29. $\sqrt{1200}$
30. $\sqrt{200}$
31. $\sqrt{300}$
32. $\sqrt{500}$
33. $\sqrt{1452}$
34. $\sqrt{1849}$
35. $\sqrt{3}\sqrt{5}\sqrt{15}$
36. $\sqrt{2}\sqrt{7}\sqrt{14}$
37. $\sqrt{64}\sqrt{12}\sqrt{3}$
38. $\sqrt{5}\sqrt{14}\sqrt{28}$
39. $\sqrt{1976}$
40. $\sqrt{1776}$

41–70. Simplify the quotients.

41. $\frac{1}{\sqrt{3}}$
42. $\frac{1}{\sqrt{2}}$
43. $\sqrt{\frac{1}{7}}$
44. $\sqrt{\frac{1}{10}}$
45. $\frac{3}{\sqrt{5}}$
46. $\frac{5}{\sqrt{6}}$
47. $\frac{10}{\sqrt{5}}$
48. $\frac{2}{\sqrt{6}}$
49. $\frac{3}{\sqrt{15}}$
50. $\frac{12}{\sqrt{6}}$
51. $\frac{1}{\sqrt{12}}$
52. $\frac{1}{\sqrt{18}}$
53. $\sqrt{\frac{1}{50}}$
54. $\sqrt{\frac{1}{32}}$
55. $\sqrt{\frac{25}{98}}$
56. $\sqrt{\frac{16}{75}}$
57. $\frac{1}{\sqrt{27}}$
58. $\frac{1}{\sqrt{125}}$
59. $\sqrt{\frac{1}{8}}$
60. $\sqrt{\frac{1}{1000}}$
61. $\frac{\sqrt{5}}{\sqrt{20}}$
62. $\frac{\sqrt{7}}{\sqrt{63}}$
63. $\frac{24}{\sqrt{24}}$
64. $\frac{45}{\sqrt{45}}$
65. $\frac{\sqrt{2}}{\sqrt{54}}$
66. $\frac{\sqrt{2}}{\sqrt{28}}$
67. $\frac{7\sqrt{5}}{\sqrt{70}}$
68. $\frac{2\sqrt{5}}{\sqrt{30}}$
69. $\frac{5\sqrt{6}}{\sqrt{10}}$
70. $\frac{7\sqrt{6}}{\sqrt{21}}$

71–80. Simplify each quotient. Each letter represents a positive real number.

71. $\frac{1}{\sqrt{a}}$
72. $\sqrt{\frac{1}{b}}$
73. $\frac{1}{\sqrt{4c^2}}$
74. $\frac{1}{\sqrt{9d^2}}$
75. $\sqrt{\frac{50}{b}}$
76. $\sqrt{\frac{45}{a}}$
77. $\sqrt{\frac{d^2}{6}}$
78. $\sqrt{\frac{c^2}{10}}$
79. $\sqrt{\frac{b^2 - 4ac}{4a^2}}$
80. $\sqrt{\frac{d^2 - 9}{9d^2}}$

6.3 Operations with Radicals

It is often possible to express a sum or difference of two radical terms as a single term. For example, since $\sqrt{5}$ is a real number, so are $3\sqrt{5}$ and $4\sqrt{5}$ and $3\sqrt{5} + 4\sqrt{5}$. Using the distributive axiom,

$$3\sqrt{5} + 4\sqrt{5} = (3 + 4)\sqrt{5} = 7\sqrt{5}$$

Square root radicals that have the same radicand are called **like radicals.** Radicals that are not like are called **unlike.** For example, $3\sqrt{5}$ and $4\sqrt{5}$ are like radicals, whereas $2\sqrt{5}$ and $4\sqrt{3}$ are unlike radicals.

Terms that contain like radicals can be combined by using the distributive axiom, but terms that contain unlike radicals cannot be so combined. Sometimes it is necessary to simplify each radical term in order to identify like radicals.

Example 1 Simplify $\sqrt{50} + \sqrt{18}$ if possible.

SOLUTION

$$\begin{aligned} &\sqrt{50} + \sqrt{18} \\ &= 5\sqrt{2} + 3\sqrt{2} && \text{(Simplifying each radical)} \\ &= (5 + 3)\sqrt{2} && \text{(Using the distributive axiom)} \\ &= 8\sqrt{2} \end{aligned}$$

Example 2 Simplify $\sqrt{48} + \sqrt{18} - \sqrt{12}$.

SOLUTION

$$\begin{aligned} &\sqrt{48} + \sqrt{18} - \sqrt{12} \\ &= 4\sqrt{3} + 3\sqrt{2} - 2\sqrt{3} && \text{(Simplifying)} \\ &= (4\sqrt{3} - 2\sqrt{3}) + 3\sqrt{2} && \text{(Collecting like radicals)} \\ &= (4 - 2)\sqrt{3} + 3\sqrt{2} && \text{(Using the distributive axiom)} \\ &= 2\sqrt{3} + 3\sqrt{2} && \text{(Simplified form)} \end{aligned}$$

Since a solution of a quadratic equation ($ax^2 + bx + c = 0$) may have the form $4 + 3\sqrt{6}$ or $\frac{6 - \sqrt{5}}{2}$, one needs to know how to evaluate quadratic polynomials for such values as these in order to check a possible solution.

Example 3 Show that $4 + 3\sqrt{6}$ is a solution of $x^2 - 8x = 38$.

SOLUTION First find the value of x^2 for $x = 4 + 3\sqrt{6}$.

$$x^2 = (4 + 3\sqrt{6})^2 = 4^2 + 2(4)(3\sqrt{6}) + (3\sqrt{6})^2$$
$$= 16 + 24\sqrt{6} + 9(6) = 70 + 24\sqrt{6}$$

Then evaluate $-8x$.

$$-8x = -8(4 + 3\sqrt{6}) = -32 - 24\sqrt{6}$$

Combining these results,

$$x^2 - 8x = 70 + 24\sqrt{6} - 32 - 24\sqrt{6} = 38$$

Thus, $x^2 - 8x = 38$ for $x = 4 + 3\sqrt{6}$ and $4 + 3\sqrt{6}$ is a solution.

Example 4 Show that $\dfrac{6 - \sqrt{5}}{2}$ is a solution of $4x^2 + 31 = 24x$.

SOLUTION

$$4x^2 + 31 = 4\left(\frac{6 - \sqrt{5}}{2}\right)^2 + 31$$
$$= 4\left(\frac{36 - 12\sqrt{5} + 5}{4}\right) + 31$$
$$= 41 - 12\sqrt{5} + 31 = 72 - 12\sqrt{5}$$
$$24x = 24\left(\frac{6 - \sqrt{5}}{2}\right) = 72 - 12\sqrt{5}$$

Since both sides of the equation result in the same value, $\dfrac{6 - \sqrt{5}}{2}$ is a solution.

Example 5 Simplify $\dfrac{8 + \sqrt{20}}{6}$.

SOLUTION First simplify the radical:

$$\sqrt{20} = \sqrt{4 \cdot 5} = 2\sqrt{5}$$

Then

$$\frac{8 + \sqrt{20}}{6} = \frac{8 + 2\sqrt{5}}{6}$$

Now factor the numerator and the denominator in order to reduce the fraction.

$$\frac{8 + 2\sqrt{5}}{6} = \frac{2(4 + \sqrt{5})}{2(3)} = \frac{4 + \sqrt{5}}{3}$$

In solving a quadratic equation by using the quadratic formula, we need to evaluate the expression $\sqrt{b^2 - 4ac}$. See Examples 6 and 7.

Example 6 Evaluate $\sqrt{b^2 - 4ac}$ for $a = 2$, $b = -6$, and $c = 3$.

SOLUTION

$$\begin{aligned}\sqrt{b^2 - 4ac} &= \sqrt{(-6)^2 - 4(2)(3)} \\ &= \sqrt{36 - 24} = \sqrt{12} = 2\sqrt{3}\end{aligned}$$

Example 7 Evaluate $\sqrt{b^2 - 4ac}$ for $a = 1$, $b = 3$, and $c = -10$.

SOLUTION

$$\begin{aligned}\sqrt{b^2 - 4ac} &= \sqrt{(3)^2 - 4(1)(-10)} \\ &= \sqrt{9 + 40} = \sqrt{49} = 7\end{aligned}$$

Exercises 6.3

In Exercises 1–24, write each expression in simplest radical form.

1. $6\sqrt{5} + 2\sqrt{5}$
2. $7\sqrt{2} - 4\sqrt{2}$
3. $5\sqrt{6} + \sqrt{6}$
4. $9\sqrt{10} - 8\sqrt{10}$
5. $3\sqrt{5} + 5\sqrt{3}$
6. $5\sqrt{3} - 4\sqrt{3} + 2\sqrt{3}$
7. $7\sqrt{3} - 5\sqrt{3}$
8. $5\sqrt{7} + 4\sqrt{7}$
9. $8\sqrt{6} - 7\sqrt{6}$
10. $2\sqrt{15} + \sqrt{15}$
11. $2\sqrt{30} + 5\sqrt{30} - 4\sqrt{30}$
12. $7\sqrt{5} - 5\sqrt{7}$
13. $6\sqrt{28} - 4\sqrt{28}$
14. $\sqrt{40} - \sqrt{10}$
15. $\sqrt{28} - 2\sqrt{7}$
16. $3\sqrt{20} + \sqrt{5}$
17. $\sqrt{3} + \sqrt{27}$
18. $5\sqrt{18} - 4\sqrt{8}$
19. $3\sqrt{24} + 2\sqrt{54}$
20. $\sqrt{32} + \sqrt{98} - 3\sqrt{2}$
21. $\sqrt{8} - \sqrt{18} - \sqrt{12}$
22. $\sqrt{20} + \sqrt{12} + \sqrt{45}$
23. $\sqrt{63} - \sqrt{28} + 2\sqrt{2}$
24. $\sqrt{54} + \sqrt{32} - \sqrt{24}$

25–40. Simplify.

25. $(5 + \sqrt{6})^2$
26. $(2 - \sqrt{7})^2$
27. $(4 - \sqrt{5})^2$
28. $(6 + \sqrt{10})^2$
29. $(-1 + \sqrt{3})^2$
30. $(-2 + \sqrt{5})^2$
31. $(-7 - \sqrt{10})^2$
32. $(-3 - \sqrt{2})^2$
33. $3(5 + \sqrt{6})$
34. $4(2 - \sqrt{7})$
35. $-2(4 - \sqrt{5})$
36. $-5(6 + \sqrt{10})$
37. $5(2 - 3\sqrt{2})^2$
38. $2(5 + 2\sqrt{6})^2$
39. $-4(-2 + 2\sqrt{3})^2$
40. $-3(-2 - 2\sqrt{5})^2$

For Exercises 41–46, find the value of the polynomial when $x = 2 + \sqrt{3}$.

41. $x^2 + 3x$
42. $x^2 - 4x$
43. $3x^2 - x$
44. $2x^2 + 5x$
45. $x^2 + 3x + 2$
46. $4x^2 - 3x + 1$

For Exercises 47–50, find the value of the polynomial when $x = 3 - \sqrt{2}$.

47. $x^2 - 3x$
48. $x^2 + 2x$
49. $3x^2 + 2x$
50. $2x^2 + 3x$

For Exercises 51 and 52, find the value of the polynomial when $x = \dfrac{-3 + \sqrt{17}}{2}$.

51. $x^2 + 3x - 2$
52. $x^2 - 3x - 2$

53–58. Find the value of the given polynomial for the given value of x.

53. $x^2 - 10x + 23$
    $x = 5 - \sqrt{2}$
54. $x^2 + x - 1$
    $x = \dfrac{1 - \sqrt{5}}{2}$
55. $x^2 - 10x + 23$
    $x = 5 + \sqrt{2}$
56. $x^2 + x - 1$
    $x = \dfrac{-1 + \sqrt{5}}{2}$
57. $x^2 - 10x + 23$
    $x = -5 + \sqrt{2}$
58. $x^2 + x - 1$
    $x = \dfrac{-1 - \sqrt{5}}{2}$

59–66. Simplify.

59. $\dfrac{8 + 2\sqrt{3}}{6}$
60. $\dfrac{4 + \sqrt{48}}{8}$
61. $\dfrac{6 - \sqrt{18}}{6}$
62. $\dfrac{12 - \sqrt{8}}{6}$
63. $\dfrac{-4 + \sqrt{20}}{2}$
64. $\dfrac{-2 - \sqrt{44}}{10}$
65. $\dfrac{14 + \sqrt{80}}{2}$
66. $\dfrac{20 - \sqrt{96}}{4}$

67–76. For the given values of a, b, and c, evaluate $\sqrt{b^2 - 4ac}$ *and simplify.*

67. $a = 1, b = 4, c = 3$
68. $a = 1, b = 6, c = 4$
69. $a = 2, b = -8, c = 5$
70. $a = 3, b = -6, c = 2$
71. $a = 1, b = 8, c = -2$
72. $a = 2, b = 1, c = -3$
73. $a = 3, b = -2, c = -5$
74. $a = 4, b = -7, c = -2$
75. $a = 2, b = -6, c = -3$
76. $a = 3, b = -6, c = -5$

6.4 Quadratic Equations: Solution by Factoring

The solution of quadratic equations by factoring, which was introduced earlier, involves four basic steps:

1. Find an equivalent quadratic equation whose right side is zero.
2. Factor the quadratic polynomial.
3. Set each factor equal to zero.
4. Solve for the variable.

This method is based on the zero-product theorem, which was introduced in Chapter 3 and is restated below for convenience.

The Zero-Product Theorem
Let r and s be any real numbers. If $rs = 0$, then $r = 0$ or $s = 0$.

In solving an equation by the factoring method, it is important to remember that the right side of the equation must be zero before the factors on the left side can be equated to zero.

Example 1 Solve for x: $x^2 - x = 6$.

SOLUTION

$$x^2 - x = 6$$

1. $x^2 - x - 6 = 0$ (Equivalent equation whose right side is zero)
2. $(x - 3)(x + 2) = 0$ (Factor the quadratic polynomial)
3. $x - 3 = 0$ or $x + 2 = 0$ (Zero-product theorem: If $rs = 0$, then $r = 0$ or $s = 0$)
4. $x = 3$ or $x = -2$

CHECK

$x^2 - x = 6$	$x^2 - x = 6$
$(3)^2 - 3 = 6$	$(-2)^2 - (-2) = 6$
$9 - 3 = 6$	$4 + 2 = 6$
$6 = 6$	$6 = 6$

Therefore, the solution set is $\{3, -2\}$.

Example 2 Solve $(x - 2)(x - 5) = 40$.

SOLUTION

$$\begin{aligned}(x-2)(x-5) &= 40\\ x^2 - 7x + 10 &= 40\\ x^2 - 7x - 30 &= 0\\ (x+3)(x-10) &= 0\end{aligned}$$

$$x + 3 = 0 \quad \text{or} \quad x - 10 = 0$$
$$x = -3 \quad \text{or} \quad x = 10$$

The solution set is $\{-3, 10\}$.

CHECK

For $x = -3$, $(x - 2)(x - 5) = (-3-2)(-3-5) = (-5)(-8) = 40$.
For $x = 10$, $(x - 2)(x - 5) = (10 - 2)(10 - 5) = (8)(5) = 40$.

The equation $X^2 = k$, where k is a positive constant, is a very special case of a quadratic equation. Since this equation can also be written as $X^2 - k = 0$ and since $k = (\sqrt{k})^2$,

$$X^2 - k = X^2 - (\sqrt{k})^2 = (X - \sqrt{k})(X + \sqrt{k})$$

Now, solving $X^2 = k$,

$$\begin{aligned}X^2 - k &= 0\\ (X - \sqrt{k})(X + \sqrt{k}) &= 0\end{aligned}$$

$$X - \sqrt{k} = 0 \quad \text{or} \quad X + \sqrt{k} = 0$$
$$X = \sqrt{k} \quad \text{or} \quad X = -\sqrt{k}$$

This result can now be stated as the following very useful theorem.

Theorem for Square Root Method
If $X^2 = k$ and $k > 0$, then $X = \sqrt{k}$ or $X = -\sqrt{k}$.

Example 3 Solve $x^2 = 49$.

SOLUTION

$x = 7$ or $x = -7$ (Using the theorem and $\sqrt{49} = 7$)

Example 4 Solve $x^2 = 75$.

SOLUTION

$x = \sqrt{75}$ or $x = -\sqrt{75}$

$\sqrt{75} = \sqrt{25(3)} = 5\sqrt{3}$ (Simplifying the radical)

$x = 5\sqrt{3}$ or $x = -5\sqrt{3}$

Example 5 Solve $3x^2 - 2 = 0$.

SOLUTION First solve for x^2.

$$x^2 = \frac{2}{3}$$

$$x = \sqrt{\frac{2}{3}} \quad \text{or} \quad x = -\sqrt{\frac{2}{3}} \quad \text{(Using the theorem)}$$

$$x = \frac{\sqrt{6}}{3} \quad \text{or} \quad x = \frac{-\sqrt{6}}{3} \quad \text{(Simplifying the radicals)}$$

$$\left(\frac{\sqrt{2}}{\sqrt{3}}\frac{\sqrt{3}}{\sqrt{3}} = \frac{\sqrt{6}}{3}\right)$$

CHECK

$$3\left(\frac{\sqrt{6}}{3}\right)^2 - 2 = 3\left(\frac{6}{9}\right) - 2 = 2 - 2 = 0$$

$$3\left(\frac{-\sqrt{6}}{3}\right)^2 - 2 = 3\left(\frac{6}{9}\right) - 2 = 2 - 2 = 0$$

Another special case occurs when $c = 0$. The equation reads

$$ax^2 + bx = 0$$

Now there is a common factor, x, for the left member of the equation.

$$ax^2 + bx = 0$$
$$x(ax + b) = 0$$
$$x = 0 \quad \text{or} \quad ax + b = 0 \quad \text{(Zero-product theorem)}$$
$$ax = -b$$
$$x = -\frac{b}{a}$$

The solution set is $\left(0, -\frac{b}{a}\right)$.

Example 6 Solve by factoring $3x^2 + 2x = 0$.

SOLUTION

$$3x^2 + 2x = 0$$
$$x(3x + 2) = 0$$
$$x = 0 \quad \text{or} \quad 3x + 2 = 0$$
$$3x = -2$$
$$x = -\frac{2}{3}$$

Therefore, the solution set is $\left\{0, -\frac{2}{3}\right\}$.

CHECK

If $x = 0$, $3(0)^2 + 2(0) = 0 + 0 = 0$.

If $x = -\frac{2}{3}$, $3x^2 + 2x = 3\left(-\frac{2}{3}\right)^2 + 2\left(-\frac{2}{3}\right)$

$$= 3\left(\frac{4}{9}\right) - \frac{4}{3} = \frac{4}{3} - \frac{4}{3} = 0$$

Example 7 Solve $4x^2 = 12x$.

SOLUTION

$$4x^2 = 12x$$
$$4x^2 - 12x = 0$$
$$4x(x - 3) = 0$$
$$4x = 0 \quad \text{or} \quad x - 3 = 0$$

Since $4 \neq 0$, $x = 0$ or $x = 3$.

CHECK

For $x = 0$, $4x^2 = 4(0)^2 = 0$ and $12x = 12(0) = 0$.
For $x = 3$, $4(3)^2 = 4(9) = 36$ and $12x = 12(3) = 36$.

Note in the preceding examples that each quadratic equation had *two* solutions. If both sides of the equation $4x^2 = 12x$ had been divided by x, this would have yielded $4x = 12$ and $x = 3$, thus losing the solution $x = 0$. Since the equation $4x^2 = 12x$ is true for $x = 0$, dividing both sides by x is the same as dividing both sides by 0, and this is not permitted. In general, division by an expression containing a variable may lose a solution, and multiplication by an expression containing a variable may introduce an extraneous solution (a number that is not a solution of the original equation).

A quadratic equation may have only one number in its solution set. Such a solution is called a **double root.**

Example 8 Solve $x^2 - 6x + 9 = 0$.

SOLUTION

$$x^2 - 6x + 9 = (x - 3)^2$$
$$x^2 - 6x + 9 = 0$$
$$(x - 3)^2 = 0$$
$$(x - 3)(x - 3) = 0$$
$$x - 3 = 0 \quad \text{or} \quad x - 3 = 0$$
$$x = 3 \quad \text{or} \quad x = 3$$

The solution set is $\{3\}$, where 3 is called a double root.

When the solutions of a quadratic equation are restricted to real numbers, there are quadratic equations that have no solution, as the following example shows.

Example 9 Solve $x^2 + 4 = 0$.

SOLUTION

$$x^2 + 4 = 0$$
$$x^2 = -4$$

Since there is no real number whose square is negative, there is no real solution to this equation. (A solution for this type of equation is provided by the set of complex numbers, which is beyond the scope of this book.)

Exercises 6.4

Solve the following equations.

1. $(x - 5)(x + 2) = 0$
2. $(x + 3)(x - 4) = 0$
3. $5x(x - 4) = 0$
4. $2x(x + 3) = 0$
5. $x^2 - 5x + 6 = 0$
6. $x^2 + x - 6 = 0$
7. $x^2 + 2x = 35$
8. $x^2 - 3x = 18$
9. $(x + 3)(x - 2) = 14$
10. $(x - 4)(x - 3) = 42$
11. $2x^2 + x = 10$
12. $3x^2 = x + 4$
13. $6x^2 = 7x - 1$
14. $5x^2 + 4 = 21x$
15. $4x^2 - 25 = 0$
16. $9x^2 - 64 = 0$
17. $4x^2 = 27$
18. $9x^2 = 32$
19. $5x^2 = 6$
20. $2x^2 = 7$
21. $5x^2 = 6x$
22. $2x^2 = 7x$
23. $4x^2 - 2x = 0$
24. $6x^2 = 24x$
25. $4x^2 - 2 = 2x$
26. $5x^2 = 20x - 15$
27. $x^2 - 12x + 36 = 0$
28. $x^2 + 25 = 10x$
29. $4x^2 + 9 = 12x$
30. $9x^2 = 60x - 100$
31. $5x^2 = 20x$
32. $4x^2 + 36x = 0$
33. $5x^2 = 20$
34. $4x^2 - 36 = 0$
35. $3x^2 - 24x + 48 = 0$
36. $5x^2 + 100x + 500 = 0$
37. $(x - 2)^2 = 1$
38. $(x + 3)^2 = 4$
39. $(x + 5)^2 = 16$
40. $(x - 7)^2 = 36$
41. $(2x - 1)^2 = 9$
42. $(3x + 2)^2 = 25$
43. $(5x + 3)^2 = 7$
44. $(4x - 5)^2 = 5$

6.5 Quadratic Equations: Completing the Square

Not all quadratic polynomials can be factored easily, and the solution of quadratic equations could be very time consuming and cumbersome if the factors had to be determined by the trial-and-error method. In order to find the solution more rapidly, other methods are available. One such method is called **completing the square.** The aim of this method is to make the left side of the equation a perfect square trinomial and the right side a constant—that is, $(x + d)^2 = k$.

By using the theorem for the square root method, for $k \geq 0$,

$$x + d = k \quad \text{or} \quad x + d = -k$$
$$x = -d + \sqrt{k} \quad \text{or} \quad x = -d - \sqrt{k}$$

If $k < 0$, then there is no solution in the set of real numbers.

Example 1 Solve $x^2 + 6x + 4 = 0$ by completing the square.

SOLUTION

1. Subtract the constant from both sides of the equation.

$$x^2 + 6x = -4$$

2. Add to both sides the number that makes $x^2 + 6x$ a perfect square trinomial.

$$x^2 + 6x + 9 = -4 + 9$$

3. Write the left side as the square of a binomial and simplify the right side.

$$(x + 3)^2 = 5$$

4. Take the square root of both sides.

$$x + 3 = \pm\sqrt{5}$$

5. Solve for x.

$$x = -3 + \sqrt{5} \quad \text{or} \quad x = -3 - \sqrt{5}$$

CHECK

If $x = -3 + \sqrt{5}$,

$$x^2 + 6x + 4 = 0$$
$$(-3 + \sqrt{5})^2 + 6(-3 + \sqrt{5}) + 4 = 0$$
$$9 - 6\sqrt{5} + 5 - 18 + 6\sqrt{5} + 4 = 0$$
$$0 = 0$$

If $x = -3 - \sqrt{5}$,

$$x^2 + 6x + 4 = 0$$

continued

$$(-3-\sqrt{5})^2+6(-3-\sqrt{5})+4=0$$
$$9+6\sqrt{5}+5-18-6\sqrt{5}+4=0$$
$$0=0$$

Therefore, the solution set is $\{-3+\sqrt{5}, -3-\sqrt{5}\}$.

It is useful to recall that the constant term of a perfect square trinomial of the form x^2+bx+c is found by taking half of the coefficient of the linear term and squaring this number. (Here the linear term is bx.)

Solving a quadratic equation by completing the square is most useful when the coefficient of x^2 is 1 and the coefficient of x is an even number. When this is not the case, fractions are involved. However, the method can still be used for the general case $ax^2+bx+c=0$, even when $a \neq 1$. The technique is to divide each term by a, the coefficient of x^2. See Examples 2 and 3.

Example 2 Solve $3x^2-6x+1=0$ by completing the square.

SOLUTION

1. Divide each term by a ($a=3$).

$$x^2-2x+\frac{1}{3}=0$$

2. Subtract the constant from each side.

$$x^2-2x=\frac{-1}{3}$$

3. Complete the square.

$$x^2-2x+1=\frac{-1}{3}+1$$
$$(x-1)^2=\frac{2}{3}$$

4. Use the theorem for $X^2=k$.

$$x-1=\pm\sqrt{\frac{2}{3}}$$

5. Simplify the radical.

$$x-1=\pm\frac{\sqrt{6}}{3}$$

6. Solve for the variable.

$$x=1\pm\frac{\sqrt{6}}{3}$$

and $$x=\frac{3\pm\sqrt{6}}{3}$$

Example 3 Solve $12x^2 - 20x + 7 = 0$ by completing the square.

SOLUTION

1. $$x^2 - \frac{20}{12}x + \frac{7}{12} = 0$$

2. $$x^2 - \frac{5}{3}x = \frac{-7}{12}$$

3. $$x^2 - \frac{5}{3}x + \left(\frac{1}{2}\cdot\frac{-5}{3}\right)^2 = \frac{-7}{12} + \frac{25}{36}$$

$$\left(x - \frac{5}{6}\right)^2 = \frac{-21 + 25}{36} = \frac{4}{36}$$

4–5. $$x - \frac{5}{6} = \pm\frac{2}{6}$$

6. $$x = \frac{5 + 2}{6} \quad \text{or} \quad x = \frac{5 - 2}{6}$$

$$x = \frac{7}{6} \quad \text{or} \quad x = \frac{1}{2}$$

CHECK

For $x = \frac{7}{6}$, $12\left(\frac{49}{36}\right) - 20\left(\frac{7}{6}\right) + 7 = \frac{49}{3} - \frac{70}{3} + 7 = \frac{-21}{3} + 7 = 0$

For $x = \frac{1}{2}$, $12\left(\frac{1}{4}\right) - 20\left(\frac{1}{2}\right) + 7 = 3 - 10 + 7 = 0$

Exercises 6.5

Solve the following equations.

1. $(x - 5)^2 = 2$
2. $(x + 3)^2 = 7$
3. $(x + 4)^2 = 45$
4. $(x - 2)^2 = 8$
5. $x^2 - 4x + 1 = 0$
6. $x^2 - 2x - 5 = 0$
7. $y^2 - 8y - 4 = 0$
8. $y^2 - 6y + 4 = 0$
9. $x^2 = 2x + 19$
10. $x^2 = 10x - 7$
11. $2 - 10t = t^2$
12. $8t = 12 - t^2$
13. $14z = 23 - z^2$
14. $z^2 + 54 = 18z$
15. $a^2 + 12(a + 2) = 6$
16. $a^2 + 16(a + 2) = 0$
17. $y^2 = 20(y - 3)$
18. $y^2 = 12(y - 1)$
19. $2x^2 + 7x = 30$
20. $3x^2 - 5x = 12$
21. $4x^2 - 20x + 17 = 0$
22. $9x^2 + 42x + 25 = 0$

23. $5x^2 - 10x - 2 = 10$
24. $6x^2 + 24x + 1 = 0$
25. $36x^2 + 12x - 3 = 0$
26. $25x^2 - 20x - 9 = 0$
27. $6a^2 - 10a + 3 = 0$
28. $9m^2 - 48m + 10 = 0$
29. $(x + 3)^2 = (2x - 5)^2$
30. $(4x - 1)^2 = (2x + 3)^2$

6.6 The Quadratic Formula

There is a formula that can be used to solve a quadratic equation by applying the method of completing the square to the general quadratic equation $ax^2 + bx + c = 0$, where $a \neq 0$.

The quadratic formula method of solution is most useful when the quadratic equation is not easily factored and when completing the square involves fractions.

Solving $ax^2 + bx + c = 0$,

1. Divide each term by a. $$x^2 + \frac{b}{a}x + \frac{c}{a} = 0$$
2. Subtract $\frac{c}{a}$ from each side. $$x^2 + \frac{b}{a}x = -\frac{c}{a}$$
3. Complete the square. $$x^2 + \frac{b}{a}x + \left(\frac{b}{2a}\right)^2 = \frac{-c}{a} + \frac{b^2}{4a^2}$$
4. Form the square. $$\left(x + \frac{b}{2a}\right)^2 = \frac{b^2 - 4ac}{4a^2}$$
5. If $b^2 - 4ac \geq 0$, take the square root of each side. $$x + \frac{b}{2a} = \frac{\pm\sqrt{b^2 - 4ac}}{2a}$$
6. Solve for x. $$x = \frac{-b \pm \sqrt{b^2 - 4ac}}{2a}$$

Since a, b, and c were chosen completely arbitrarily, except $a \neq 0$ and $b^2 - 4ac \geq 0$, the set of equations

$$\left\{x = \frac{-b + \sqrt{b^2 - 4ac}}{2a}, \quad x = \frac{-b - \sqrt{b^2 - 4ac}}{2a}\right\}$$

can be used as a formula for the solution of a quadratic equation.

The Quadratic Formula

The quadratic equation $ax^2 + bx + c = 0$, with $a \neq 0$, has the solutions

$$x = \frac{-b + \sqrt{b^2 - 4ac}}{2a}, \quad x = \frac{-b - \sqrt{b^2 - 4ac}}{2a}$$

where $b^2 - 4ac \geq 0$.

Example 1 Solve $2x^2 + 3x + 1 = 0$ by using the quadratic formula.

SOLUTION Comparing with $ax^2 + bx + c = 0$,
$2x^2 + 3x + 1 = 0$
it is seen that $a = 2$, $b = 3$, and $c = 1$. Therefore

$$x = \frac{-b \pm \sqrt{b^2 - 4ac}}{2a}$$

yields

$$x = \frac{-3 + \sqrt{9 - 4(2)(1)}}{4} = \frac{-3 + 1}{4} = \frac{-2}{4} = -\frac{1}{2}$$

or

$$x = \frac{-3 - \sqrt{9 - 4(2)(1)}}{4} = \frac{-3 - 1}{4} = \frac{-4}{4} = -1$$

CHECK

If $x = -\frac{1}{2}$,

$$2x^2 + 3x + 1 = 0$$
$$2\left(-\frac{1}{2}\right)^2 + 3\left(-\frac{1}{2}\right) + 1 = 0$$
$$\frac{1}{2} - \frac{3}{2} + 1 = 0$$
$$0 = 0$$

If $x = -1$,

$$2x^2 + 3x + 1 = 0$$
$$2(-1)^2 + 3(-1) + 1 = 0$$
$$2 - 3 + 1 = 0$$
$$0 = 0$$

Therefore, $\left\{-\frac{1}{2}, -1\right\}$ is the solution set.

Note that this equation ($2x^2 + 3x + 1 = 0$) could also be solved by factoring.

$$2x^2 + 3x + 1 = 0$$
$$(2x + 1)(x + 1) = 0$$
$$2x + 1 = 0 \quad \text{or} \quad x + 1 = 0$$
$$x = -\frac{1}{2} \quad \text{or} \quad x = -1$$

This result agrees with the answers obtained by using the quadratic formula.

Example 2 Solve $2x^2 - 3x = x^2 - 1$ by using the quadratic formula.

SOLUTION First, an equivalent equation of the form $ax^2 + bx + c = 0$ must be found:

$$2x^2 - 3x = x^2 - 1$$

continued

$$x^2 - 3x + 1 = 0$$
$$a = 1,\ b = -3,\ c = 1$$
$$x = \frac{-(-3) \pm \sqrt{(-3)^2 - 4(1)(1)}}{2}$$
$$x = \frac{3 + \sqrt{5}}{2} \quad \text{or} \quad x = \frac{3 - \sqrt{5}}{2}$$

The solution set is $\left\{\frac{3 + \sqrt{5}}{2}, \frac{3 - \sqrt{5}}{2}\right\}$.

Note that when b is negative, then $-b$ is positive. In this case, $-(-3) = +3$.

Example 3 Solve $2x^2 - 2x - 3 = 0$.

SOLUTION $a = 2,\ b = -2,\ c = -3$

$$x = \frac{-(-2) \pm \sqrt{(-2)^2 - 4(2)(-3)}}{2(2)}$$
$$= \frac{2 \pm \sqrt{4 + 24}}{4} = \frac{2 \pm \sqrt{28}}{4} = \frac{2 \pm 2\sqrt{7}}{4} = \frac{2(1 \pm \sqrt{7})}{2(2)}$$
$$x = \frac{1 \pm \sqrt{7}}{2}$$

CHECK $x = \frac{1 + \sqrt{7}}{2}$

$$2\left(\frac{1 + 2\sqrt{7} + 7}{4}\right) - 2\left(\frac{1 + \sqrt{7}}{2}\right) - 3 = \frac{8 + 2\sqrt{7}}{2} - 1 - \sqrt{7} - 3$$
$$= 4 + \sqrt{7} - 1 - \sqrt{7} - 3 = 0$$

The check for the other root is similar.
Note that when c is negative, then $-4ac$ is positive. In this case, $-4ac = -4(2)(-3) = +24$.

Example 4 Solve $9x^2 + 30x + 25 = 0$.

$$x = \frac{-30 \pm \sqrt{(30)^2 - 4(9)(25)}}{2(9)}$$
$$= \frac{-30 \pm \sqrt{900 - 900}}{2(9)} = \frac{-30}{6(3)} = \frac{-5}{3}$$

The solution set is $\left\{-\frac{5}{3}\right\}$.

CHECK For $x = -\frac{5}{3}$,

continued

$$9x^2 + 30x + 25 = 9\left(\frac{-5}{3}\right)^2 + 30\left(\frac{-5}{3}\right) + 25$$

$$= 9\left(\frac{25}{9}\right) + 10(-5) + 25 = 25 - 50 + 25 = 0$$

In this case, $-\frac{5}{3}$ is a double root.

The Discriminant

The expression $b^2 - 4ac$ that occurs under the radical sign in the quadratic formula

$$x = \frac{-b \pm \sqrt{b^2 - 4ac}}{2a}$$

plays an important role in determining the kind of solutions the equation has. This expression, $b^2 - 4ac$, is called the **discriminant** of the equation $ax^2 + bx + c = 0$.

If $b^2 - 4ac = 0$, then the roots are equal; that is, the equation has a double root.

If $b^2 - 4ac = n^2$ (n being an integer other than zero), and if a, b, and c are integers, then the roots are unequal and rational, in which case the quadratic polynomial $ax^2 + bx + c$ can be factored over the integers.

If $b^2 - 4ac$ is positive but not a perfect square, then the roots are unequal and irrational. In other words, the roots involve radicals.

When a, b, and c are integers and the roots are irrational, then the roots occur in **conjugate pairs** having the forms $r + t\sqrt{s}$ and $r - t\sqrt{s}$. In checking, this means that only one solution needs to be checked by substitution. The other solution must be the conjugate.

As examples,

If $3 + 2\sqrt{5}$ is one root, the other root must be $3 - 2\sqrt{5}$

If $\frac{1 - \sqrt{7}}{2}$ is one root, the other root must be $\frac{1 + \sqrt{7}}{2}$

The quadratic equation also has solutions when $b^2 - 4ac$ is negative, but these solutions are not real numbers and are beyond the scope of this text.

Example 5 Find the value of the discriminant, $b^2 - 4ac$. State if the roots are equal or unequal, rational or irrational. If the roots are rational, factor the polynomial.

a. $5x^2 - 8x + 2 = 0$ b. $5x^2 + 9x - 2 = 0$

c. $25x^2 - 20x + 4 = 0$

continued

SOLUTION

a. For $5x^2 - 8x + 2 = 0$, $a = 5$, $b = -8$, and $c = 2$

$$b^2 - 4ac = (-8)^2 - 4(10) = 64 - 40 = 24$$

Since 24 is not a perfect square, the roots are **irrational and unequal.**

b. For $5x^2 + 9x - 2 = 0$, $a = 5$, $b = 9$, and $c = -2$

$$b^2 - 4ac = 9^2 - 4(-10) = 81 + 40 = 121$$

Since $121 = (11)^2$, the roots are **rational and unequal.** Factoring, $5x^2 + 9x - 2 = (5x - 1)(x + 2)$.

c. For $25x^2 - 20x + 4 = 0$, $a = 25$, $b = -20$, $c = 4$

$$b^2 - 4ac = (-20)^2 - 4(100) = 400 - 400 = 0$$

The roots are **rational and equal.** Moreover, the polynomial is a perfect square.

$$25x^2 - 20x + 4 = (5x - 2)^2$$

Exercises 6.6

1–20. Solve by using the quadratic formula. Check each solution.

1. $x^2 - 3x + 1 = 0$
2. $x^2 + x - 1 = 0$
3. $x^2 + 3x + 1 = 0$
4. $x^2 - x - 1 = 0$
5. $2x^2 + 5x + 2 = 0$
6. $3x^2 + 5x + 1 = 0$
7. $2x^2 - 5x - 2 = 0$
8. $3x^2 + 5x + 2 = 0$
9. $x^2 - 5x - 25 = 0$
10. $5x^2 - 3x - 9 = 0$
11. $5x^2 - 2x - 1 = 0$
12. $3x^2 - 6x + 2 = 0$
13. $4y^2 + 8y + 3 = 0$
14. $2y^2 + 7y - 3 = 0$
15. $4y^2 - 8y - 3 = 0$
16. $2y^2 - 7y + 3 = 0$
17. $5t^2 + 3 = 9t$
18. $3t^2 + 4 = 12t$
19. $3t^2 = 4t + 2$
20. $5t^2 = 6t + 1$

In Exercises 21–40, find the value of the discriminant, $b^2 - 4ac$. State if the roots are equal or unequal, rational or irrational. If the roots are rational, factor the polynomial.

21. $x^2 + 3x + 2 = 0$
22. $x^2 - 3x - 2 = 0$
23. $5x^2 + x - 2 = 0$
24. $5x^2 + x - 4 = 0$
25. $9x^2 - 60x + 100 = 0$
26. $2x^2 - 3x - 2 = 0$
27. $3x^2 - 7x - 3 = 0$
28. $9x^2 - 30x + 25 = 0$
29. $5x^2 - 9x + 3 = 0$
30. $5x^2 - 9x - 2 = 0$

31. $x^2 + 8x + 12 = 0$

32. $4x^2 - 10x - 25 = 0$

33. $4x^2 + 4x + 1 = 0$

34. $x^2 - 10x + 25 = 0$

35. $9x^2 - 12x + 4 = 0$

36. $x^2 - 10x - 25 = 0$

37. $8x^2 + 5x - 2 = 0$

38. $3x^2 - 2x - 8 = 0$

39. $9x^2 - 4 = 0$

40. $25x^2 - 20x + 4 = 0$

Use the quadratic formula to solve the equations in Exercises 41–50 for the indicated variable. Treat any letter other than the variable you are solving for as a constant.

41. $x^2 + px + q = 0$ for x

42. $x^2 + 2px + q = 0$ for x

43. $3x^2 + 8xy - 3y^2 = 0$ for x

44. $4x^2 - 9xy + 2y^2 = 0$ for x

45. $W^2 = L(L - W)$ for W

46. $W^2 = L(L - W)$ for L

47. $P = EI + RI^2$ for I

48. $s = (k - s)^2$ for s

49. $px^2 + qx + r = 0$ for x

50. $am^2 + bm + c = 0$ for m

6.7 Verbal Problems

There are many practical applications that involve solving a quadratic equation. When you are solving verbal problems that involve the application of quadratic equations, it is especially important to check each root of the equation in the statement of the problem to see if the necessary conditions are met. Often the equation will have two roots, but only one may apply to a given problem. For example, lengths of sides of rectangles, triangles, and the like are always positive numbers; ages of individuals are positive numbers; digits in a numeral cannot be fractions; and the number of people at a gathering cannot be fractional or negative.

Example 1 One leg of a right triangle is 1 foot longer than the other leg and 8 feet shorter than the hypotenuse. Find the lengths of the sides of the right triangle.

SOLUTION If the lengths of the legs of the triangle are designated by a and b, where b is the shorter leg, and the length of the hypotenuse by c, the following relationships are given:

1. $a^2 + b^2 = c^2$ (Theorem of Pythagoras)
2. $a = c - 8$ (Given)
3. $a = b + 1$ (Given)

continued

From (2) and (3) we obtain the equivalent equations $c = a + 8$ and $b = a - 1$. Substituting $(a + 8)$ for c and $(a - 1)$ for b in equation (1) yields the equivalent equation

$$a^2 + (a - 1)^2 = (a + 8)^2$$
$$a^2 + a^2 - 2a + 1 = a^2 + 16a + 64$$

4. $a^2 - 18a - 63 = 0$

Equation (4) is a quadratic equation that can be solved by factoring:

$$(a - 21)(a + 3) = 0$$
$$a - 21 = 0 \quad \text{or} \quad a + 3 = 0$$
$$a = 21 \quad \text{or} \quad a = -3$$

Since the verbal problem contains the assumption $a > 0$, we disregard the solution $a = -3$ and accept $a = 21$. From (2) we know that $a = c - 8$; therefore, $c = 29$. From (3) we know that $b = a - 1$; therefore, $b = 20$. The sides of the triangle are 20 feet, 21 feet, and 29 feet.

Example 2 An 18-foot pole and a 30-foot pole are 15 feet apart. What length of wire is needed to join the tops of the two poles?

SOLUTION First make a sketch such as the one in Fig. 6.3. Find the lengths of the legs of the right triangle. The length of the hypotenuse determines the length of the wire.

$$x^2 = (15)^2 + (12)^2$$
$$x^2 = 225 + 144$$
$$x^2 = 369$$
$$x = 19.2 \text{ feet, correct to the nearest tenth}$$

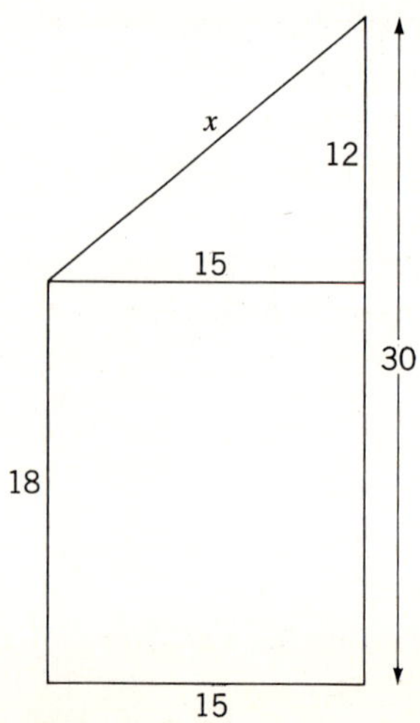

Figure 6.3

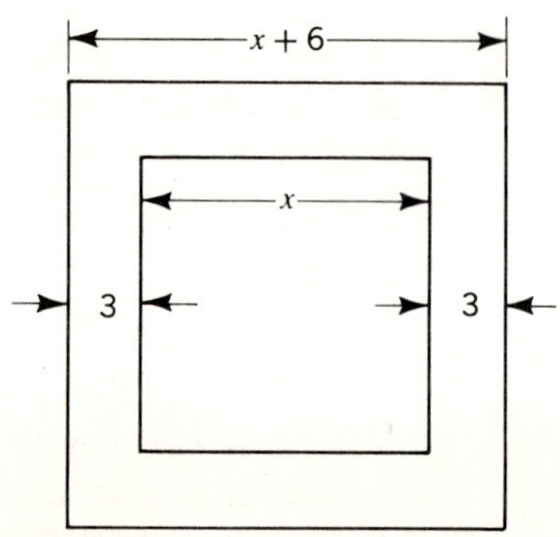

Figure 6.4

Al-Khowarizmi's Solution of a Quadratic Equation

Around 825 the Arab al-Khowarizmi showed two methods for solving quadratic equations, both based on the geometric methods of the Greeks. To solve $x^2 + 10x = 39$, he drew a square like the one illustrated and then gave the solution that follows.

> You halve the number of roots, yielding 5 in this case.
> You multiply this number by itself; the product is 25.
> Add this to 39; the sum is 64.
> Now take the root of this, which is 8.
> From 8 subtract half the number of roots, 5 in this case.
> The remainder is 3.
> Three is the root of the square you wanted; the square itself is 9.

Comparing the solution of $x^2 + 10x = 39$ to the solution of $x^2 + px = q$, that is,

$$\sqrt{\left(\frac{10}{2}\right)^2 + 39} - \frac{10}{2} \quad \text{to} \quad \sqrt{\left(\frac{p}{2}\right)^2 + q} - \frac{p}{2}$$

we note that his method is similar to our "completing the square" method. Note also that the negative solution -13 was ignored.

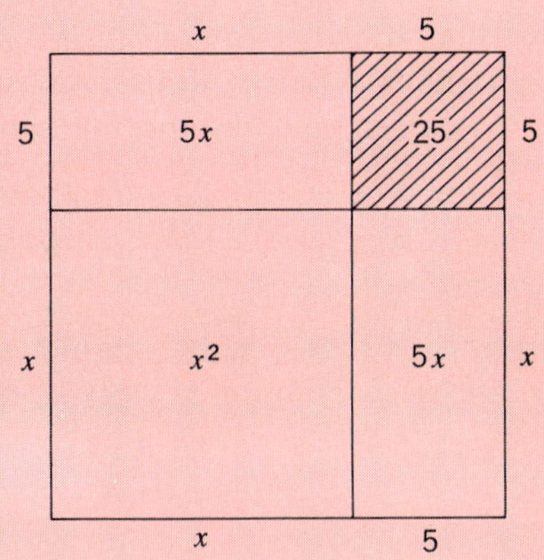

Example 3 A square flower bed has a 3-foot walk surrounding it. If the walk were to be replaced and planted with flowers, the new flower bed would have 4 times the area of the original bed. What is the length of one side of the original bed?

SOLUTION

Let x = length of a side of original flower bed
Then x^2 = area of original bed
$x + 6$ = length of a side of new flower bed
and $(x + 6)^2$ = area of new flower bed

continued

$$
\begin{aligned}
(x+6)^2 &= 4x^2 \\
x^2 + 12x + 36 &= 4x^2 \\
3x^2 - 12x - 36 &= 0 \\
3(x^2 - 4x - 12) &= 0 \\
x^2 - 4x - 12 &= 0 \\
(x+2)(x-6) &= 0 \\
x + 2 = 0 \quad &\text{or} \quad x - 6 = 0 \\
x = -2 \quad &\text{or} \quad x = 6
\end{aligned}
$$

Again we discard the negative answer, and the length of the original side is 6 feet.

CHECK

$$
\begin{aligned}
\text{Area of original bed} &= x^2 = 6^2 = 36 \text{ square feet} \\
\text{Area of new bed} &= (x+6)^2 = (6+6)^2 = 12^2 \\
&= 144 \text{ square feet} \\
\text{Area of new bed} &= 4 \text{ times area of old bed} \\
144 &= 4 \times 36 \\
144 &= 144
\end{aligned}
$$

Example 4 Working alone, a carpenter can make a set of cabinets in 3 hours less time than his helper can. Working together, they can make the set of cabinets in 6 hours. Find the time (correct to the nearest minute) that each requires to make the set alone.

SOLUTION Let x = time it takes helper alone
Then $x - 3$ = time it takes carpenter alone

FORMULAS

$tr = w$, and w of first + w of second = w of both = 1, where t = time, r = rate, and w = amount of work done.

	Working Alone			***Working Together (Whole Job)***		
Formula	t ·	r	= w	t ·	r	= w
Carpenter	$x - 3$	$\frac{1}{x-3}$	1	6	$\frac{1}{x-3}$	$\frac{6}{x-3}$
Helper	x	$\frac{1}{x}$	1	6	$\frac{1}{x}$	$\frac{6}{x}$

EQUATION

Work of helper + work of carpenter = whole job

$$\frac{6}{x} + \frac{6}{x-3} = 1$$

with restrictions $x > 0$ and $x - 3 > 0$

continued

$$6(x - 3) + 6x = x(x - 3)$$
$$12x - 18 = x^2 - 3x$$
$$x^2 - 15x + 18 = 0$$

Since $x^2 - 15x + 18$ is not readily factorable and since $\frac{b}{a} = -15$ is not an even integer, the method using the quadratic formula is recommended for solving the equation.

SOLUTION OF EQUATION
$ax^2 + bx + c = 0$ if and only if

$$x = \frac{-b \pm \sqrt{b^2 - 4ac}}{2a}$$

If $x^2 - 15x + 18 = 0$, then $a = 1$, $b = -15$, $c = 18$. Thus

$$x = \frac{-(-15) \pm \sqrt{(-15)^2 - 4(18)}}{2} = \frac{15 \pm \sqrt{225 - 72}}{2}$$

$$x = \frac{15 + \sqrt{153}}{2} \quad \text{or} \quad x = \frac{15 - \sqrt{153}}{2}$$

Approximating (by using the tables), $\sqrt{153} = 12.37$. Thus, approximately to the nearest hundredth,

$$x = \frac{15 + 12.37}{2} \quad \text{or} \quad x = \frac{15 - 12.37}{2}$$

$$x = \frac{27.37}{2} \quad \text{or} \quad x = \frac{2.63}{2}$$

$$x = 13.68 \quad \text{or} \quad x = 1.32$$

and $x - 3 = 10.68$ or $x - 3 = -1.68$. (This solution is rejected, since $x - 3 > 0$.) Thus $x = 13.68$ and $x - 3 = 10.68$ are correct to the nearest hundredth. Since there are 60 minutes in an hour, 0.68 hours = 0.68 (60) minutes = 40.8 minutes. Therefore, correct to the nearest minute,

x = 13 hours, 41 minutes, time of helper alone
$x - 3$ = 10 hours, 41 minutes, time of carpenter alone

Example 5 A plane flies 300 miles with a tailwind of 10 mph and returns against a wind of 20 mph. What is the speed of the plane in still air if the total flying time is 4 hours?

SOLUTION Let x represent the speed of the plane in still air.
Then $x + 10$ = speed with the tailwind
and $x - 20$ = speed against the wind

continued

Formula	r	$\cdot$	t	$=$	d
With wind	$x + 10$		$\frac{300}{x + 10}$		300
Against wind	$x - 20$		$\frac{300}{x - 20}$		300

EQUATION

Time going + time returning = total time
= 4 hours

$$\frac{300}{x + 10} + \frac{300}{x - 20} = 4$$

Multiplying both sides of this equation by the LCD $(x + 10)(x - 20)$,

$$300(x - 20) + 300(x + 10) = 4(x + 10)(x - 20)$$

Dividing both sides by 4,

$$\begin{aligned} 75(x - 20) + 75(x + 10) &= (x + 10)(x - 20) \\ 150x - 750 &= x^2 - 10x - 200 \\ x^2 - 160x + 550 &= 0 \end{aligned}$$

Solving this equation by completing the square, since the coefficient of x^2 is 1 and since the coefficient of x is even,

$$\begin{aligned} x^2 - 160x + (-80)^2 &= -550 + 6400 \\ (x - 80)^2 &= 5850 \\ x - 80 &= \pm\sqrt{5850} \\ &= \pm 76.5 \text{ to the nearest tenth} \end{aligned}$$

Therefore $x = 80 + 76.5$ or $x = 80 - 76.5$
$x = 156.5$ or $x = 3.5$

Since speed is always a positive number, $x \neq 3.5$, because $x - 20$ would then be a negative number. Therefore, the speed of the plane in still air is 156.5 mph, correct to the nearest tenth.

Exercises 6.7

Solve the following exercises.

1. The square of a number is 91 more than 6 times the number. Find the number.

2. If 36 is subtracted from 20 times a number, the result is the square of the number. Find the number.

3. The sum of a number and its reciprocal is 2. Find the number.

4. The difference between a number and its reciprocal is $\frac{15}{4}$. Find the number.

5. Five times the square of a number added to 10 times the number equals 315. Find the number.

6. If 4 times a number is added to 3 times its square, the sum is 95. Find the number.

7. The hypotenuse of a right triangle is 13 inches. If one leg is 7 inches longer than the other, how long are the legs of the triangle?

8. The hypotenuse of a right triangle is 8 units, and the legs are equal in length. Find the length of a leg of the triangle.

9. The equation $1600y = 1200x - x^2$ represents the orbit of a certain projectile initially fired at 200 feet per second.
   a. Find x for $y = 0$. (The largest value gives the range.)
   b. Find x for $y = 200$. (This gives the horizontal distance the projectile has traveled when its vertical height is 200 feet.)

10. The equation $-16t^2 + 2000t - 18{,}400 = 0$ is used to find the time a certain projectile is in flight. Find t.

11. In chemistry, $x^2 = 4(1 - x)^2$ gives the number of moles, x, that react when 1 mole of pure ethyl alcohol is mixed with 1 mole of acetic acid. Find x. (By chemistry considerations, x is positive and less than or equal to 1.)

12. The equation $P = EI - RI^2$ is a formula for electric power. Find the amperage I if $P = 360$ watts, $E = 110$ volts, $R = 5$ ohms.

13. The equation $s^2 = 8rh - 4h^2$ gives the span, s, of a circular arch whose height is h and whose radius is r. Find h if $r = 50$ and $s = 80$ and $h \leq 50$.

14. The equation $D = k(L - 2c)^2$ is a formula that gives the deflection, D, for a certain beam of length, L. Find L if $D = \frac{529}{40}$ and $k = \frac{1}{10}$ and $c = \frac{1}{4}$.

15. The length of a rectangle is 1 inch more than 3 times its width. The area is 52 square inches. Find the dimensions of the rectangle.

16. The length of a rectangle is 7 meters less than twice its width. The area of the rectangle is 660 square meters. Find the width and length of the rectangle.

17. A wire is stretched from the top of a 4-foot fence to the top of a 20-foot vertical pole. If the fence and the pole are 30 feet apart, find the length of the wire.

18. A 36-inch length of copper tubing is bent to form a right triangle with a 15-inch hypotenuse. Find the lengths of the other two sides of the triangle.

19. A concrete walk of uniform width extends around a rectangular lawn having dimensions 20 feet by 80 feet. Find the width of the walk if the area of the walk is 864 square feet.

20. A rectangular piece of sheet metal is twice as long as it is wide. From each of its four corners a square piece 2 centimeters on a side is cut out. The flaps are then turned up to form an uncovered metal box. If the volume of this box is 320 cubic centimeters, find the dimensions of the original piece of sheet metal.

21. It takes John 3 hours longer to do a certain job than it does his brother Bob. For 3 hours they worked together; then John left and Bob finished the job in 1 hour. How many hours would it have taken Bob to do the whole job by himself?

22. A new copier takes 5 minutes less time to make 150 copies than an older copier. When both machines are used, 150 copies can be made in 6 minutes. Find the time for the new copier, working alone, to make 150 copies.

23. One of two outlets can fill a swimming pool in 6 hours. The time for the other outlet to fill the pool is 2 hours longer than the two outlets together. Find the time it takes for the two outlets together to fill the tank, correct to the nearest minute.

24. One inlet pipe takes 12 minutes longer than another inlet pipe to fill a tank. An outlet pipe can empty the tank in 45 minutes. When all three pipes are open, it takes 15 minutes to fill the tank. Find the time it takes to fill the tank if only the larger inlet pipe is open.

25. A fisherman trolled upstream in a motorboat to a spot 8 miles from his campsite and then returned to camp. If the rate of the current was 2 mph and if the round trip took 3 hours, find the rate of the motorboat in still water.

26. A jet plane flying against a headwind of 20 mph takes 20 minutes longer to fly a distance of 2610 miles than a plane with the same still air speed flying in the opposite direction. Find the still air speed of the plane.

27. An airplane flew 720 kilometers with a tailwind. The return flight was made against the same wind. The total flying time was 7 hours. The speed of the plane in still air was 210 kph. Find the speed of the wind.

28. A ship took 1 hour longer to travel 120 nautical miles against a current than it did to travel the same distance with the same current. The speed of the current was 5 knots (nautical miles per hour). Find the speed of the ship in still water.

29. In economics, a supply law relates the number of units, x, of a commodity that can be supplied and the price, P, of one unit. For the supply law,

$$P = x^2 + 8x + 60$$

find the number of units that can be supplied when the price per unit is 80.

30. In economics, a demand law relates the number of units, x, of a commodity demanded and the price, P, of one unit. For the demand law,

$$100P = 756 - 24x - x^2$$

find the number of units that are demanded when the price is 5.

Review Exercises

Approximate the square roots for Exercises 1–8 correct to the nearest hundredth.

1. $\sqrt{27}$
2. $\sqrt{39}$
3. $(\sqrt{14})^2$
4. $\sqrt{14^2}$
5. $\sqrt{87}$
6. $-\sqrt{44}$
7. $-\sqrt{8^2}$
8. $\sqrt{(-8)^2}$

Write the radicals for Exercises 9–20 in simplest form.

9. $\sqrt{5}\sqrt{20}$
10. $\sqrt{20}$
11. $\sqrt{160}$
12. $\sqrt{500}$
13. $\sqrt{20} - \sqrt{45}$
14. $\sqrt{28} + 2\sqrt{63}$
15. $\dfrac{24 - 10\sqrt{8}}{6}$
16. $\dfrac{14 + \sqrt{60}}{4}$
17. $\dfrac{-8 - \sqrt{56}}{8}$
18. $(3 + 2\sqrt{5})^2$
19. $(2 - 3\sqrt{7})^2$
20. $\sqrt{5}(\sqrt{10} - \sqrt{5} - \sqrt{45})$

Solve the equations in Exercises 21–26 by completing the square.

21. $x^2 + 2x - 1 = 0$
22. $x^2 - 12x - 9 = 0$
23. $x^2 - 1 = 4x$
24. $x^2 + 10x + 1 = 0$
25. $x^2 + 3x = 2$
26. $2x^2 - 12x = 7$

Solve the equations in Exercises 27–40 either by factoring or by using the quadratic formula. Use the discriminant to see if the polynomial can be factored.

27. $3x^2 + 6x + 2 = 0$
28. $3x^2 - 6x + 2 = 0$
29. $(2x - 3)(x + 2) = 0$
30. $(3x + 4)(2x - 1) = 0$
31. $(x - 4)(x + 2) = 4$
32. $(3x + 1)(x - 1) = 1$
33. $(2y - 5)^2 = 4$
34. $(3y + 2)^2 = 6$
35. $2x^2 - 8x - 1 = 0$
36. $3x^2 = 4x + 3$
37. $12z^2 = 48z$
38. $a^2 + 18a + 81 = 0$
39. $6x^2 - 10x + 2 = 0$
40. $9x^2 = 2(3x + 1)$

41. John can mow his lawn in 20 minutes less time with his power mower than with his hand mower. One day his power mower broke down 15 minutes after he started mowing, and he had to complete the job with his hand mower. It took him 25 minutes to finish mowing by hand. How long does it take John to do the complete job with the power mower?

42. Find a number whose square is equal to the number itself.

43. The sum of the reciprocals of two consecutive integers is $\frac{11}{30}$. Find the numbers.

44. The perimeter of a rectangle is 36 feet, and its area is 45 square feet. Find the dimensions of the rectangle.

45. A pilot left a Chicago airport and flew 200 miles south to a town T with a tailwind of 20 mph. From T he flew back to Chicago against a headwind of 30 mph. If his total flying time was $2\frac{1}{3}$ hours, what was the average speed of the plane in still air?

46. One leg of a right triangle is 9 inches longer than the other leg. The hypotenuse is 45 inches long. Find the lengths of the legs of the triangle.

47. The height H of a projectile at the end of t seconds is given by

$$H = cvt - \frac{1}{2}gt^2$$

Solve for t.

48. The total surface area T of a right circular cylinder of radius r and height h is given by $T = 2\pi r(r + h)$. Solve for r.

Answers

Answers to Exercises 1.1

1. 24
2. 72
3. 2
4. 5
5. 15
6. 13
7. 8
8. 11
9. 25
10. 9
11. 64
12. 8
13. 9
14. 4
15. 49
16. 64
17. 729
18. 64
19. 23
20. 56
21. 6
22. 1
23. 200
24. 100
25. 100
26. 1400
27. 0
28. 1
29. 20
30. 0
31. 1
32. 25
33. 100
34. 40
35. 94
36. 27
37. 216
38. 975
39. 1000
40. 729
41. $x + 8$
42. $y + 5$
43. $8x$
44. $4t$
45. n^2
46. y^2
47. t^3
48. c^3
49. $y - 4$
50. $x - 1$
51. $\frac{x}{6}$
52. $\frac{y}{5}$
53. st
54. yz
55. c^2
56. y^2
57. z^3
58. n^3
59. $x + y$
60. $a + b$
61. xy
62. ab
63. $y - x$
64. $b - a$
65. $\frac{x}{y}$
66. $\frac{a}{b}$
67. $7 - x$
68. $10 - x$
69. $\frac{5}{y}$
70. $\frac{10}{x}$
71. $x^2 + y^2$
72. $a^2 + b^2$
73. $25 - t^2$
74. $n^2 - 36$
75. $\frac{a + b}{ab}$
76. $\frac{xy}{x + y}$
77. $5n^3$
78. $7k^3$
79. $4y - 16$
80. $36 - 6x$

Answers to Exercises 1.2

1. {18, 20, 22, 24}
2. {36, 38, 40, 42, 44}
3. {57, 59, 61}
4. {71, 73, 75, 77, 79}
5. {30, 33, 36, 39, 42, 45, 48}
6. {32, 36, 40, 44, 48}
7. {10, 20, 30, 40, 50, 60, 70, 80, 90}
8. {12, 24, 36, 48, 60, 72, 84, 96}
9. {1, 2, 4, 7, 14, 28}
10. {1, 2, 3, 4, 6, 8, 12, 24}
11. {1, 2, 3, 5, 6, 10, 15, 30}
12. {1, 2, 13, 26}
13. {1, 2, 5, 10, 25, 50}
14. {1, 2, 3, 6, 9, 18}
15. {2, 3, 5, 7, 11, 13, 17, 19}
16. {31, 37, 41, 43, 47}
17. {53, 59, 61, 67, 71, 73, 79}
18. {83, 89, 97}
19. {20, 21, 22, 24, 25, 26}
20. {72, 74, 75, 76, 77, 78, 80, 81, 82}
21. {2, 7}
22. {2, 3}
23. {2, 3, 5}
24. {2, 13}
25. {2, 5}
26. {2, 3}
27. {2, 4, 6, 12, 18, 36}
28. {1, 3, 7, 21}

29. {1}
30. ∅
31. ∅
32. {4, 6, 8, 12, 16, 24, 48}
33. {15, 30, 45}
34. {12, 24, 36, 48}
35. {36, 72, 108, 144, 180}
36. {260}
37. {1, 2, 3, 6}
38. {1, 3, 5, 15}
39. $3 \cdot 5$
40. $2 \cdot 7$
41. $2 \cdot 3^2$
42. $2^3 \cdot 5$
43. $2 \cdot 3 \cdot 11$
44. $2^2 \cdot 13$
45. $2 \cdot 3 \cdot 13$
46. $2^2 \cdot 3 \cdot 5$
47. $2 \cdot 7^2$
48. $3 \cdot 5^2$

Answers to Exercises 1.3

1. 15
2. 21
3. 15
4. 21
5. 15
6. 20
7. 9
8. 40
9. 60
10. 84
11. 180
12. 168
13. 12
14. 52
15. 3
16. 7
17. 80
18. 200
19. 400
20. 1600
21. 16
22. 34
23. 1000
24. 1000
25. 17
26. 37
27. 35
28. 72
29. 5
30. 15
31. 20
32. 40
33. 54
34. 30
35. 22
36. 6
37. 4
38. 21
39. 12
40. 48
41. 29
42. 15
43. 9
44. 32
45. 16
46. 6
47. 66
48. 720
49. 9
50. 7
51. 1
52. 13
53. 4
54. 5
55. 8
56. 9
57. 72
58. 60
59. 26
60. 27

Answers to Exercises 1.4

1. 74
2. 96
3. 180
4. 73
5. 0
6. 36
7. 12
8. 2
9. 180
10. 16
11. 45
12. 10
13. 120
14. 40
15. 64
16. 75
17. 34
18. 35
19. 25
20. 49
21. 196
22. 1
23. 2
24. 420
25. 213
26. 2
27. 19
28. 75
29. 19
30. 7
31. 125
32. 98
33. 4
34. 64
35. 19
36. 6
37. 122
38. 5
39. 240
40. 108
41. 49
42. 150
43. 14
44. 169
45. 60
46. 405
47. 4
48. 2
49. 45
50. 13
51. 180 square feet
52. 48 centimeters
53. 63 square inches
54. 60 m^3
55. 120
56. 12 ohms
57. 100°
58. 68°
59. $740
60. 20
61. 490 meters
62. 10 lumens
63. 20,100
64. $200
65. 75
66. 4

Answers to Exercises 1.5

1. Positive
2. Signed
3. Signed
4. Signed
5. Signed
6. Signed
7. Positive
8. Positive
9. Signed
10. Signed
11. Signed
12. Positive
13. Positive
14. Signed
15. Signed
16. Signed
17. Positive
18. Signed
19. Signed
20. Positive
21. 35 degrees south latitude
22. 250 B.C.
23. 20° below 0°F
24. Loss of 35 points
25. Force of 20 pounds away
26. A deceleration of 32 ft/sec^2
27. Loss of 150 BTU
28. A decrease of $5
29. Clockwise rotation of 40 degrees
30. 30 centimeters behind a mirror
31. 80 degrees west longitude
32. $50 loss
33. 70 feet below sea level
34. A loss of 5 yards
35. 15 mph north wind
36. A 2800 mph downward velocity
37. 6 steps downward
38. 8 hours earlier
39. 35 pounds underweight
40. A decrease of 20 cubic centimeters
41. 3
42. -2
43. -4
44. -1
45. 4
46. 0
47. 1
48. 5
49. -5
50. 6
51. 5
52. -3
53. -4
54. $1\frac{1}{2}$
55. $-2\frac{1}{2}$
56. 4
57. -2
58. -5
59. 2
60. -1

61. 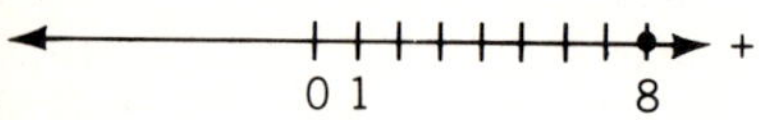

62. 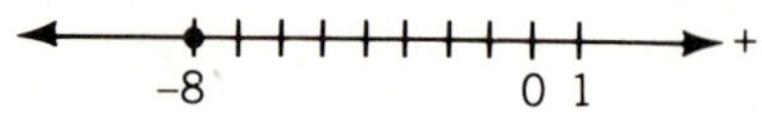

63. 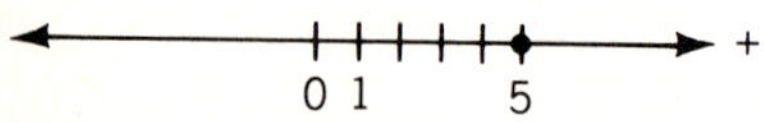

64. 

65. 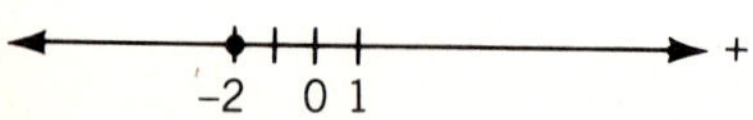

66. 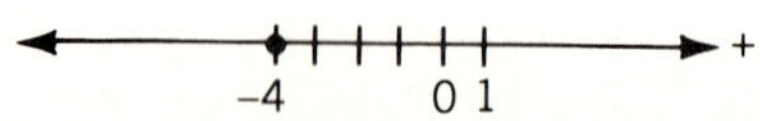

67. 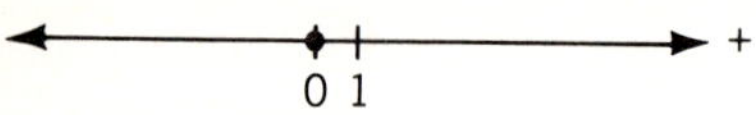

68. 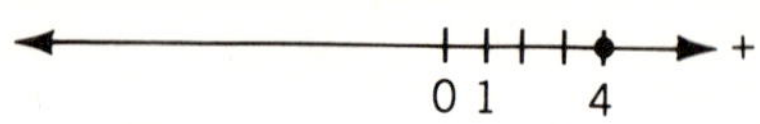

69. 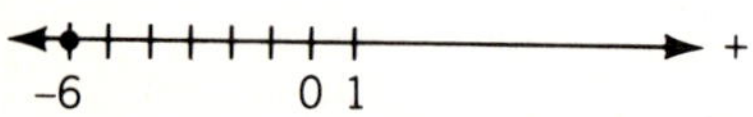

70. 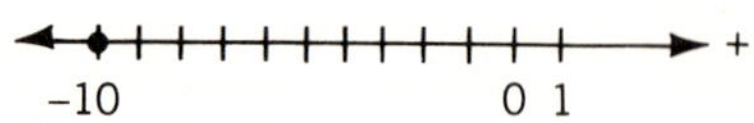

71. 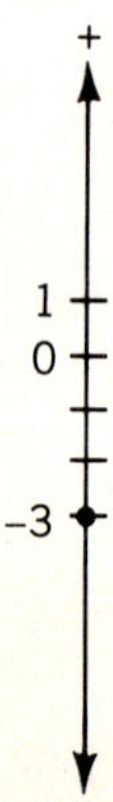

72. 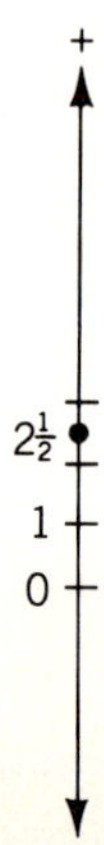

73. 

74. 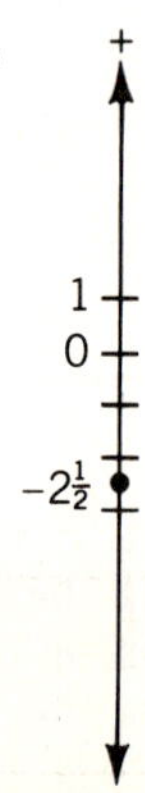

75. 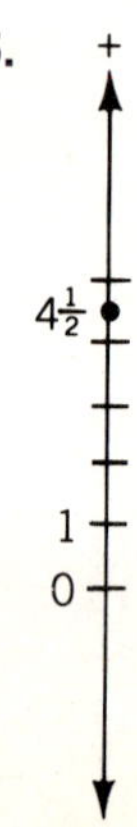

76. 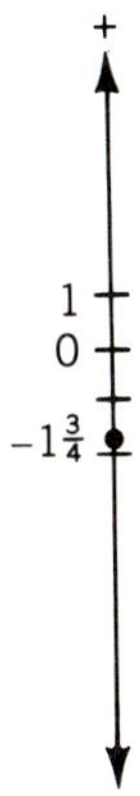

77. 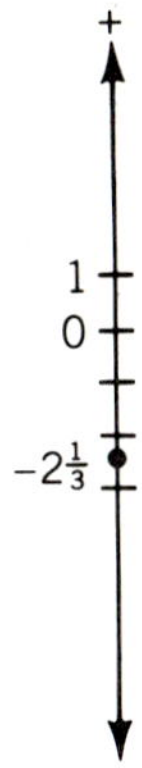

78. 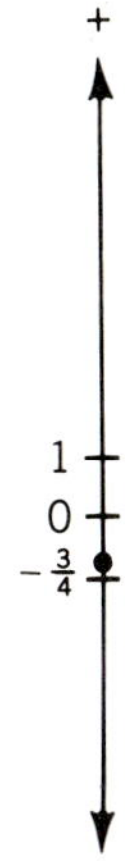

79. 

80. 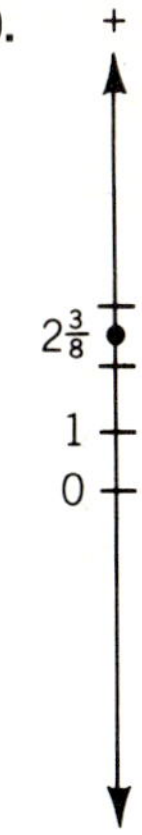

Answers to Exercises 1.6

1. 3
2. 3
3. $2\frac{1}{2}$
4. -3
5. $\frac{1}{3}$
6. 5
7. 1
8. 0
9. 10
10. 3
11. 8
12. 6
13. 6
14. 4
15. 0
16. 6
17. -5
18. -12
19. 12
20. 8
21. 2
22. 16
23. 0
24. 3
25. $1\frac{3}{4}$
26. $7\frac{1}{2}$
27. 12
28. 9
29. 11
30. 15
31. 24
32. 7
33. 7
34. 16
35. 0
36. 4
37. $>$
38. $<$
39. $<$
40. $>$
41. $>$
42. $<$
43. $>$
44. $<$
45. $>$
46. $>$
47. $<$
48. $>$
49. $<$
50. $<$
51. $=$
52. $=$
53. $<$
54. $>$
55. $>$
56. $>$
57. $>$
58. $<$
59. $>$
60. $<$
61. $<$
62. $>$

63. 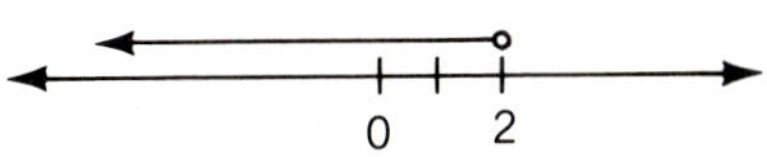

64. 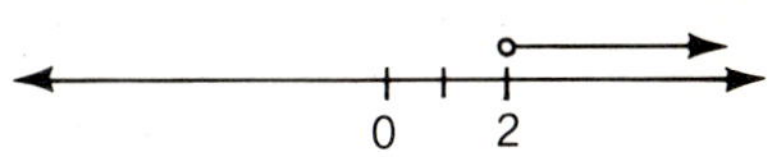

65. 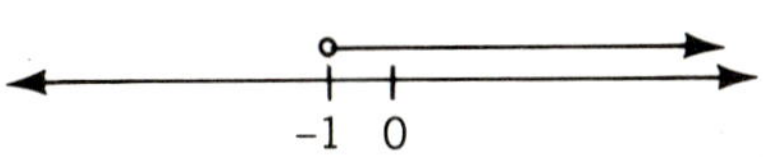

66. 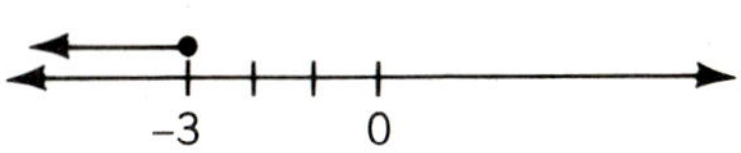

67. 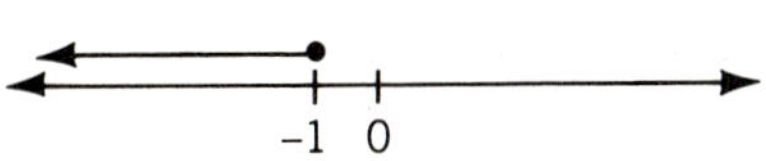

68. 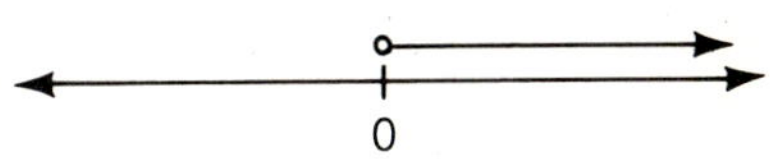

69. 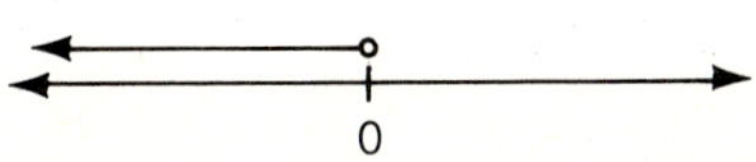

70. 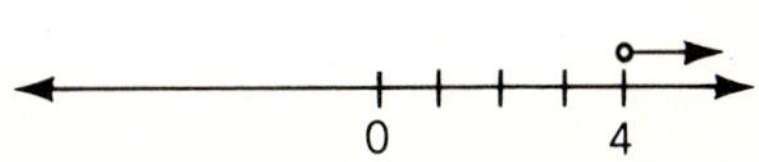

71. 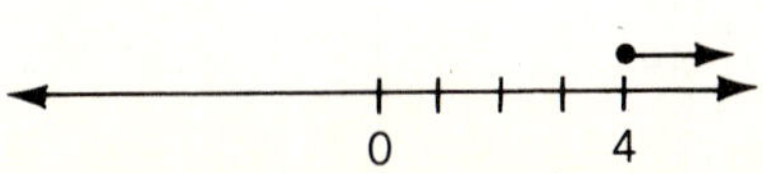

72. 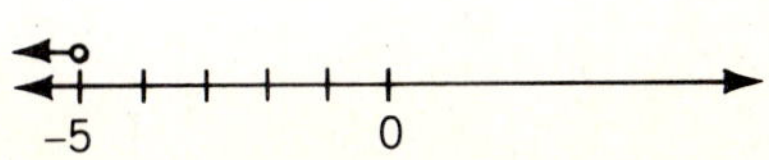

73. 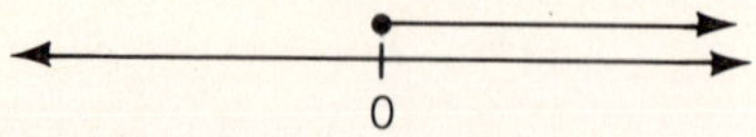

74. 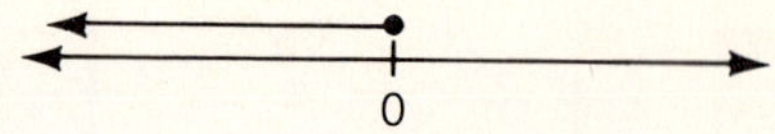

75. 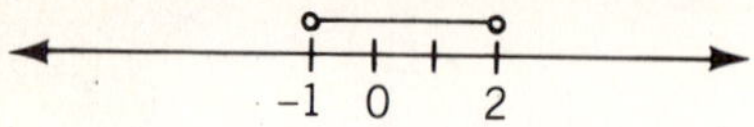

76. 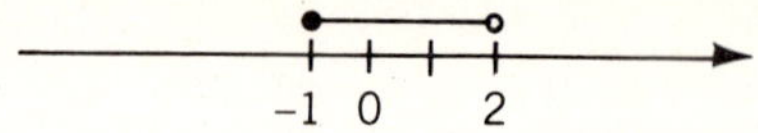

77. 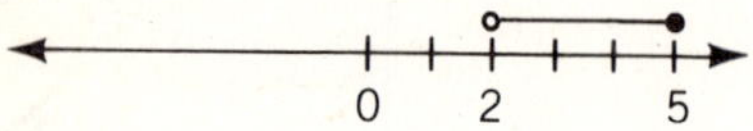

78. 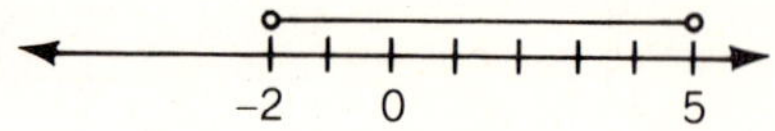

79. 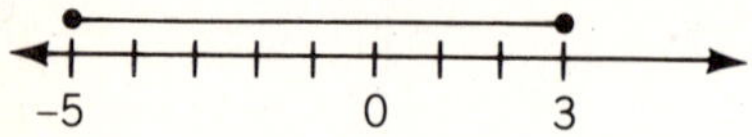

80. 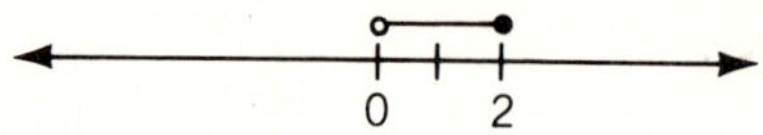

Answers to Exercises 1.7

1. −4
2. −3
3. 4
4. 3
5. 12
6. 15
7. −20
8. −30
9. 0
10. 0
11. 12
12. 14
13. −12
14. −14
15. 6
16. −6
17. −6
18. 6
19. −5
20. −8
21. −7
22. −9
23. 0
24. 0
25. 0
26. 0
27. 18
28. 0
29. 0
30. 6
31. −30
32. −40
33. 6
34. −6
35. 5
36. 6
37. −10
38. −15
39. −100
40. −100
41. 5
42. −15
43. −25
44. 4
45. −50
46. −105
47. −8
48. −5
49. 0
50. 0
51. 0
52. 0
53. −20
54. −10
55. 70
56. −55
57. −9
58. 8
59. 0
60. 0
61. −4
62. −4
63. −10
64. −15
65. −40
66. −50
67. −7
68. 0
69. 8
70. 14
71. −15
72. −20
73. −20
74. −8
75. −4
76. −40
77. −27
78. 20
79. −1
80. 0
81. 0
82. 49
83. $404
84. −9
85. 66 kilograms
86. 7 yards
87. 50 mph
88. $32
89. −5, (5 centimeters below C)
90. −2
91. −1
92. 1
93. 1
94. −1
95. a. $C = 4$ b. $N = -3$
c. $Al = 3$
96. $55.25

Answers to Exercises 1.8

1. 6
2. −13
3. 14
4. 4
5. 6
6. 7
7. −7
8. −11
9. −6
10. 5
11. 1
12. −4
13. 35
14. 65
15. 0
16. 12
17. 14
18. 0
19. 3
20. −3
21. 4
22. −4
23. 10
24. −10

25. −10
26. −2
27. 0
28. 0
29. −14
30. 16
31. −28
32. −5
33. 10
34. −2
35. −2
36. 5
37. 7
38. −7
39. −23
40. −19
41. 11
42. 16
43. −7
44. −4
45. −40
46. −160
47. 10
48. 20
49. 8
50. 12
51. −8
52. −12
53. 0
54. 0
55. 0
56. 0
57. −74
58. −38
59. 26
60. 56
61. −9
62. −2
63. −10
64. 6
65. 59
66. −26
67. 46
68. −9
69. 50
70. 3
71. 15
72. 5
73. −9
74. −8
75. 35
76. −40
77. −5
78. −9
79. −9
80. −5
81. 6°
82. 13°
83. 16°
84. 14°
85. 5°
86. 2°
87. 51 degrees
88. 13 degrees
89. 6 degrees
90. 8 degrees
91. 19,757 feet
92. 14,777 feet
93. 154 feet
94. 74 feet
95. 30,320 feet
96. 1250 years
97. 250 years
98. 612 − 1 = 611 years
99. 925 − 1 = 924 years
100. 4380 − 1 = 4379 years

Answers to Exercises 1.9

1. 35
2. −35
3. 35
4. −16
5. 16
6. −100
7. −1000
8. 100
9. −1000
10. 105
11. 1000
12. −88
13. −88
14. 221
15. 1225
16. −8000
17. −250
18. 2500
19. 0
20. 0
21. 0
22. 70
23. 90
24. −80
25. −700
26. 900
27. 100
28. −500
29. −36
30. 0
31. −48
32. 460
33. −54
34. −64
35. −125
36. 36
37. −32
38. 36
39. −686
40. 6
41. −15
42. 4
43. 0
44. 10
45. −36
46. 0
47. 24
48. 24
49. 32
50. −24
51. −40
52. −40
53. 35
54. 35
55. 16
56. 1000
57. 1000
58. −84
59. −6
60. 3024
61. 3000
62. −2000
63. −560
64. 720
65. 7500
66. −1
67. −3
68. −4
69. 1
70. −1

Answers to Exercises 1.10

1. 7
2. −7
3. −7
4. 7
5. −8
6. 125
7. −20
8. 0
9. Undefined
10. 8
11. −8
12. −14
13. −3
14. −3
15. 3
16. 3
17. 0
18. Undefined
19. −7
20. −9

21. 5
22. −6
23. −7
24. 9
25. −11
26. −13
27. 25
28. −125
29. 5
30. Undefined
31. 0
32. Undefined
33. −1
34. 2
35. −60
36. 0
37. −10
38. 4
39. −1
40. 1
41. 6
42. 0
43. −2
44. Undefined
45. −1
46. −1
47. −5
48. 25
49. 5
50. −20
51. −5
52. −12
53. −3
54. 0
55. Undefined
56. −10
57. −4
58. −1
59. −27
60. −3
61. −20
62. −1
63. −1
64. −8
65. −2
66. −1
67. −2
68. 19
69. 75
70. −4
71. −1
72. −8
73. −9
74. 4
75. 14
76. −1
77. −4
78. 1
79. 1
80. −2

Answers to Exercises 1.11

1. 0
2. 7
3. −4
4. −3
5. 4
6. 30
7. 6
8. 4
9. 10
10. −10
11. 2
12. 4
13. −6
14. −2
15. 30
16. −35
17. 0
18. −6
19. −4
20. 0
21. −40
22. 0
23. 46
24. −15
25. 0
26. −6
27. −7
28. 17
29. 9
30. −17
31. 7
32. 127
33. 35
34. 3
35. −4
36. 1
37. 4
38. 81
39. −2
40. 21
41. 20
42. −6000
43. −22
44. 29
45. 28
46. −23
47. 25
48. 60
49. 0
50. −19
51. −20°
52. 5°
53. 25
54. −6
55. 60 centimeters
56. −6 feet
57. a. 6 b. −50 c. 575 d. −20 e. 340
58. a. 3 b. −1
59. a. −500 b. 300
60. 500

Review Exercises for Chapter 1

1. 12
2. 6
3. −24
4. −4
5. −7
6. −25
7. 0
8. Undefined
9. −1
10. −4
11. $x + y$
12. xy
13. $5x - 4$
14. $\dfrac{2ab}{a + b}$
15. $x^3 + x^2$
16. {1, 2, 5, 7, 10, 14, 35, 70}
17. {2, 10, 14, 70}
18. {1, 5, 7, 35}
19. {2, 5, 7}
20. {10, 14, 35, 70}
21. a. 14 b. −6 c. 24
22. a. −16 b. 0 c. −1
23. a. 6 b. 4 c. −1
24. a. 9 b. −7 c. −15
25. a. 5 b. −2 c. −3
26. a. 12 b. −4 c. 36
27. a. −56 b. 0 c. 20
28. a. 13 b. −29
29. a. 9 b. 1
30. a. −25 b. −10
31. a. 700 b. −24

32. a. -4 b. 2
33. True
34. False
35. True
36. True
37. True
38. False
39. False
40. True
41. True
42. True

43. 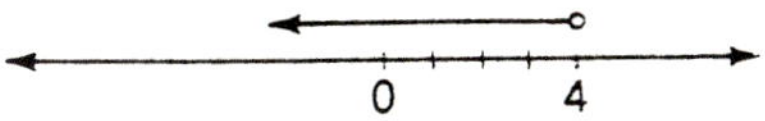

44. 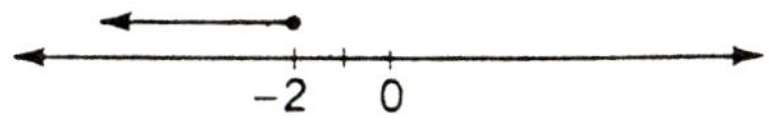

45. 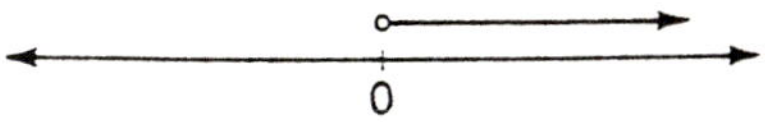

46. 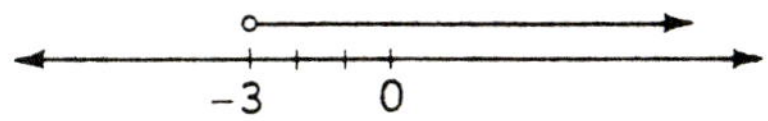

47. 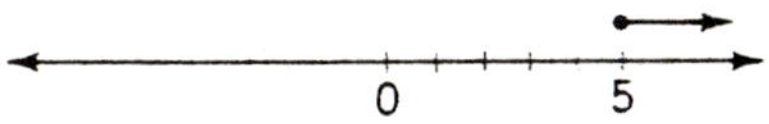

48. 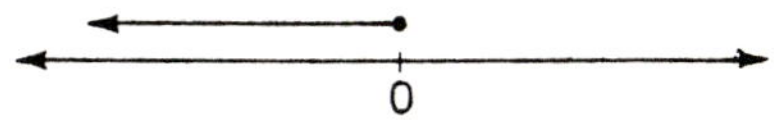

49. 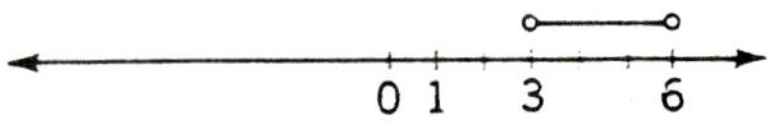

50. 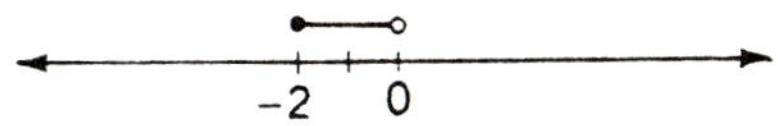

51. 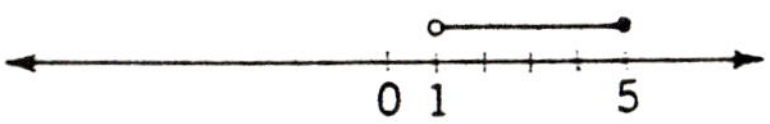

52. 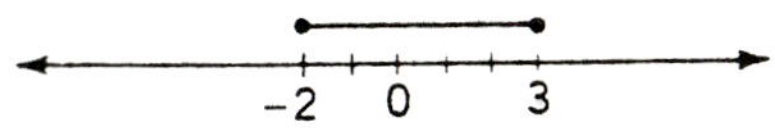

53. -1
54. -550
55. 80
56. 2

Answers to Exercises 2.1

1. $x + y + 4$
2. $x - 3$
3. $-x$
4. $x + y - 4$
5. $36x^2y^2$
6. -5
7. $2ab - 4$
8. $30a - 12b$
9. $m + n - 7$
10. $m - n - 1$
11. $-15x^2$
12. $27xy$
13. $x + y - 9$
14. $-30x^2y$
15. 3
16. $x - y + 1$
17. $24abc$
18. $5ab - 4c + d$
19. $-abcn^3$
20. $-35ab^2k^2$
21. $2x + 10$
22. $-3y - 6$
23. $5n - 35$
24. $-12x + 18$
25. $-x + 5$
26. $-3x - 4$
27. $y - 6$
28. $3x + 3y - 6$
29. $-x + 3y$
30. $ax + bx$
31. $6ay - 3by$
32. $a - b + c$
33. $x - y - 5$
34. $8x - 12y + 16$
35. $a - 2b + 2c - 6$
36. $2x + y - 4z + 10$
37. $3a - 6b + 4x + 8y$
38. $-c + d$
39. $4x^3 - 4x^2 - x + 1$
40. $ax + 5a - 2x - 10$
41. $5x$
42. $10x$
43. x
44. 0
45. $-2xy$
46. z
47. $2x + 5$
48. $8k$
49. $-7t + 6$
50. $5x^2 - 3x - 2$
51. $3z^2 + 5z + 4$
52. $p - q$
53. $6x^2 + 7xy - 10y^2$
54. $x^2 - y^2$
55. $3z^2 + 2z$
56. $-3k + 7$
57. $x^2 - x + 5$
58. $5x - 2$
59. $-x + 6$
60. $4m + 6n$
61. $5s + t$
62. $5p + 3q$
63. $3x^2 - xy - 10y^2$
64. $xy - 8xz$
65. $9x^2 - 16y^2$
66. $-3ab + ac$
67. $3x + 1$
68. $y + 3$
69. $3x^2 - 3x + 3$
70. $3y^2 - 4y + 2$
71. $8x + 8y$
72. $-8c - 8d$
73. $a - b$
74. $-7x$
75. $2xy - 12x + 15y$
76. $-x + 5$
77. $k - 4$
78. $ab - 16a + 6b$

79. $4x + 5$
80. $5k - 4$
81. $4y^2 - 8y + 9$
82. $4x^2 - x + 17$
83. $3x + 3y$
84. $-9a + 9b$
85. $4y$
86. $15xy - 12x + 3y$
87. $-24y$
88. $5x + 4$
89. $-x - 6$
90. $4ab + a - 15b$
91. $11k - 4$
92. $7x - 27$
93. $-3m + 12n$
94. $4n + 21$
95. $5x - 8$
96. $-3c - 12d$

Answers to Exercises 2.2

1. 2
2. 8
3. -2
4. 2
5. -7
6. -5
7. 7
8. 17
9. -7
10. 7
11. 0
12. 0
13. -26
14. 26
15. 3
16. 5
17. -2
18. 15
19. 2
20. 4
21. -2
22. -8
23. 3
24. 5
25. -2
26. -1
27. -2
28. 2
29. -7
30. $\varnothing$
31. 4
32. -5
33. $\varnothing$
34. 3
35. 1
36. 5
37. -8
38. 5
39. 8
40. -3
41. 0
42. -8
43. 20
44. 15
45. 5
46. -6
47. R (identity)
48. 27
49. 11
50. R
51. -8
52. 13
53. R (identity)
54. $\varnothing$
55. 1
56. 1
57. -3
58. -6
59. $\varnothing$
60. R (identity)

Answers to Exercises 2.3

1. -5
2. -4
3. 1
4. -7
5. -3
6. -1
7. 10
8. -20
9. 4
10. 0
11. $\varnothing$
12. $\varnothing$
13. -1
14. -30
15. 1
16. 9
17. 9
18. R
19. R
20. 3
21. 4
22. -5
23. -5
24. 10
25. -8
26. 3
27. -2
28. R
29. 2
30. 3
31. 2
32. 5
33. 5
34. 7
35. R
36. 4
37. 2
38. 5
39. 24
40. 3
41. $\varnothing$
42. -2
43. 7
44. $\varnothing$
45. -2
46. 4
47. 2
48. -1
49. R
50. 4
51. 3
52. 1

Answers to Exercises 2.4

1. $x > 2$

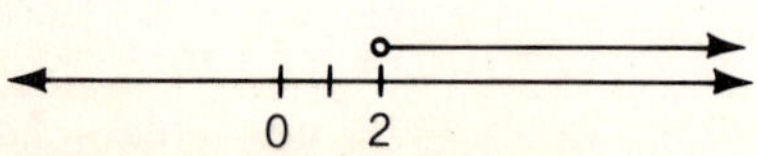

2. $x > -2$

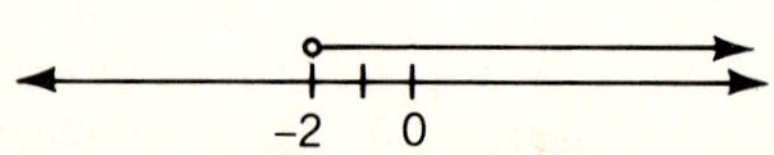

3. $x > 3$

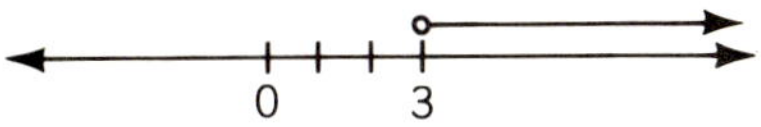

4. $x < 1$

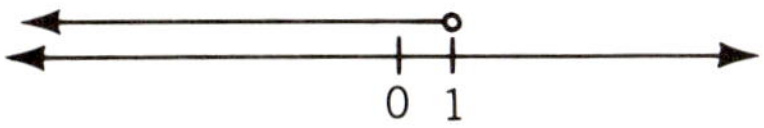

5. $x < -6$

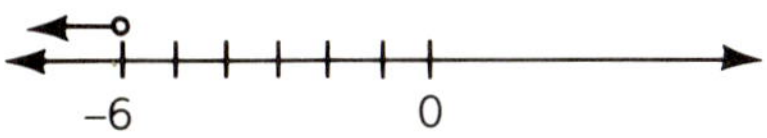

6. $x < -4$

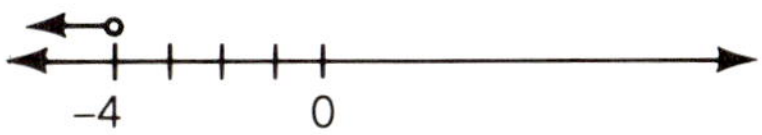

7. $x \geq 2$

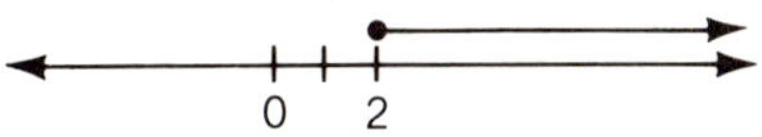

8. $x \leq 4$

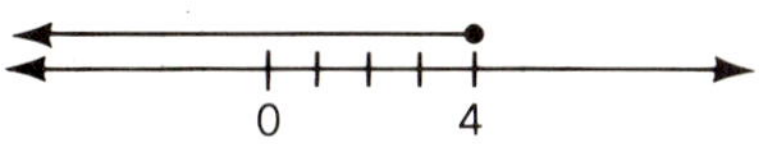

9. $x \leq 5$

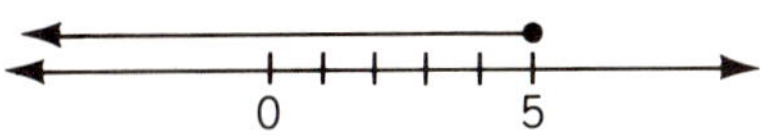

10. $x \geq 5$

11. $x > -1$

12. $x \geq 2$

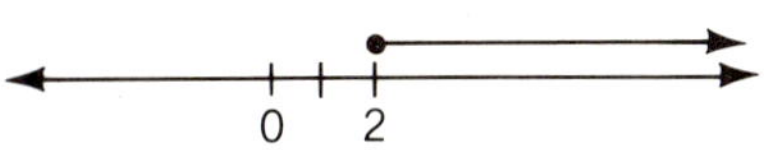

13. $x < -3$

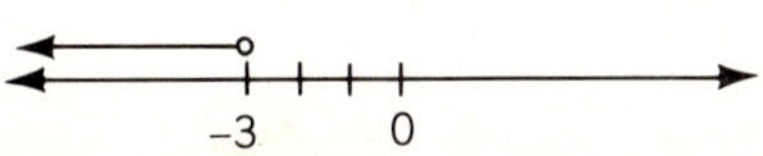

14. $x \geq -3$

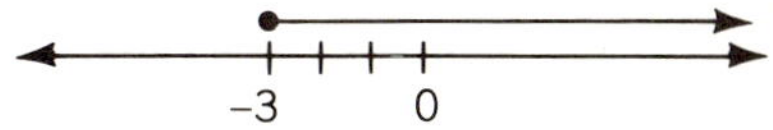

15. R, the set of all real numbers

16. $\varnothing$

17. $x \geq 0$

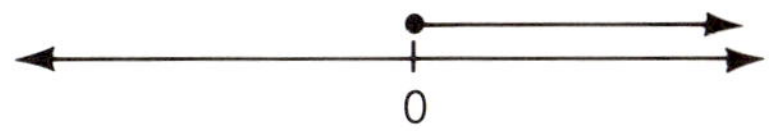

18. $x \leq 2$

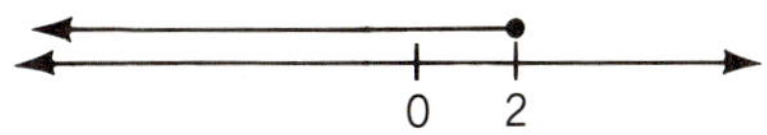

19. $x \leq -1$

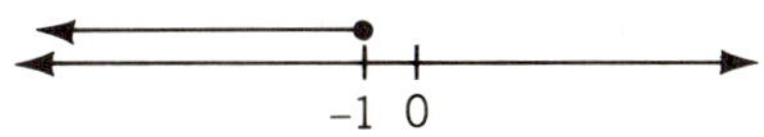

20. $x < -1$

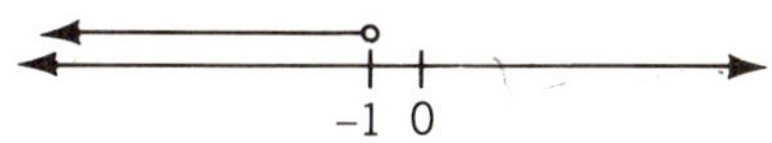

21. $x \geq 3$

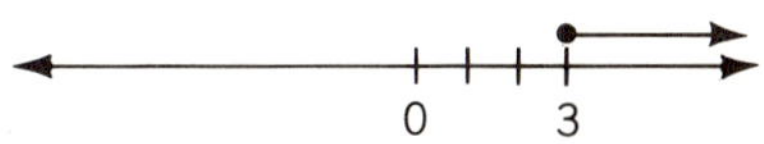

22. $x \geq 12$

23. $x \leq 12$

24. $x < -3$

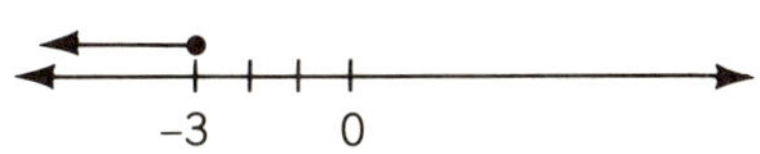

25. $x \geq -3$

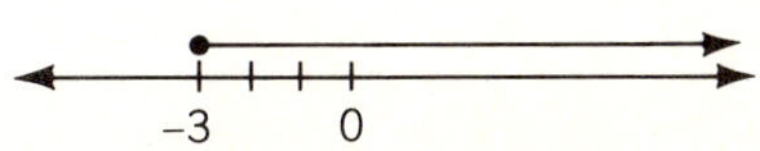

26. $x \leq \frac{1}{6}$

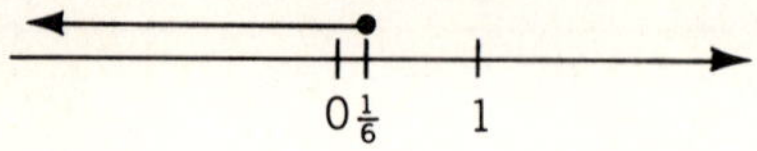

27. $x > 8$

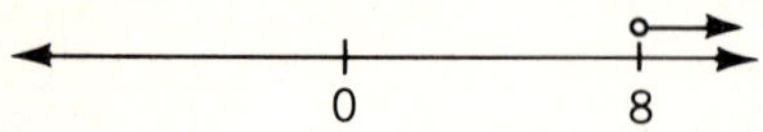

28. $x < -34$

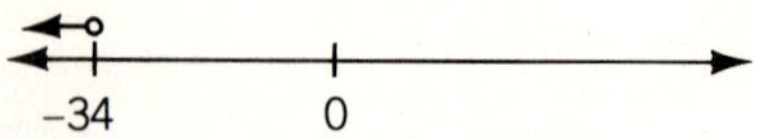

29. $x > 46$

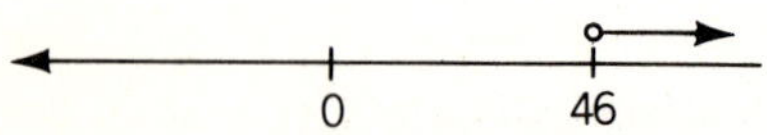

30. $x \geq 1$

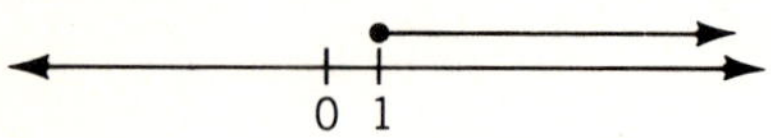

31. $x < 0$

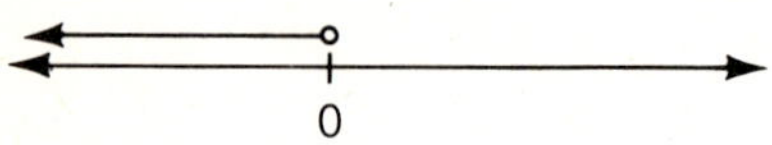

32. $x > 0$

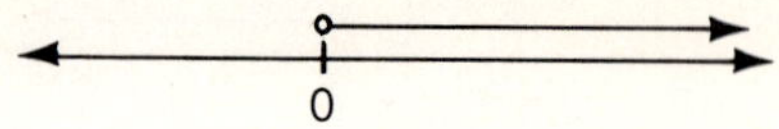

33. $\varnothing$

34. True for all real numbers.

35. $x < 2$

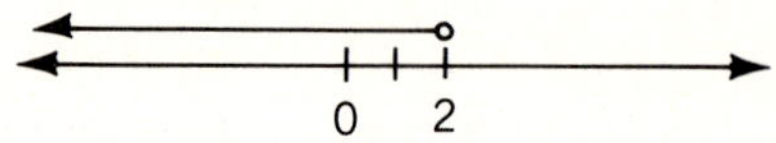

36. True for all real numbers

37. $x < 2$

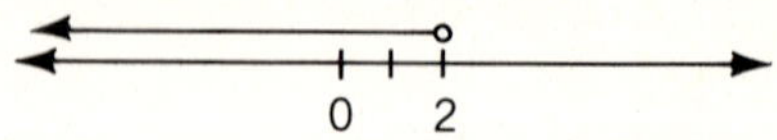

38. $x \geq 13$

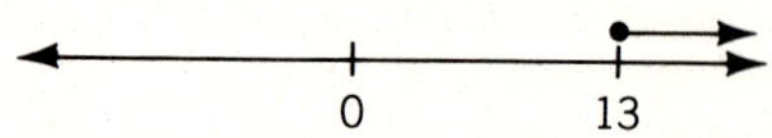

39. $\varnothing$

40. $x \geq 13$

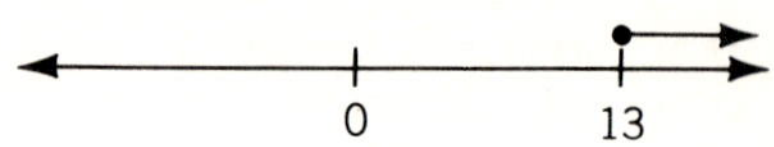

Answers to Exercises 2.5

1. $L = \frac{A}{W}$

2. $W = \frac{V}{LH}$

3. $h = \frac{V}{\pi r^2}$

4. $L = \frac{P - 2W}{2}$

5. $t = \frac{D}{r}$

6. $x = \frac{y - b}{m}$

7. $h = \frac{2A}{b}$

8. $a = \frac{2A}{h} - b$

9. $r = \frac{A - P}{Pt}$

10. $d = \frac{W}{F}$

11. $r = \frac{C}{2\pi}$

12. $b = \frac{3V}{h}$

13. $s = \frac{A - a^2}{2a}$

14. $r = \frac{P}{I^2}$

15. $C = \frac{5}{9}(F - 32)$

16. $h = \frac{2A}{a + b}$

17. $B = 180 - A - C$

18. $b = P - a - c$

19. $n = \frac{D}{A} + 1$

20. $A = \frac{D}{n - 1}$

21. $C = \frac{100M}{Q}$

22. $O = I - EI$

23. $n = \frac{C - S}{R}$

24. $S = C - Rn$

25. $d = \frac{SD}{s}$

26. $D = \frac{ds}{S}$

27. $t = \frac{pvT}{PV}$

28. $P = \frac{pvT}{tV}$

29. $p = \frac{PVt}{vT}$

30. $V = \frac{pvT}{Pt}$

31. $W = \frac{CL}{100}$

32. $L = \frac{100W}{C}$

33. $a = \frac{r}{c} - x$

34. $x = \frac{r}{c} - a$

35. $A = \frac{C(y + 12)}{y}$
36. $W = \frac{150C}{A}$
37. $f = \frac{s - ps}{p}$
38. $n = \frac{r - d}{dr}$
39. $G = E - \frac{rE}{R}$
40. $R = \frac{rE}{E - G}$
41. $y = 8 - x$
42. $y = 5 + 2x$
43. $y = 3x - 6$
44. $y = 5x - 20$
45. $x = 9 + 2y$
46. $x = 4 - 7y$
47. $x = 2y + 4$
48. $x = 8 - 4y$
49. $x = \frac{7y + 8}{2}$
50. $y = \frac{20 - 5x}{4}$
51. $y = 10 + 5x$
52. $y = 12 - 4x$
53. $x = 12 - 4y$
54. $x = 18 + 6y$
55. $x = -y - 1$
56. $x = 2y - 6$
57. $y = \frac{6 - 3x}{2}$
58. $x = \frac{5y + 10}{2}$
59. $y = \frac{-ax - c}{b}$
60. $x = \frac{-by - c}{a}$

Answers to Exercises 2.6

1. $2(x + 5) = x + 13$; 3
2. $2x + 3 = 3x - 7$; 10
3. $4x - 24 = 5x - 18$; -6
4. $3x + 19 = 4(x + 6)$; -5
5. $5(x - 4) = x + 16$; 9
6. $5x = 4(28 - x) - 4$; 12, 16
7. $2(L + 4) = 3L - 14$; 26 to 22
8. $3N + (3N + 6) + N = 62$;
   8 nickels, 24 dimes, 30 pennies, \$3.10
9. $(2x + 3) + x + 7 = 100$;
   37 singles, 63 doubles
10. $x + 2x - 9 = 165$;
    58 dune buggies, 107 bicycles
11. $x + 2x - 10 = 80$;
    30 meters and 50 meters
12. $x + 2(x - 2000) = 20{,}000$;
    daughter, \$8000; son, \$12,000
13. $x + (x + 1) = 75$; 37, 38
14. $3x = 2(x + 1) + 10$; 12, 13
15. $x + (x + 2) = 90$; 44, 46
16. $x + (x + 2) = 48$; 23, 25
17. $4(x + 2) = 5x - 9$; 17, 19
18. $2(x + 2) = 3x - 4$; 8, 10
19. $x + (x + 1) + (x + 2) = 36$; 11, 12, 13
20. a. $x - 1$ and $x + 1$
    b. $(x - 1) + x + (x + 1) = 159$;
    52, 53, 54
21. $(x - 1) + (x + 1) = x + 10$; 9, 10, 11
22. $x + (x + 1) + (x + 2) = 5x - (x + 2)$;
    5, 6, 7
23. $x + (x + 2) + (x + 4) = 114$; 36, 38, 40
24. $x + (x + 2) + (x + 4) = 129$; 41, 43, 45
25. $2(x + 4) + 3x = 73$; 13, 15, 17
26. $3(x + 4) + 2x = 132$; 24, 26, 28
27. $3x = x + 12$; 6 years
28. $3x + 7 = 2(x + 7)$; 7, 21
29. $(x - 4) + (x - 3 - 4) = 45$; 25, 28
30. $(x + 2 - 7) = 2(x - 7)$; 9, 11
31. $x + 5x + (5x + 2) = 68$; 6, 30, 32
32. $x + (x - 3) + 6x = 53$; 7, 4, 42

Answers to Exercises 2.7

1. $2W + 2(3W) = 80$; 10 meters, 30 meters
2. $2W + 2(W + 3) = 30$;
   6 inches, 9 inches
3. $2W + 2(2W - 5) = 62$; 12 feet, 19 feet
4. $2(W + 8) + 2(3W - 20) = 296$;
   40 centimeters, 120 centimeters
5. $4S = 32$; $S = 8$ meters, $A = 8^2 = 64$ square meters
6. $4S = 40$; $S = 10$ inches, $A = 100$ square inches
7. a. $2W - 8 = W + 8$; 16 yards, 32 yards
   b. 96 yards c. 512 square yards
   d. 96 yards e. 576 square yards
8. a. $4W - 50 = 2W$; 25 meters, 100 meters
   b. 250 meters c. 2500 square meters
   d. 200 meters e. 2500 square meters
9. $x + (x + 5) + 2x - 5 = 60$; 15 inches, 20 inches, 25 inches
10. $x + (x + 5) + (3x - 7) = 48$;
    10 centimeters, 15 centimeters,
    23 centimeters
11. $x + 2x + 3x = 180$; 30°, 60°, 90°

12. $x + x + 2x = 180$; 45°, 45°, 90°
13. $10x + 10(x + 1) + 10(x + 2) = 180$; 50°, 60°, 70°
14. $10x + 10(x + 2) + 10(x + 4) = 180$; 40°, 60°, 80°
15. $40x + 40(x + 20) = 5600$; 60 inches, 80 inches
16. $2.5(x + 1) + 3x = 24.5$; 5 meters and 4 meters
17. $2(12 + 2x) + 2(18 + 2x) = 84$; 3 inches
18. $20(18 - 2x) = 200$; 4 centimeters
19. $2x + 2(2x - 6) = 78$; 15 feet, 24 feet
20. $2x + 2(x + 1) = 18$; 4 meters, 5 meters
21. $2x = 6$; 3 feet, 2 feet
22. $4(x - 2) = 12$; 5 feet, 3 feet old; 9 feet, 3 feet new
23. $x + (x - 3) + x + x + (x - 3) = 34$; 8 inches
24. $\frac{12(x + 2)}{2} = \frac{10x}{2} + 18$; 6 inches, 8 inches

Answers to Exercises 2.8

1. $40x + 35(300) = 37(x + 300)$; 200 liters
2. $2.8(45) + 3.6x = 3.0(x + 45)$; 15 pounds
3. $50x + 90(100 - x) = 60(100)$; 75 pounds at 50¢, 25 pounds at 90¢
4. $72x + 12(120 - x) = 36(120)$; 48 pounds of cement, 72 pounds of sand
5. $100x + 10(50x) + 25(10x + 60) = 10{,}000$; 10 dollars, 500 dimes, 160 quarters
6. $25(D + 4) + 5(2D + 2) + 10D = 740$; 14 dimes, 18 quarters, 30 nickels
7. $25x + 20(36 - x) = 795$; 15 at 25¢, 21 at 20¢
8. $5x + 3(420 - x) = 1480$; 110 at \$5, 310 at \$3
9. $0.60x + 0.40(3x) = 90$; 50 kilograms of 60% ore, 150 kilograms of 40% ore
10. $0.12x + 0.10(2x) = 4800$; \$15,000 at 12%, \$30,000 at 10%
11. $6 + 2x = 1.5(x + 6)$; 6 grams
12. $0.30(10) + 0.80x = 0.6(x + 10)$; 15 liters
13. $0.12x + 0.16x + 0.18(100{,}000 - 2x) = 15{,}000$; \$37,500 at 12% and 16%, \$25,000 at 18%
14. $0.10(20{,}000) + 0.15x + 0.08(40{,}000 - x) = 7300$; \$30,000
15. $0.75(100 - x) = 0.30(100)$; 60 liters
16. $0.01(60) = 0.03(60 - x)$; 40 kilograms
17. $0.40(100 + x) = 0.50(100)$; 25 pints
18. $x + 0.75(20 - x) = 0.90(20)$; 12 ounces
19. $3(2x) + 2x + (2x - 10) = 290$; 60 pounds of beef, 30 pounds of ham, 50 pounds of chicken
20. $20x + 40(x + 2) + 30(2)(x + 2) = 680$; 4 pepper, 6 melon, 12 tomato

Answers to Exercises 2.9

1. $55x + 50x = 210$; 2 hours
2. $65x + 55(x - 3) = 315$; $x = 4$ hours, 12 noon
3. $3x + 3(x + 15) = 465$; 70 mph
4. $4x + 4(x + 22) = 712$; 78 kph, 100 kph
5. $4x = 5(x - 100)$; 500 kph
6. $4x = 5(x - 10)$; 50 kph
7. $30x = 20(5 - x)$; $x = 2$ hours, $d = 60$ miles
8. $12x = 60(3 - x)$; $x = \frac{5}{2}$; $d = 30$ miles
9. $36x = 60(x - 2)$; 1 P.M.
10. $30x = 10(x + 1)$; $\frac{1}{2}$ hour
11. $60x = 1 + 40x$; $\frac{1}{20}$ hour = 3 minutes
12. $90x = 3 + 80x$; $\frac{3}{10}$ hour = 18 minutes
13. $3x + 3(x + 80) = 2190$; 325 mph, 405 mph
14. $6x + 6(x - 18) = 972$; 90 kph, 72 kph
15. $96x = 32(x + 6)$; $x = 3$ hours, 288 nautical miles
16. $5(x + 50) = 4x + 700$; 450 kph, 500 kph
17. $4x = 3(x + 15)$; 45 mph
18. $570x = 475(11 - x)$; 5 hours
19. $75x + 90(3 - x) = 240$; $x = 2$ hours, 150 kilometers
20. $4x + 15 = 3(x + 20)$; $x = 45$, 180 miles

Review Exercises for Chapter 2

1. 5
2. 5
3. -7
4. $\varnothing$
5. -2
6. 2
7. 4
8. 4
9. -5
10. 0
11. -12
12. 1
13. -5
14. $\frac{3}{2}$
15. 8
16. 6
17. -7
18. 4
19. 1
20. $\varnothing$
21. R
22. 4
23. $\frac{1}{2}$
24. -2
25. -5
26. 5
27. -3
28. 7
29. $\varnothing$
30. 3
31. 0
32. -9
33. 2
34. -3
35. R
36. -6
37. 3
38. -5
39. 2
40. 4

41. $y = 3x - 5$
42. $x = \dfrac{5 + y}{3}$
43. $y = -2x - 5$
44. $x = \dfrac{-y - 5}{2}$
45. $x = \dfrac{10 + 2y}{5}$
46. $y = \dfrac{5x - 10}{2}$
47. $x = \dfrac{y - 2}{3}$
48. $y = \dfrac{x - 2}{3}$
49. $y = \dfrac{c - ax}{b}$
50. $x = \dfrac{c - by}{a}$
51. $d = \dfrac{C}{\pi}$
52. $t = \dfrac{A - P}{Pr}$
53. $y = P - x - z$
54. $b = \dfrac{S - 2a}{2}$
55. $t = \dfrac{RT - g}{R}$
56. $d = \dfrac{M - n^2}{2n}$
57. $A = 30 - 2B$
58. $H = \dfrac{W + 190}{5}$
59. $B = \dfrac{20 - A + 3C}{2}$
60. $C = \dfrac{A + 2B - 20}{3}$

61. $6 + 2x = 5(x - 3)$; 7
62. $3(2x - 15) - 3x = 135$; bus, 60 kph; train, 105 kph
63. $x + 0.4(60) = 0.5(x + 60)$; 12 grams
64. $2x + 2(2x - 2) = 56$; $x = 10$; $\dfrac{3(10)(18)}{360} = 1\frac{1}{2}$ gallons
65. $40x + 45(2x) + 35(2x - 1) = 765$; 4 pounds of peaches, 8 pounds of plums, 7 pounds of bananas
66. $\frac{1}{2}x(x + 11) - 2310 = \frac{1}{2}x(x - 10)$; $x = 220$ feet
67. $x + (x + 3) + \frac{x}{2} + 17 = 70$; 20 years
68. $1.5(2x + 5 - 40) = 28.5$; 27 kilograms, 32 kilograms
69. $65x + 50(x + 1) = 280$; $x = 2$ hours, 2 hours after 11 A.M. = 1 P.M.
70. $x > -4$

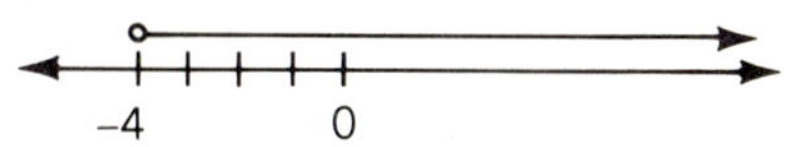

71. $x \geq 5$

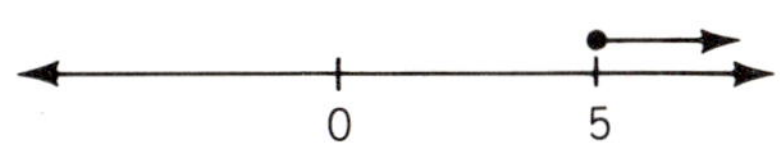

72. $x < 3$

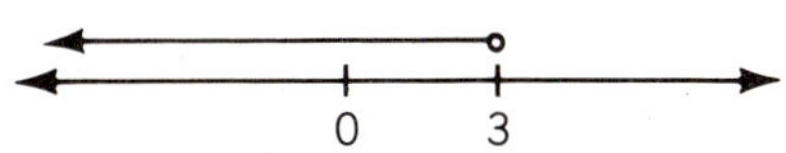

73. True for all real numbers
74. $\varnothing$
75. $x \leq 0$

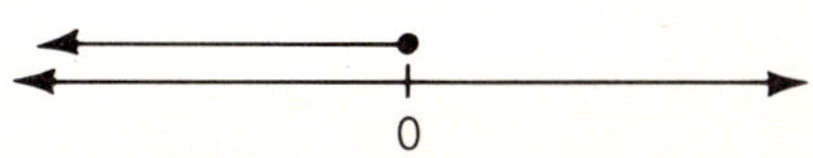

Answers to Exercises 3.1

1. $2x^3$; 3
2. $5y^2$; 2
3. $-2t^4$; 4
4. $-8x^1$; 1
5. y^1; 1
6. $4t^5$; 5
7. xy^2; 3
8. x^3y; 4
9. $-3x^2y^3$; 5
10. $-7x^3y^3$; 6
11. $6x^2$
12. $28y^2$
13. $5y^2$
14. $6x^2$
15. $-24x^4$
16. $-12y^4$
17. $-15x^2y^2$
18. $-10x^2y^2$
19. x^4y^4
20. $6x^4y^4$
21. x^5
22. y^6
23. $12y^4$
24. $8x^6$
25. $5x^4$
26. $4y^5$
27. $35z^6$
28. $6z^4$
29. $-6x^6$
30. $-5x^8$
31. $2y^8$
32. $3y^7$
33. $-24t^4$
34. $-36t^6$
35. $15x^2$
36. $8y^2$
37. $16x^4$
38. $9y^4$
39. $25t^6$
40. $36t^6$
41. $-x^6$
42. $-z^5$
43. $24y^4$
44. $50y^6$
45. $-18x^2$
46. $-75y^2$
47. $20t^5$
48. $-100t^7$
49. $-8x^6$
50. $100x^6$
51. $-x^3y^2$
52. $-2x^3y^3$
53. $3x^2y^3$
54. $20x^3y^2$
55. $-r^2s^2$
56. $-u^2v^2$
57. $-x^3y^3$
58. r^3t^3
59. $u^2v^3w^2$
60. $-r^2s^3t^2$

Answers to Exercises 3.2

1. $5x^2 + 3x - 2$
2. $x^2 - x - 7$
3. $7x - 4$
4. $8y - 36$
5. $y^3 + y^2 - 5y + 3$
6. $y^3 - 3y^2 + 3y - 1$
7. $8x^2 - 2y^2$
8. $9x^2$
9. $x + y$
10. $4x - 3y - 3z$
11. $7x^2 + 5x + 1$
12. $y^3 + 2y^2 - y - 3$
13. $6y^2 + 3y - 5$
14. $8x^2 - 4x - 4$
15. $5x - 10$
16. $3x - y$
17. $t^3 - 6t^2 + 12t - 8$
18. $t^3 - 9t^2 + 27t - 27$
19. $2y - 4z$
20. $-3a + b$
21. $-6y - 8$
22. $a - 2b - 2c$
23. $2s - 2t$
24. $x + 5y - 12$
25. $10x + 5$
26. $16x^2 - 16xy$
27. $-x - 1$
28. $y^2 + 3y - 14$
29. $2y^2$
30. $5x^4 + 2x^2 - 1$
31. $4x^2 + 1$
32. $-a^2 + 4b^2 + 5$
33. $2t - 9$
34. $-5c^3 - 6c^2 + 3c - 1$
35. $x^3 - 2x^2 - 2x - 1$
36. $2x^2 + x + 18$
37. $-a^2 - 3ab + 4b^2$
38. $3x^2 - 2x + 9$
39. $a + 5b - c$
40. $-6y^3 + 2y^2 + 3y - 13$
41. $-5x^2 - 2x + 10$
42. $2y^2 + 4xy - x^2$
43. $6x^4 + x^3 - x^2 + 21x - 6$
44. $6ab + b^2$
45. $-3x^4 + 11x^3 - 11x^2 + x - 1$
46. $x^3 - 7x^2y + 13xy^2 + 17y^3$
47. $-2a^3 + 3b^3 + a^2b - 4ab^2$
48. $-x^4 - 4x^2 - 2x + 1$
49. $3x^3 - 10x^2y - 2xy^2 + y^3 + 4$
50. $3a^3 + 8a^2b + 18ab^2 - 10b^3$

Answers to Exercises 3.3

1. $x^2 - 4x$
2. $x^2 + 3x$
3. $6x^2 + 10x$
4. $10x^2 - 20x$
5. $-12y^2 + 3y$
6. $-8y^2 - 12y$
7. $-x^4 - 4x^2$
8. $-x^5 + 16x^3$
9. $8x^4 - 4x^2y^2$
10. $18x^4 - 30x^3y$
11. $10x - 15y + 30$
12. $12x + 8y - 20$
13. $-x^3 - 2x^2 - x$
14. $-y^3 + 4y^2 - 4y$
15. $-12y^3 + 6y^2 + 18y$
16. $-18x^3 + 2x^2 + 2x$
17. $5x^3y + 10x^2y^2 - 5xy^3$
18. $3x^3y - 9x^2y^2 + 18xy$
19. $-2x^4 + 6x^3 + 8x^2$
20. $-5x^4 + 10x^3 + 40x^2$
21. $x^3 - 3x^2 + 3x - 1$
22. $x^3 + 6x^2 + 12x + 8$
23. $y^3 - 125$
24. $y^3 + 216$
25. $16x^3 + 36x^2y + 12xy^2 + y^3$
26. $27x^3 - 54x^2y + 36xy^2 - 8y^3$
27. $343r^3 + 8s^3$
28. $1000r^3 - 729t^3$
29. $x^4 - x^2 + 2x - 1$
30. $x^4 - 5x^2 + 4$
31. $x^2 + 2xy + y^2 - 25$
32. $x^2 - 2xy + y^2 - 16$
33. $a^2 + b^2 + c^2 + 2ab + 2ac + 2bc$
34. $a^2 + b^2 + c^2 - 2ab - 2ac + 2bc$
35. $x^2 + 8x + 15$
36. $y^2 + 9y + 14$
37. $x^2 - 8x + 15$
38. $y^2 - 9y + 14$
39. $x^2 - 2x - 15$
40. $y^2 - 5y - 14$
41. $x^2 + 2x - 15$
42. $y^2 + 5y - 14$
43. $2x^2 - 5xy - 3y^2$
44. $3x^2 + 10xy - 8y^2$
45. $9x^2 - 18xy + 8y^2$
46. $25x^2 - 20xy + 3y^2$
47. $x^4 + 5x^2 - 36$
48. $x^4 - 24x^2 - 25$
49. $x^4 - 37x^2 + 36$
50. $x^4 - 116x^2 + 1600$
51. $y^4 - 625$
52. $y^4 - 81$
53. $9x^2 - 4y^2$
54. $16x^2 - 25y^2$
55. $6x^2 - 5xy - 6y^2$
56. $20x^2 + 9xy - 20y^2$
57. $xy + 6x + 4y + 24$
58. $xy + 8x + 3y + 24$
59. $2xy - x - 10y + 5$
60. $3xy - 2x - 21y + 14$
61. $x^3 + x^2 - x - 1$
62. $x^3 + 4x^2 - 9x - 36$
63. $y^3 - 6y^2 + 4y - 24$
64. $y^3 - 8y^2 + y - 8$
65. $16rs - 4r - 4s + 1$
66. $64uv + 24u - 24v - 9$
67. $x^5 + 2x^3 - 2x^2 - 4$
68. $y^5 - 8y^3 + 8y^2 - 64$
69. $64x^2 + 48xy + 9y^2$
70. $49x^2 - 28xy + 4y^2$
71. $2x^3 - 3x^2 - 3x + 2$
72. $x^3 + 4x^2 + x - 6$
73. $-3x^3 + 10x^2 - 6x + 1$
74. $16x^3 + 4x^2 - 6x + 1$
75. $x^4 + x^3 - 6x^2 - 4x + 8$
76. $-x^4 - x^3 - 9x + 81$
77. $x^4 - 3x^3 + 3x - 1$
78. $x^4 - 2x^3 + 3x^2 + 2x - 4$
79. $x^4 - 1$
80. $-x^4 + 1$
81. $x^2 - 2x + 1$
82. $x^2 - 6x + 9$
83. $25x^2 - 40x + 16$
84. $4x^2 - 4x + 1$
85. $x^4 + 2x^3 + 3x^2 + 2x + 1$
86. $x^4 + 2x^3 - x^2 - 2x + 1$
87. $-x^3 + 1$
88. $x^3 + 1$
89. $x^3 + 3x^2 + 3x + 1$
90. $-x^3 + 3x^2 - 3x + 1$

Answers to Exercises 3.4

1. $4(x + 3)$
2. $5(x - 4)$
3. $3(2x - 1)$
4. $7(2x + 1)$
5. $3(x^2 + 2x + 3)$
6. $5(x^2 + 3xy + y^2)$
7. $4(2ab - 3a + 1)$
8. $3(2x + y - 1)$
9. $4y^2(y - 25)$
10. $2x(2x + 1)$
11. $a(y - 1)$
12. $by(y + 1)$
13. $x^3(x - 1)$
14. $x^3(1 - x)$ or $-x^3(x - 1)$
15. $x^4(1 - x)$ or $-x^4(x - 1)$
16. $x^4(x - 1)$
17. $x(24x^2 - 30x + 1)$
18. $n(n^2 - 3n + 1)$
19. $cn^2(c + n)$
20. $xy^2(x - 3)$
21. $a(x + y - z)$
22. $x(x^2 - x + 1)$
23. $-5x(x + 5)$
24. $-9x(4x + 1)$
25. $6(4x^2 + 2x + 1)$
26. $4(5y^2 - 2y + 3)$
27. $c(c^2 - c - 1)$
28. $x(x + y - 1)$
29. $2y(3y^2 - y + 5)$
30. $3t(t^2 + t - 4)$
31. $-5x(5x^2 + 3)$
32. $-4x^2(2x + 5)$
33. $-xy(y^2 + y + 1)$
34. $-x^2(x + y + xy)$
35. $-x^2(x^2 - 2x + 6)$
36. $-5x^2(x^2 + 2x - 5)$
37. $4xy(x - y - 1)$
38. $9xy(x - 2y - 2)$
39. $-6ab(a^2 + b^2 - ab)$
40. $-8an(n^2 + 5n - 4)$
41. $(x + 3)(x + 6)$
42. $(x + 5)(x - 4)$
43. $(x - 7)(2x - 3)$
44. $(x + 2)(3x - 4)$
45. $(x + 1)(5x - 1)$
46. $(x - 5)(4x - 1)$
47. $(x - 4)(x^2 + 4)$
48. $(x + 8)(x^2 - 2)$
49. $(t + 1)(t^2 - 5)$
50. $(t - 1)(t^2 + 6)$

Answers to Exercises 3.5

1. $x^2 - 16$
2. $x^2 - 49$
3. $y^2 - 25$
4. $y^2 - 36$
5. $x^2 - 9y^2$
6. $x^2 - 4y^2$
7. $64x^2 - 1$
8. $81x^2 - 1$
9. $100x^2 - 81y^2$
10. $64x^2 - 25y^2$
11. $x^4 - 1$
12. $x^4 - 25$
13. $81x^4 - 1$
14. $16x^4 - 1$
15. $x^4 - y^4$
16. $x^4 - 9y^4$
17. $625x^4 - 16$
18. $100x^4 - 81$
19. $49a^4 - 36b^4$
20. $400a^4 - 49b^4$
21. $(x - 9)(x + 9)$
22. $(x - 3)(x + 3)$
23. $(y - 2)(y + 2)$
24. $(y - 4)(y + 4)$
25. $(5x - 1)(5x + 1)$
26. $(8x - 1)(8x + 1)$
27. $(7y - 1)(7y + 1)$
28. $(6y - 1)(6y + 1)$
29. Prime
30. Prime
31. Prime
32. Prime
33. $(6x - 5)(6x + 5)$
34. $(2x - 7)(2x + 7)$
35. $(8x - 7y)(8x + 7y)$
36. $(9x - 8y)(9x + 8y)$
37. $(x^2 - 3)(x^2 + 3)$
38. $(x^2 - 5)(x^2 + 5)$
39. $(5y^2 - 4)(5y^2 + 4)$
40. $(4y^2 - 11)(4y^2 + 11)$
41. $(7x^2 - 12y^2)(7x^2 + 12y^2)$
42. $(10x^2 - 3y^2)(10x^2 + 3y^2)$
43. Prime
44. Prime
45. Prime
46. Prime
47. $3x^2 - 108$
48. $5x^2 - 80$
49. $x^3 - 49x$
50. $x^3 - 25x$
51. $2x^3 - 162x$
52. $4x^3 - 36x$
53. $y^4 - 64y^2$
54. $y^4 - 100y^2$
55. $4x^4 - 100x^2$
56. $9x^4 - 9x^2$
57. $-3y^4 + 48y^2$
58. $-5y^4 + 320y^2$
59. $16x^4 - 16$
60. $25x^4 - 100$
61. $4(x - 3)(x + 3)$
62. $5(x - 4)(x + 4)$
63. $x(5x - 2)(5x + 2)$
64. $x(8x - 3)(8x + 3)$
65. $4(x^2 + 9)$
66. $4(x^2 + 4)$
67. $5x(3x - 1)(3x + 1)$
68. $4x(5x - 1)(5x + 1)$
69. $x^2(6x - 7y)(6x + 7y)$
70. $y^2(2x - 9y)(2x + 9y)$
71. $25(x^2 - 2)(x^2 + 2)$
72. $9(x^2 - 3)(x^2 + 3)$
73. $25x^2(x - 2)(x + 2)$
74. $9x^2(x - 3)(x + 3)$
75. $25x^2(x^2 + 4)$
76. $9x^2(x^2 + 9)$

Answers to Exercises 3.6

1. $x^2 + 9x + 18$
2. $x^2 + 9x + 14$
3. $x^2 - 9x + 18$
4. $x^2 - 9x + 14$
5. $x^2 + 3x - 18$
6. $x^2 - 5x - 14$
7. $x^2 - 3x - 18$
8. $x^2 + 5x - 14$
9. $x^2 + 4xy - 5y^2$
10. $x^2 - 5xy - 6y^2$
11. $x^2 - 11xy + 24y^2$
12. $x^2 - 9xy + 20y^2$
13. $x^4 - 29x^2 + 100$
14. $x^4 - 10x^2 + 9$
15. $x^4 - 21x^2 - 100$
16. $x^4 - 8x^2 - 9$
17. $y^4 + 7y^2 - 144$
18. $y^4 + 4y^2 - 32$
19. $x^4 - x^2y^2 - 2y^4$
20. $x^4 + 2x^2y^2 - 3y^4$
21. $x + 3$
22. $x + 6$
23. $x - 3$
24. $x - 6$
25. $x + 3$
26. $x + 6$
27. $x - 1$
28. $x - 1$
29. $x + y$
30. $x + y$
31. $(x + 1)(x + 5)$
32. $(x + 1)(x + 7)$
33. $(x - 1)(x - 5)$
34. $(x - 1)(x - 7)$
35. $(x + 1)(x - 5)$
36. $(x + 1)(x - 7)$
37. $(x - 1)(x + 5)$
38. $(x - 1)(x + 7)$
39. Prime
40. Prime
41. $(y + 6)(y - 1)$
42. $(y + 3)(y + 5)$
43. $(y - 6)(y - 1)$
44. $(y - 1)(y - 15)$
45. $(y - 3)(y + 2)$
46. $(y - 3)(y + 5)$
47. $(y - 2)(y - 3)$
48. $(y - 15)(y + 1)$
49. Prime
50. Prime
51. $(x - 1)(x - 2)$
52. $(x + 3)(x - 1)$
53. $(x - 3y)(x + y)$
54. $(x - y)(x - 2y)$
55. $(x - 2y)(x + 5y)$
56. $(x + 5y)(x - 7y)$
57. $(x - 3y)(x - 7y)$
58. $(x + 4y)(x - 2y)$
59. $(x + 14y)(x - y)$
60. $(x + 9y)(x - 3y)$
61. Prime
62. $(x + y)^2$
63. $(x - 2y)^2$
64. Prime
65. $(x - 3)^2$
66. Prime
67. Prime
68. $(x - 5)^2$
69. $-(x + 1)(x - 2)$
70. $-(x - 1)(x + 4)$
71. $(x^2 + 1)(x^2 + 2)$
72. $(x^2 + 1)(x^2 + 7)$
73. $(x^2 - 2)(x^2 + 7)$
74. $(x^2 - 21)(x^2 + 1)$
75. $(x^2 - 8)(x^2 + 1)$
76. $(x^2 + 6)(x^2 - 2)$
77. $(x^2 - 6)^2$
78. $(x^2 - 8)^2$
79. $(y^2 - 3)(y^2 + 4)$
80. $(y^2 + 5)(y^2 - 6)$
81. $3(x - 1)(x - 2)$
82. $10(x + 5)(x - 7)$
83. $150x^2(x - 2)(x - 5)$
84. $4ay(a - 3)(a + 4)$
85. $20(y^2 + 7y - 6)$
86. $-5b(b - 1)(b - 6)$
87. $4(x + 3)(x + 5)$
88. $5(x - 1)(x + 7)$
89. $100x(x + 2)(x - 4)$
90. $6y^2(y + 2)(y - 6)$
91. $15(y^2 + 2y + 3)$
92. $-7(a + 3)(a - 4)$
93. $-cd(d - 4)(d + 5)$
94. $30(u^2 + 5u - 4)$
95. $2x(x - 6y)(x + 10y)$
96. $-uv(u^2 + v^2 - 5)$
97. $-nk^2(k - 4)(k + 7)$
98. $xy(x + 5y)(x - 7y)$
99. $u^2(u - v)(u + 3v)$
100. $18(d^2 + dx + 2x^2)$

Answers to Exercises 3.7

1. $x^2 + 6x + 9$
2. $x^2 + 8x + 16$
3. $x^2 - 4x + 4$
4. $x^2 - 10x + 25$
5. $y^2 + 12y + 36$
6. $y^2 - 14y + 49$
7. $y^2 - 18y + 81$
8. $y^2 + 16y + 64$
9. $16x^2 - 8x + 1$
10. $9x^2 + 6x + 1$
11. $25x^2 + 10x + 1$
12. $4x^2 - 4x + 1$
13. $4x^2 + 28x + 49$
14. $36x^2 + 60x + 25$
15. $9x^2 - 48x + 64$
16. $81x^2 - 72x + 16$
17. $x^2 + 2xy + y^2$
18. $x^2 - 2xy + y^2$
19. $100x^2 - 140xy + 49y^2$
20. $16x^2 - 24xy + 9y^2$
21. $x^4 - 20x^2 + 100$
22. $x^4 + 6x^2 + 9$
23. $x^4 + 2x^2 + 1$
24. $x^4 - 8x^2 + 16$
25. $x^6 - 16x^3 + 64$
26. $x^6 - 2x^3 + 1$
27. $16x^4 - 72x^2y^2 + 81y^4$
28. $625x^4 - 200x^2y^2 + 16y^4$
29. $144x^4 - 24x^2y^2 + y^4$
30. $10{,}000x^4 - 200x^2y^2 + y^4$

31. $12x$
32. $20x$
33. $14t$
34. $10t$
35. $8y$
36. $40y$
37. $18y$
38. $22y$
39. $4x^2$
40. $2x^2$
41. $16y^2$
42. $6y^2$
43. $60xy$
44. $80xy$
45. $2xy$
46. $12xy$
47. $12xy$
48. $40xy$
49. $140t$
50. $144t$
51. No
52. No
53. $(y + 4)^2$
54. No
55. No
56. No
57. No
58. $(a + 9)^2$
59. $(x - 5)^2$
60. $(x - 3)^2$
61. $(x + 2)^2$
62. $(x + 9)^2$
63. $(x - 1)^2$
64. $(x - 5)^2$
65. $(3y + 1)^2$
66. $(4y + 1)^2$
67. $(6t - 1)^2$
68. $(7t - 1)^2$
69. $(4x - 5)^2$
70. $(3x - 8)^2$
71. $(y - 1)(y - 16)$
72. $(y - 1)(y - 9)$
73. Prime
74. $(x - 2)(x + 2)$
75. $(x - 7)(x + 7)$
76. Prime
77. $(2x + 3y)^2$
78. $(5x + 2y)^2$
79. $(t - 4)^2$
80. $(t - 6)^2$
81. $(8x - 5y)^2$
82. $(7x - 3y)^2$
83. Prime
84. $(u + 2)(u + 8)$
85. $(u + 1)(u + 9)$
86. Prime
87. $(x^2 - 10)^2$
88. $(x^2 + 30)^2$
89. $(x^2 + 6)^2$
90. $(x^2 - 8)^2$
91. $(x^2 + 4)(x^2 + 9)$
92. $(x^2 - 2)(x^2 - 32)$
93. $(11y^2 - 1)^2$
94. $(12y^2 - 1)^2$
95. Prime
96. $(t - 2)(t - 50)$
97. $(t - 3)(t - 27)$
98. $(t - 5)(t + 20)$
99. $(t + 81)(t - 1)$
100. Prime

Answers to Exercises 3.8

1. $2x + 11$
2. $2x - 5$
3. $3x - 2$
4. $3y - 5$
5. $3y - 5$
6. $4a - 7$
7. $5x - 1$
8. $5x + 1$
9. $5x + 1$
10. $5x - 1$
11. $3x - 4$
12. $4a + 1$
13. $y - 4$
14. $3x + 2$
15. $5p - 2$
16. $2x - 5$
17. $x - 6$
18. $x + 6$
19. $x + 6$
20. $x - 6$
21. $(2x - 5)(x - 3)$
22. $(5x - 3)(x - 5)$
23. $(3x + 2)(2x - 5)$
24. $(4x - 3)(2x + 3)$
25. $(6x - 1)(5x + 2)$
26. $(6x - 5)(3x + 2)$
27. $(8x - 5)(3x + 2)$
28. $(3x + 5)(2x - 3)$
29. $(7x + 4)(5x - 3)$
30. $(3x + 2)(4x - 3)$
31. $(2x + 1)(x + 1)$
32. $(3x + 1)(x + 1)$
33. $(3x - 2)(x - 1)$
34. $(2x - 1)(x - 3)$
35. $(5x - 2)(x + 1)$
36. $(7x - 1)(x + 1)$
37. $(7x + 1)(x - 2)$
38. $(5x + 1)(x - 3)$
39. Prime
40. Prime
41. $(3y + 7)(y - 1)$
42. $(5y - 1)(y + 2)$
43. $(3y + 1)(y - 7)$
44. $(5y + 2)(y - 1)$
45. Prime
46. $(5y + 1)(y + 2)$
47. $(3y - 1)(y - 7)$
48. Prime
49. $(4t - 1)(t - 5)$
50. $(2t - 1)(3t + 7)$
51. $(2t + 3)^2$
52. $(6t - 7)(t + 1)$
53. $(2t + 1)(2t - 5)$
54. Prime
55. Prime
56. $(3t - 4)^2$
57. $(2u + 1)(u - 1)$
58. $(5u - 4)(u + 1)$
59. $(3x + 8y)(3x - y)$
60. $(4x - 3y)(x - 2y)$
61. $(2x + 3)(5x - 2)$
62. $(3x + 2)(5x - 4)$
63. $(7x + 4)(3x - 2)$
64. $(2x - 1)(7x + 4)$
65. $(4x - 1)(3x - 5)$
66. $(2x - 1)(6x - 5)$
67. $(5x - 3)^2$
68. $(4x - 5)^2$
69. $(25x - 9y)(x - y)$
70. $(16x - 25y)(x - y)$
71. $(4x^2 + 1)(x^2 + 4)$
72. $(x^2 + 9)(4x^2 + 1)$
73. $(3x^2 - 5)(2x^2 + 1)$
74. $(2x^2 - 1)(x^2 + 1)$
75. $(3x^2 - 7)(x^2 + 1)$
76. $(2x^2 - 1)(3x^2 + 5)$
77. $(6t^2 - 1)^2$
78. $(8t^2 - 5)^2$
79. $(2t^2 - 9)(t^2 + 5)$
80. $(2t^2 - 7)(4t^2 + 3)$
81. $2(3x + 1)(x + 1)$
82. $3(2y - 1)^2$
83. $5(y - 1)(4y + 5)$
84. $8(x - 1)(2x + 3)$

85. $9(3t - 2)^2$
86. $8(2x - 1)(x - 7)$
87. $x^2(5x - 1)(2x - 1)$
88. $3x^2(2x^2 + 4x - 7)$
89. $-6x^2(x^2 + 2x - 5)$
90. $-2x^2(6x + 1)(5x - 1)$
91. $8(2x + y)(x - 3y)$
92. $10(5x - y)(x - 5y)$
93. $2xy(x + 5y)(x - 2y)$
94. $4xy(x - 4y)(x + 5y)$
95. $4a^2(a^2 - ab + 2b^2)$
96. $25a^2(a^2 - 4ab + b^2)$
97. 4, −4, 5, −5
98. 1, −1, 4, −4, 11, −11
99. 3, −3, 9, −9
100. 14, −14, 18, −18, 46, −46

Answers to Exercises 3.9

1. $3(x - 2)(x + 2)$
2. $5(x - 4)(x + 4)$
3. $9(x - 2)(x + 1)$
4. $2(2x - 1)(x - 1)$
5. $9(2y - 1)^2$
6. $4(3y + 1)^2$
7. $5(t - 1)(3t - 4)$
8. $6(t - 7)(t + 2)$
9. $-2x^2(2x - 3)(2x + 3)$
10. $-4x^2(x^2 + 25)$
11. $-9y^2(y^2 + 4)$
12. $-9y^2(2y - 1)(2y + 1)$
13. $x(x + 4)(x - 9)$
14. $5x(x - 3)(x + 7)$
15. $9x^2(2x - 9)(2x - 1)$
16. $4x^2(3x - 2)(3x - 8)$
17. $2y(3y + 5)(2y - 1)$
18. $y(3y - 4)(2y + 3)$
19. $4t(5t + 4)^2$
20. $3t(3t + 2)^2$
21. $(x + 2y)(x - 2y)(x^2 + 4y^2)$
22. $(x - 3y)(x + 3y)(x^2 + 9y^2)$
23. $(x - 2)^2(x + 2)^2$
24. $(x - 3)^2(x + 3)^2$
25. $(x - 4)(x + 4)(x^2 + 1)$
26. $(x - 1)(x + 1)(x^2 + 81)$
27. $(5x - 1)(5x + 1)(x^2 + 16)$
28. $(2x^2 - 9)^2$
29. $(5x^2 - 4)^2$
30. $(2x - 3)(2x + 3)(x^2 + 9)$
31. $7(x - 10)(x + 10)(x^2 + 100)$
32. $x^2(x^2 + 25)$
33. $x^2(x^2 + 1)$
34. $100(x - 5)(x + 5)(x^2 + 25)$
35. $2(x - 2)(x + 2)(x - 3)(x + 3)$
36. $x(x - 4)^2(x + 4)^2$
37. $49(x - 1)^2(x + 1)^2$
38. $x(x - 1)(x + 1)(x - 6)(x + 6)$
39. $36(t^2 + 2t - 1)$
40. $49(t^2 - t + 1)$
41. $2y^2(5y - 1)^2$
42. $4y^2(y - 6)^2$
43. $-3t(t + 3)(t - 3)$
44. $-2t(7t - 2)^2$
45. $-5y(3y - 8)^2$
46. $-y^2(y - 8)(y + 8)$
47. $-2x^2(2x + 1)^2$
48. $-3x^2(4x + 1)^2$
49. $6(3x^2 - 4)(x^2 + 5)$
50. $15(2x^2 - 7)(x^2 + 4)$
51. $(x^2 + x + 1)(x^2 - x + 1)$
52. $(x^2 + 2x + 2)(x^2 - 2x + 2)$
53. $(x + 5)(x - 2)(x + 2)$
54. $(x - 8)(x - 1)(x + 1)$
55. $(a - b + c)(a - b - c)$
56. $(x - y + z)(x - y - z)$
57. $(x - 2)(x - 3)(x + 2)(x + 3)$
58. $(x - 2)(x + 2)(x - 4)(x + 4)$
59. $(x - 1)(x + 5)(x^2 + 4x + 5)$
60. $(x + 1)^2(x^2 + 2x - 1)$

Answers to Exercises 3.10

1. $\{-2, 3\}$
2. $\{-4, 1\}$
3. $\{0, 5\}$
4. $\{-6, 6\}$
5. $\{-4, 4\}$
6. $\{0, -7\}$
7. $\left\{-8, \frac{1}{2}\right\}$
8. $\left\{3, -\frac{2}{3}\right\}$
9. $\{-6, -9\}$
10. $\{-5, -8\}$
11. $\{1, 2\}$
12. $\{3, 4\}$
13. $\{6, -4\}$
14. $\{0, 3\}$
15. $\{0, 2\}$
16. $\{-5, 8\}$
17. $\{-6, 6\}$
18. $\{-7, 7\}$
19. $\{-9, 9\}$
20. $\{-8, 8\}$
21. $\{-3, 5\}$

22. $\{-6, 2\}$
23. $\{7\}$
24. $\{10\}$
25. $\{4, -3\}$
26. $\{-2, 3\}$
27. $\{-6, 1\}$
28. $\{-2, 3\}$
29. $\{-4, 4\}$
30. $\{-5, 5\}$
31. $\{0, 3, -3\}$
32. $\{0, 1, -1\}$
33. $\{-2, 8\}$
34. $\{-1, 9\}$
35. $\{-4, 6\}$
36. $\{-6, 10\}$
37. $\left\{-2, \frac{1}{5}\right\}$
38. $\left\{-4, \frac{1}{4}\right\}$
39. $\{0, 6\}$
40. $\{0, -4\}$
41. $\{0\}$
42. $\{0\}$
43. $\{0, 2, 3\}$
44. $\{0, 4, -6\}$
45. $\{0, 5, -4\}$
46. $\{0, -3, -6\}$
47. $\{1, -1, 2, -2\}$
48. $\{1, -1, 9, -9\}$
49. $\{2, -2\}$
50. $\{9, -9\}$
51. $\{5, -5\}$
52. $\{3, -3\}$
53. $\{1, -1\}$
54. $\{10, -10\}$
55. $\{3, -3\}$
56. $\{7, -7\}$
57. $\{2, -2, 6, -6\}$
58. $\left\{2, -2, \frac{1}{2}, -\frac{1}{2}\right\}$
59. $\{3, 5, 7\}$
60. $\{-4, 6, 8\}$
61. 1591
62. 7224
63. 396
64. 484
65. 1156
66. 1209
67. 81
68. 7225
69. 621
70. 172
71. 1596
72. 2475
73. 1575
74. 8091
75. 7221
76. 121
77. 1764
78. 256
79. 2024
80. 2601

Review Exercises for Chapter 3

1. x^5
2. y^3
3. x^4
4. p^9
5. $-12x^4$
6. $-2a^8$
7. $-10n^3x^4$
8. $3a^6$
9. $12x^3$
10. $-50x^3$
11. $x^3 + 3x^2 + 2x$
12. $x^4 + 2x^2 - x$
13. $6x^3 - 8x^2 - 2x$
14. $4ax^3 - 4bx^2 + 4cx$
15. $x^2 - 5x + 6$
16. $y^2 - 3y - 4$
17. $6a^2 + 7a - 3$
18. $a^2 + 10a + 25$
19. $25x^2 - 1$
20. $4y^2 - 12y + 9$
21. $2x^3 - 3x^2 - 14x$
22. $2x^5 + x^4 - 2x^3$
23. $-3x^7 - 2x^5 + 4x^4$
24. $6x^2 - 29xy - 16y^2$
25. $2x^3 + 3x^2 - 20x$
26. $18x^3 + 24x^2 + 8x$
27. $x^3 + 8x^2 + 21x + 18$
28. $x^3 - x^2 - 10x - 8$
29. $6x^3 - x^2 + 8x + 3$
30. $x^4 + 2x^3 - 8x^2 - 13x + 6$
31. $3x(x - 2 + 8y)$
32. $(t + 3)(t + 8)$
33. $(x + 5)^2$
34. $(7y^2 - 1)(7y^2 + 1)$
35. $xy(x - 4y)(x + 4y)$
36. $16(x^2 + 4)$
37. $-3x(x^2 - 9) = -3x(x - 3)(x + 3)$
38. $(3a + 1)(2a - 3)$
39. $3x(12 + x^2 - x^4) = -3x(x^4 - x^2 - 12)$
$= -3x(x^2 - 4)(x^2 + 3)$
$= -3x(x - 2)(x + 2)(x^2 + 3)$
40. $25(4p^2 - q^2) = 25(2p - q)(2p + q)$
41. $4y(16y^2 - 9) = 4y(4y - 3)(4y + 3)$
42. $(x - 2)(x + 2)(x^2 + 4)$
43. $2y(6x^2 - 11x - 10) = 2y(2x - 5)(3x + 2)$
44. $(a + b)(8a - 3b)$
45. $2y(y - 3)(y + 3)(y^2 + 9)$
46. $3ax^4(5x^2 + 14x - 3) = 3ax^4(x + 3)(5x - 1)$
47. $(5r + 1)^2$
48. $(6x - 1)^2$
49. $(2p - 5q)^2$
50. $(x - 3)^2(x + 3)^2$
51. $\left\{-2, \frac{7}{2}\right\}$
52. $\{0, -7\}$

53. $\{5, -12\}$
54. $\{0, 3\}$
55. $\{2, 9\}$
56. $\{-1\}$
57. $\{-1\}$
58. $\{1, 6\}$
59. $\{0, 3\}$
60. $\{8, -8\}$

Answers to Exercises 4.1

1. $\frac{1}{3}$
2. $\frac{-1}{3}$
3. $\frac{-3}{5}$
4. $\frac{2}{5}$
5. $\frac{-3}{4}$
6. $\frac{-2}{3}$
7. y
8. a
9. $\frac{1}{3x^2}$
10. $3x^2$
11. m^2
12. $\frac{1}{2y}$
13. $4x^2y$
14. $5ab$
15. -1
16. $-3xy$
17. $\frac{1}{3xy}$
18. $\frac{-1}{xz}$
19. $\frac{1}{ac}$
20. $\frac{a + b}{4x}$
21. $\frac{x + 3}{x}$
22. $\frac{1}{y - 2}$
23. $\frac{x - 4}{x - 2}$
24. $\frac{y}{y + 3}$
25. $\frac{x + 2}{x}$
26. $\frac{y - 3}{y - 9}$
27. $\frac{y - 8}{y}$
28. $\frac{3}{x + 3y}$
29. $\frac{x - y}{x + y}$
30. $\frac{y}{5}$
31. $\frac{5}{x}$
32. $\frac{y + 5}{y - 5}$
33. $\frac{r}{4}$
34. $\frac{1}{3}$
35. $\frac{c - d}{6}$
36. $\frac{uv}{u + v}$
37. $4(t + 2)$
38. $9(t - 3)$
39. $\frac{a + b}{a - b}$
40. -1
41. -1
42. $\frac{x - y}{x + y}$
43. $2y - 3$
44. $x + 2$
45. $\frac{m^2 + 1}{m + 1}$
46. $\frac{a^2 + 1}{a + 1}$
47. $\frac{y}{3}$
48. $\frac{y}{3}$
49. $a - b$
50. $\frac{1}{(r + 2s)^3}$
51. $\frac{2a}{b}$
52. $\frac{1}{3}$
53. $3(5n - 1)$
54. $\frac{1}{4(1 - 3t)}$
55. $\frac{x + 1}{x + 4}$
56. $\frac{x + 4}{x + 2}$
57. $\frac{z - 5}{z + 4}$
58. $\frac{2(t + 2)}{3(t - 3)}$
59. $\frac{1}{2t + 1}$
60. $\frac{3}{x - 2}$
61. -1
62. -1
63. $-(3x + 2y)$
64. $\frac{1}{6 - x}$
65. $3(5n - 1)$
66. $\frac{-1}{5x + 8y}$
67. $\frac{x + 1}{x + 7}$
68. $\frac{n - 8}{n - 1}$
69. $\frac{2}{y - 4}$
70. $\frac{y(2y - 1)}{3}$
71. $\frac{a + b}{a - b}$
72. $\frac{x - y}{x + y}$
73. $\frac{4a}{a + b}$
74. $\frac{x + y}{xy}$
75. $\frac{2(n + 1)}{n - 2}$
76. $\frac{a - 4}{2(a + 4)}$
77. $\frac{-1}{x + 7}$
78. $\frac{-1}{x + 5}$
79. $\frac{7n + 2}{7n + 5}$
80. $\frac{5n - 2}{5n + 2}$
81. $\frac{m - n + 1}{m + n - 1}$
82. $\frac{x}{z}$
83. $\frac{a}{b}$
84. $\frac{r + s - t}{r - s + t}$
85. -2
86. $\frac{-1}{2}$
87. $\frac{s - r}{t - s} = \frac{r - s}{s - t}$
88. -1
89. $\frac{x + 2}{x + 3}$
90. $\frac{t + 1}{t - 1}$

Answers to Exercises 4.2

1. $\frac{8}{9}$
2. $\frac{2}{9}$
3. $\frac{1}{4}$
4. $\frac{7}{150}$
5. $\frac{1}{2}$
6. $\frac{3}{2}$
7. 3
8. $\frac{24}{25}$
9. $\frac{25}{54}$
10. 1
11. $\frac{6}{xy}$
12. $\frac{20}{rt}$
13. $\frac{x}{2}$
14. $\frac{1}{2x}$
15. $\frac{1}{2y}$
16. $\frac{3}{11b}$
17. $8x$
18. $15t$
19. $\frac{2a}{3b}$
20. $\frac{5y^2}{x^2}$
21. $\frac{6xy}{(x-3)(y-2)}$
22. $\frac{x^2-y^2}{16}$
23. $\frac{4x}{9ay}$
24. $\frac{-b}{ay}$
25. $\frac{-1}{10t}$
26. $\frac{4x^2}{9y^2}$
27. $\frac{4(x-2)}{x+2}$
28. $\frac{x}{x-1}$
29. $\frac{4xy}{x-y}$
30. $\frac{r+s}{rs}$
31. $\frac{t-2}{t}$
32. $\frac{y}{y+3}$
33. $\frac{2a(a-5)}{a-1}$
34. $\frac{x-2y}{2x^2}$
35. $\frac{x+3}{3x}$
36. $\frac{3(x-2)}{x+2}$
37. $\frac{(a-b)^2}{a}$
38. $\frac{-5y^4(y+1)}{24}$
39. $\frac{2(n-2)}{n(n+3)}$
40. $\frac{x^2}{5(x-2)}$
41. 1
42. 1
43. $\frac{2x}{x-3}$
44. $\frac{7}{5}$
45. $\frac{x(x-1)}{x-9}$
46. $\frac{1}{y-2}$
47. 1
48. $\frac{(5y+2)(y+1)}{3y(3y-2)}$
49. $5(3t-5)$
50. $4y^2-25$
51. x^2-36
52. $t(t+4)$
53. $\frac{1}{3}$
54. 5
55. $\frac{3}{25xy}$
56. $\frac{1}{12ab}$
57. $\frac{4}{9}$
58. $\frac{18}{25}$
59. $(x+3)(x+9)$
60. $(y-4)(y-5)$
61. $\frac{x^2}{y^2}$
62. $\frac{2x^2}{(x+2y)(2x+3y)}$
63. $\frac{7x(x+7)}{(x-7)^3}$
64. $\frac{12y}{y+4}$
65. $\frac{(x-1)(x-3)}{(x+1)(x+3)}$
66. $\frac{(x+6)(x+1)}{(x-6)(x-1)}$
67. $\frac{1}{b+9}$
68. $\frac{1}{a-8}$
69. $\frac{(x+7)^2}{(x+6)^2}$
70. $\frac{(x-4)^2}{(x-5)^2}$
71. $\frac{y}{x}$
72. -1
73. -1
74. $\frac{-4}{t^2+4}$
75. $(y+5)^2$
76. y^2-49
77. $\frac{-1}{3}$
78. $\frac{-9}{25}$
79. $\frac{1}{3(y+3)^2}$
80. $\frac{1}{(5y+1)^2}$

Answers to Exercises 4.3

1. $\frac{20}{36}$
2. $\frac{21}{56}$
3. $\frac{12}{21}$
4. $\frac{15}{25}$

5. $\frac{-28}{42}$

6. $\frac{-175}{35}$

7. $\frac{16}{36}$

8. $\frac{-8}{20}$

9. $\frac{-18}{42}$

10. $\frac{-40}{-10}$

11. $\frac{3x}{45}$

12. $\frac{12x^2}{16x^2}$

13. $\frac{6xy}{15xy}$

14. $\frac{-27x}{45x^2}$

15. $\frac{6x^3}{45x^2}$

16. $\frac{-15x^2y}{21xy^3}$

17. $\frac{-28x^3yz}{42x^2z^2}$

18. $\frac{4a^4b}{4a^2b^4}$

19. $\frac{-15x^3}{40x^2}$

20. $\frac{-12x^2y}{30xy^3}$

21. $\frac{-18x^2y^2}{42xyz}$

22. $\frac{5a^2b}{5ab^3}$

23. $\frac{10}{2x + 2}$

24. $\frac{3x}{6x - 15}$

25. $\frac{5x - 15}{5x + 20}$

26. $\frac{2x^2}{x^2 - 2x}$

27. $\frac{2x^2}{2x^2 - 10x}$

28. $\frac{2y + 4}{y^2 + 2y}$

29. $\frac{3y^2 - 9y}{y^2 - 9}$

30. $\frac{t^2 + 6t + 9}{t^2 + t - 6}$

31. $\frac{x^2 + 3x + 2}{x + 1}$

32. $\frac{-1}{3 - 4x}$

33. $\frac{x}{x - 7}$

34. $\frac{-x}{2x^2 - 5x}$

35. $\frac{-y^2 - 4y}{y^2 - 16}$

36. $\frac{x^2 + 5x + 6}{x^2 + 6x + 8}$

37. $\frac{a^2 + 2a + 1}{a^2 - 2a - 3}$

38. $\frac{6x^3 - 6x^2}{6x^2 - 6x}$

39. $\frac{t^2 + 5t - 14}{14 + 9t + t^2}$

40. $\frac{a^2 - b^2}{(a + b)^2}$

41. $\frac{4x^2 + 2x}{4x^2 - 1}$

42. $\frac{6t^2 + 6t}{18t^2 + 6t}$

43. $\frac{r^2 - 3r}{r^2 + r - 12}$

44. $\frac{u^2 - 3u - 10}{u^2 - 4u - 12}$

45. $\frac{6x^2 - 17x + 10}{x - 2}$

46. $\frac{-1}{6y - 5}$

47. $\frac{t}{t - 1}$

48. $\frac{-x^2 - 5x}{x^2 - 25}$

49. $\frac{2r + 7}{4r^2 - 49}$

50. $\frac{2x^3 + x^2 - 6x}{2x^3 + 5x^2 - 12x}$

51. $\frac{y^2 - 7y + 6}{12 - 8y + y^2}$

52. $\frac{12x^2 - 60x}{2x^2 + 2x - 60}$

53. $\frac{5x^2 + 10x}{5x + 10}$

54. $\frac{4x^2 - 9y^2}{(2x - 3y)^2}$

55. $\frac{x^3 - 25x^2}{3x^2 - 75x}$

56. $\frac{3 - 9y}{1 - 9y^2}$

57. $\frac{12a^3 + 18a^2}{10a^2 + 13a - 3}$

58. $\frac{7r^4 - 7r^3 - 14r^2}{7r^4 - 28r^2}$

59. $\frac{-2a^3 + 5a^2}{4a^2 - 10a}$

60. $\frac{9x + 27}{(x - 3)(x + 3)^2}$

61. $\frac{15n^3 + 30n^2 + 15n}{3n^3 - 3n}$

62. $\frac{ab}{(b - a)^2}$

Answers to Exercises 4.4

1. $\frac{5}{7}$

2. $\frac{7}{9}$

3. $\frac{5}{x}$

4. $\frac{7}{y}$

5. $\frac{2x - 6}{7}$

6. $\frac{2y + 7}{9}$

7. $\frac{1}{2}$

8. $\frac{3}{4}$

9. $\frac{2}{x}$

10. $\frac{3}{y}$

11. $\frac{3}{5}$

12. $\frac{1}{3}$

13. 3

14. 5

15. $\frac{t + 1}{2t}$

16. $\frac{t - 1}{2t}$

17. $\frac{3}{2}$

18. $\frac{7}{6}$

19. $\frac{3}{2t}$

20. $\frac{7}{6t}$

21. $\frac{x}{10y}$

22. $\frac{a}{6b}$

23. $\frac{1}{3y^2}$

24. $\frac{-1}{4x^2}$

25. $\frac{2x+3}{x^2}$

26. $\frac{5-2y}{y^2}$

27. $\frac{x}{y(y+1)}$

28. $\frac{3}{x(x+1)}$

29. $\frac{-10}{x(x+2)}$

30. $\frac{-9}{y(y+3)}$

31. $\frac{x^2-1}{x}$

32. $\frac{y^2+1}{y}$

33. $\frac{x+y}{xy}$

34. $\frac{b-a}{ab}$

35. $\frac{3y+5}{(y+1)(y+2)}$

36. $\frac{3x}{(x-3)(x+6)}$

37. $\frac{x+3}{x+1}$

38. $\frac{x(x-1)}{x+1}$

39. $\frac{-1}{x-2}$

40. -1

41. $\frac{5a-13}{6(a+1)}$

42. $\frac{3x+10}{15(x-1)}$

43. $\frac{2t}{(t+4)^2}$

44. $\frac{17x-4}{(3x-1)^2}$

45. $\frac{6}{(2x-3)^2}$

46. $\frac{25}{(t-5)^2}$

47. $\frac{2}{(y+2)(y-2)}$

48. $\frac{6}{(x-3)(x+3)}$

49. $\frac{5x+17}{(x-5)(x+2)}$

50. $\frac{-3(3x-2)}{(x+6)(x+1)}$

51. $\frac{t(5t+2)}{9(t+2)(t-2)}$

52. $\frac{5t}{(t-3)(3t+1)}$

53. $\frac{-2x(x-2)}{3(x+4)(x-4)}$

54. $\frac{3x+1}{4x(x-7)}$

55. $\frac{x-7}{2(3x+2)}$

56. $\frac{1}{x(x+2)}$

57. $\frac{x-4}{x-5}$

58. $\frac{x-6}{x+1}$

59. $\frac{3}{t+1}$

60. $\frac{13x}{2(x-4)(x+6)}$

61. $\frac{2}{(a-1)(a+1)^2}$

62. $\frac{-3(x+1)}{(3x-1)(3x+1)^2}$

63. $\frac{1}{(x+3)(x+4)}$

64. $\frac{6}{(x-3)(x+3)}$

65. $\frac{y+15}{(4y+3)(5y-1)}$

66. $\frac{y}{(2y-1)(3y-1)}$

67. $\frac{4y^2-y+9}{(y-2)(y-3)(2y+1)}$

68. $\frac{-8x-15}{(3x+2)(x+1)(x-6)}$

69. $\frac{1}{y(y-5)}$

70. $\frac{x^2-28x-49}{(x+7)^2(x-7)^2}$

71. $\frac{x+1}{x}$

72. $\frac{4a^2}{2a+5}$

73. $\frac{1}{6(x-4)}$

74. $\frac{1}{x(x-2)}$

75. $\frac{x-y}{x+y}$

76. -2

Answers to Exercises 4.5

1. $\frac{22}{31}$

2. $\frac{10}{17}$

3. $\frac{6}{5}$

4. $\frac{32}{73}$

5. $\frac{11}{20}$

6. $\frac{20}{11}$

7. $\frac{y}{y-1}$

8. $2t$

9. $\frac{3 - x}{3}$

10. $\frac{ab}{a + b}$

11. $\frac{1}{t}$

12. $\frac{2x + 3}{3x + 2}$

13. $\frac{10x + 2y^2}{xy^2}$

14. $b - a$

15. $\frac{a}{a^2 - 3}$

16. $\frac{y}{3y + 1}$

17. $\frac{x - 2}{x - 5}$

18. $\frac{y - 2}{y - 4}$

19. $\frac{y + 3}{y - 4}$

20. $\frac{n + 3}{3n - 1}$

21. $\frac{xy}{x + y}$

22. $\frac{ab}{2b + 3a}$

23. $\frac{x - 1}{x + 1}$

24. $\frac{x + 5}{x - 5}$

25. $\frac{a + 1}{a(a + 3)}$

26. $\frac{-3}{5}$

27. $\frac{13}{8}$

28. $\frac{3}{4}$

29. $1 - x$

30. $\frac{y^2 + 1}{y^3}$

Answers to Exercises 4.6

1. $x^2 + 3x - 2 + \frac{4}{x}$

2. $4x^2 - 2x + 5 - \frac{2}{x}$

3. $3y - 2 + \frac{3}{2y}$

4. $5y^2 + 3y - 1 - \frac{1}{3y}$

5. $6x - 4 + \frac{5x + 2}{x^2}$

6. $12x^2 - 9 + \frac{4x - 5}{x^2}$

7. $x - 5 + \frac{3x^2 + 2}{x^3}$

8. $1 + \frac{-5x^2 + 3}{x^4}$

9. $3x^2y + 2x - \frac{2}{3}$

10. $x^2 - 3xy + 6 - \frac{2}{xy}$

11. $2x^2 - 3x + 9 - \frac{29}{x + 4}$

12. $5x^3 - 9x^2 + 14x - 26 + \frac{49}{x + 2}$

13. $2x^2 - 3x - 2$

14. $3x^2 - 11x - 4$

15. $2x^2 - 2x - 1 - \frac{7}{2x + 7}$

16. $4x^3 + 4x^2 + x + 8 + \frac{7}{x - 1}$

17. $2x^2 - x - 1 + \frac{4}{x - 1}$

18. $3x^2 - 13x + 4$

19. $x^2 + 5 - \frac{8}{x + 4}$

20. $x^2 + 2 + \frac{1}{x + 3}$

21. $y^2 - 4$

22. $y^2 - y - 2$

23. $a^2 + 9a + 81$

24. $a^2 - 4a + 16$

25. $4t^2 - 2t + 1$

26. $y^2 + 5y + 25$

27. $5x^2 + \frac{2x - 1}{x^2 - 6}$

28. $2x^2 - \frac{5}{x^2 - 7}$

29. $x^4 + 1$

30. $3x^2 + x - 6$

31. $x^3 + x^2 + x + 1$

32. $x^3 - x^2 + x - 1$

33. $x - 2 + \frac{8}{x + 2}$

34. $x + 3 + \frac{18}{x - 3}$

35. $2x^3 + 4 + \frac{7}{x - 3}$

36. $t^2 - 3t + 2$

37. $x^2 - 2x + 2$

38. $y^2 + 4y + 8$

39. $y^2 + y + 3$

40. $y^2 + y - 5$

41. $t - 1 - \frac{3t}{t^2 + 4}$

42. $t^2 - t + 5 - \frac{6t - 30}{t^2 - 5}$

Answers to Exercises 4.7

1. $\{-4\}$
2. $\{-3\}$
3. $\{4\}$
4. $\{12\}$
5. $\{-5\}$
6. $\varnothing$
7. $\varnothing$
8. $\{0\}$
9. $\{0\}$
10. $\{6\}$
11. $\{3\}$
12. $\{2\}$
13. $\left\{\frac{1}{4}\right\}$
14. $\left\{\frac{1}{5}\right\}$
15. $\{5\}$
16. $\{0\}$
17. $\left\{\frac{1}{3}\right\}$
18. $\{1\}$
19. $\{10\}$
20. $\left\{\frac{1}{2}\right\}$
21. $R, x \neq 4$
22. $R, x \neq -3$
23. $\left\{\frac{1}{6}\right\}$
24. $\left\{\frac{1}{10}\right\}$
25. $\{13\}$
26. $\{-10\}$
27. $\{3\}$
28. $\{-9\}$
29. $\{12\}$
30. $\{4\}$
31. $\{7\}$
32. $\{4\}$
33. $\{9\}$
34. $\{-2\}$
35. $\{2\}$
36. $\{8\}$
37. $\{-3\}$
38. $\{-1\}$
39. $\{12\}$
40. $\{6\}$
41. $\left\{\frac{1}{2}\right\}$
42. $\left\{\frac{1}{2}\right\}$
43. $R, x \neq 0, x \neq -5$
44. $R, x \neq 0, x \neq 7, x \neq -7$
45. $\{2\}$
46. $\{8\}$
47. $\{-1\}$
48. $\{6\}$
49. $\left\{\frac{1}{8}\right\}$
50. $\left\{\frac{1}{5}\right\}$
51. $\{-2\}$
52. $\left\{\frac{3}{2}, -1\right\}$
53. $\{3, -2\}$
54. $\{5\}$
55. $x = 2a - 3b$
56. $x = \frac{a^2 + b^2}{b}$
57. $x = \frac{2ab}{a + b}$
58. $x = 0$
59. $y = \frac{a + b}{2}$
60. $y = \frac{a}{2}$

Answers to Exercises 4.8

1. 35 centimeters by 56 centimeters
2. $88\frac{8}{9}$ grams of oxygen
   $11\frac{1}{9}$ grams of hydrogen
3. 20 cups; $1\frac{1}{3}$ cups
4. \$1125 and \$675
5. 13
6. 3750
7. 175 pounds of sodium
   115 pounds of chlorine
8. 48 inches; 36 inches
9. \$1.23
10. \$4.32
11. \$12,000; \$14,400
12. 18 feet by 20 feet
13. $56\frac{1}{4}$ mph
14. 40 kph
15. \$18
16. \$7.20
17. 24 gallons
18. \$5.85
19. 25%
20. 15%
21. 43
22. 21
23. \$6; \$7.35
24. \$238.24; \$197.74
25. 4 minims
26. 2 grains
27. 8 grams
28. 50 milliliters
29. 12 cubic feet of cement
    36 cubic feet of sand
    60 cubic feet of rock
30. \$24,000; \$14,400; \$9,600

Answers to Exercises 4.9

1. $\frac{9}{x} + \frac{9}{3x} = 1;$

   12 hours and 36 hours

2. $\frac{1}{2} \cdot \frac{1}{x} + \frac{1}{2} \cdot \frac{1}{5x} = 1$

   $5x = 3$ hours

   $x = \frac{3}{5}$ hour $= 36$ minutes

3. $\frac{x}{6} + \frac{x}{12} = 1$; 4 days

4. $\frac{x}{10} + \frac{x}{15} = 1$; 6 hours

5. $\frac{x}{12} + \frac{x}{15} + \frac{x}{20} = 1;$

   $x = 5$ minutes
   $1000x = 5000$ minutes
   $= 83\frac{1}{3}$ hours

6. $\frac{x}{18} + \frac{x}{30} + \frac{x}{45} = 1;$

   9 minutes

7. $\frac{6}{15} + \frac{6}{x} = 1$; 10 days

8. $\frac{20}{36} + \frac{20}{x} = 1$; 45 hours

9. $\frac{x + 10}{40} + \frac{x}{60} = 1;$

   18 minutes

10. $\frac{15}{70} + \frac{15 + x}{28} = 1;$

    7 minutes

11. $\frac{x}{15} + \frac{x}{20} - \frac{x}{12} = 1;$

    30 hours

12. $\frac{6}{8} + \frac{6}{12} - \frac{6}{x} = 1;$

    24 hours

13. $\frac{1}{x} + \frac{1}{4x} = \frac{3}{4};$

    $\frac{5}{3}, \frac{20}{3}$

14. $x - \frac{1}{x} = \frac{3}{2}$

    $2x^2 - 2 = 3x$
    $(2x + 1)(x - 2) = 0$

    $x = \frac{-1}{2}$ and $-\frac{1}{x} = 2$

    or $x = 2$ and $-\frac{1}{x} = -\frac{1}{2}$

15. $\frac{1}{8} = \frac{1}{4b} + \frac{1}{b}$

    where $a = 4b$;
    $a = 40$ inches

16. If $x = R_1$, then $3x = R_2$

    $\frac{1}{15} = \frac{1}{x} + \frac{1}{3x};$

    20 ohms and 60 ohms

Review Exercises for Chapter 4

1. $\frac{2x^2z}{3y^2}$
2. $\frac{5a^3y}{2b}$
3. -1
4. $\frac{9(x + 2y)}{x - 2y}$
5. $\frac{2x - 15}{2x - 3}$
6. $\frac{-4x^3(x - 1)}{3(x + 1)}$
7. 1
8. $\frac{x - 1}{3}$
9. $\frac{a^2}{(a - 4)^2}$
10. $\frac{x + 1}{x + 3}$
11. $\frac{5 - x}{x + 3}$
12. $\frac{6x^2 - 3x}{x^2 - 9}$
13. $\frac{4x - 15}{(x - 5)(x - 1)}$
14. $\frac{4}{x(x + 1)(x - 1)}$
15. $\frac{x^2 + 14x + 4}{2x(x + 2)}$
16. $\frac{1}{(x + 3)(x + 4)}$
17. $\frac{16}{(x - 2)^2(x + 2)^2}$
18. $\frac{xy + 3}{xy - 3}$
19. $\frac{-1}{3(x + 3)}$
20. $\frac{5(5x + 2)}{12(3x + 1)}$
21. $x^2 - x - \frac{1}{x - 3}$
22. $x^2 - 2 + \frac{4}{x^2 + 3}$
23. $3x^2 - 2x + 4 - \frac{1}{5x + 10}$
24. $a^2 + a + 1$
25. $5x^2 - 2x - 3 - \frac{7}{2x + 1}$
26. $\{9\}, x \neq 6, -6$
27. $\{0\}, x \neq -2$
28. $\{9\}, x \neq 3, -3$
29. $\left\{\frac{1}{4}\right\}, x \neq -1, 3$
30. $\{4\}, x \neq 0, 3, -5$
31. $R, x \neq 4, -4$
32. $\varnothing$

33. $\{-5\}$, $x \neq 3$

34. $x = \frac{a-b}{2}$; $x \neq a, -b$;

$a \neq -b$

35. $x = \frac{p-1}{c+1}$; $c \neq -1$, $x \neq 0$

36. $\frac{(x-3)-2}{x+1} = \frac{1}{4}$; $\frac{4}{7}$

37. Original fraction: $\frac{3x}{5x}$;

$\frac{3x-20}{5x+10} = \frac{1}{2}$

$\frac{3x}{5x} = \frac{150}{250}$

38. $\frac{x}{14} + \frac{x}{35} = 1$; 10 hours

39. $\frac{3}{4} + \frac{3}{x} = 1$;

12 hours

40. $\frac{20}{9} = \frac{120}{x}$; 54 kilograms

41. $\frac{1}{3.8} = \frac{5}{x}$; 19 liters

42. $\frac{30.5}{1} = \frac{x}{6}$; 183 centimeters

43. $5x + 8x = 65$
40 centimeters, 25 centimeters

Answers to Exercises 5.1

1. $(-3, -1)$
2. $(2, -2)$
3. $(2, 4)$
4. $(0, 0)$
5. $(3, -5)$
6. $(-4, -4)$
7. $(-4, 3)$
8. $(0, 4)$
9. $(3, 0)$
10. $(-4, 0)$

11. 13. 15. 17. 19. 21.

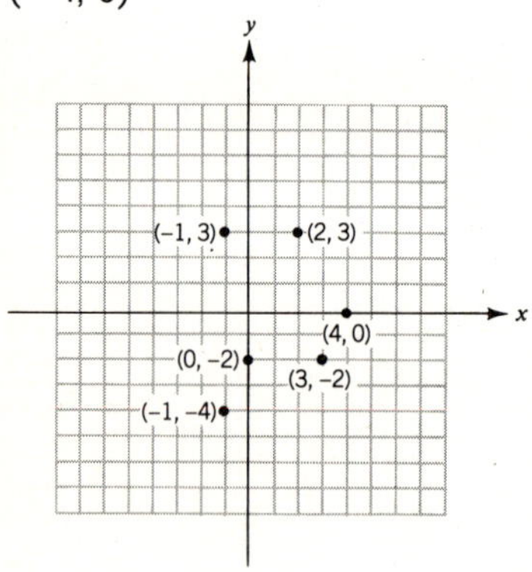

12. 14. 16. 18. 20. 22.

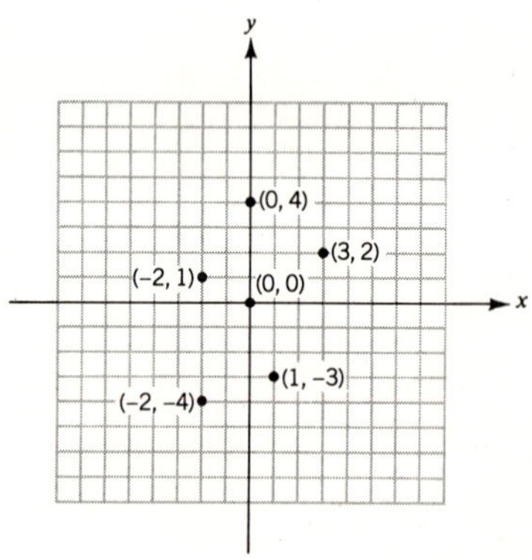

23. (1, 0)

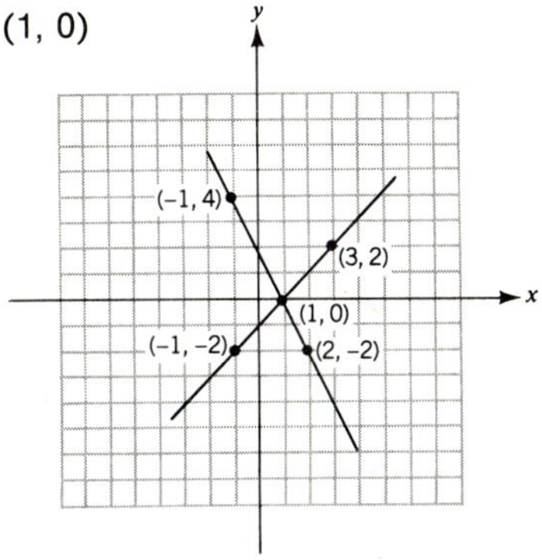

24. (5, −1)

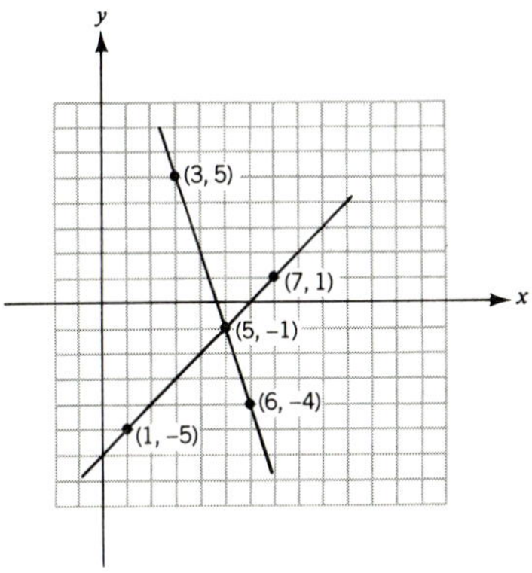

25. Yes
26. Yes
27. No
28. No
29. Yes
30. No
31. Yes
32. No
33. No
34. Yes
35. Yes
36. No
37. No
38. Yes
39. Yes
40. No
41. No
42. Yes
43. (0, 5), (3, 0)
44. (5, 3), (3, 0)
45. (3, 5), (5, −3)
46. (3, 5)
47. (3, 5), (0, 5)
48. (3, 5), (3, 0)
49. (−2, 6)
50. (−2, 0), (0, 6)
51. (2, 6), (6, −2)
52. (−2, 0)
53. (−2, 6), (−2, 0)
54. (2, 6), (−2, 6), (0, 6)

55.

x	y
0	−6
2	0
1	−3
−2	−12
1	−3

Straight line

56.

x	y
0	3
5	0
−5	6
15	−6
10	−3

Straight line

57.

x	y
0	−8
8	0
−2	−10
4	−4
3	−5

Straight line

58. $3x - 2y = 12$

x	y
0	−6
4	0
2	−3
−4	−12
8	6

Straight line

Answers to Exercises 5.2

1. x-intercept = 3,
y-intercept = −6
2. x-intercept = −2,
y-intercept = 6
3. x-intercept = 4,
y-intercept = 4
4. x-intercept = 4,
y-intercept = −4
5. x-intercept = 2,
y-intercept = −3
6. x-intercept = 6,
y-intercept = −4

7.

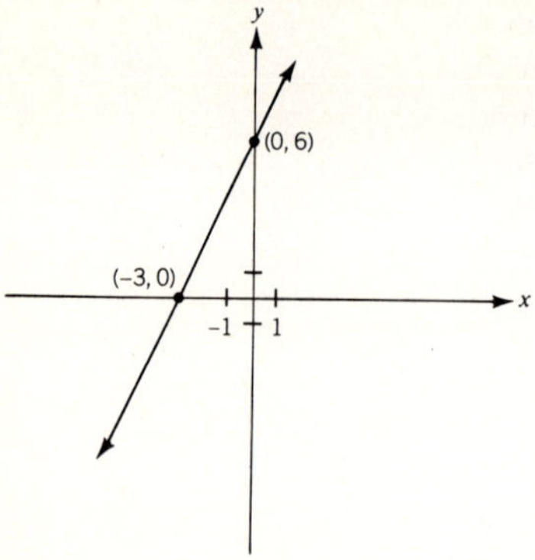

8.

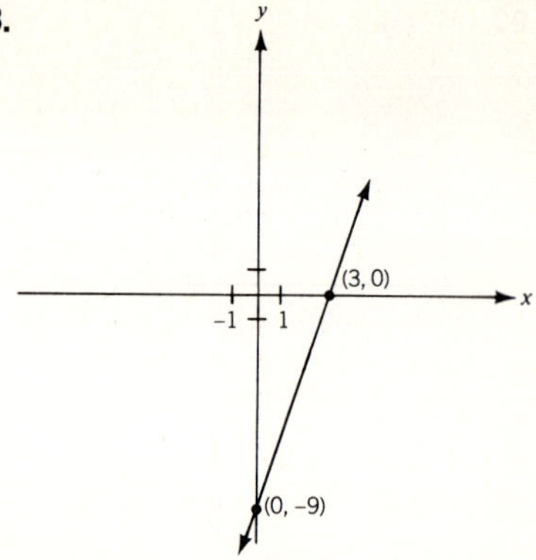

9.

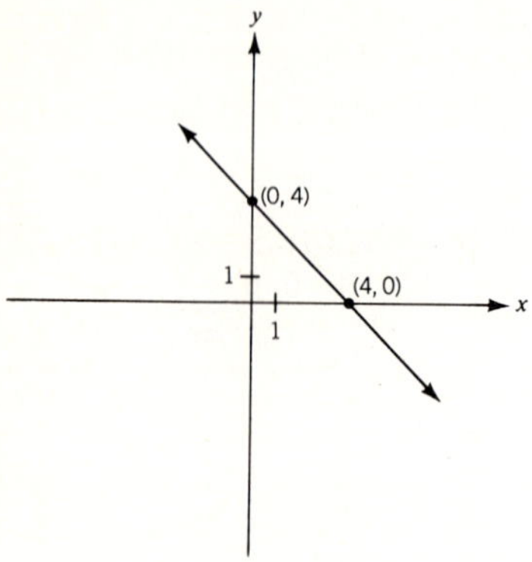

10.

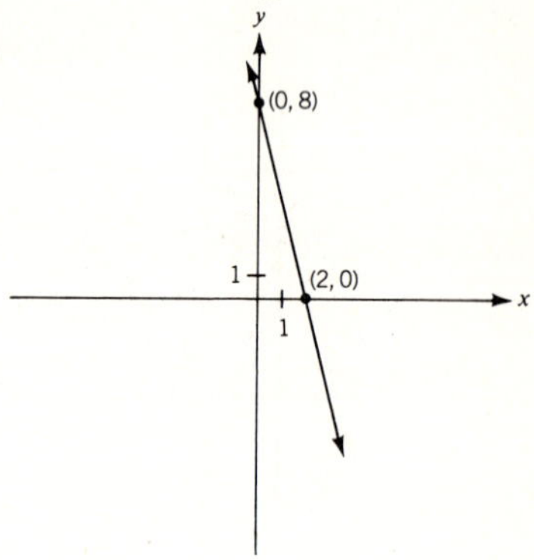

11.

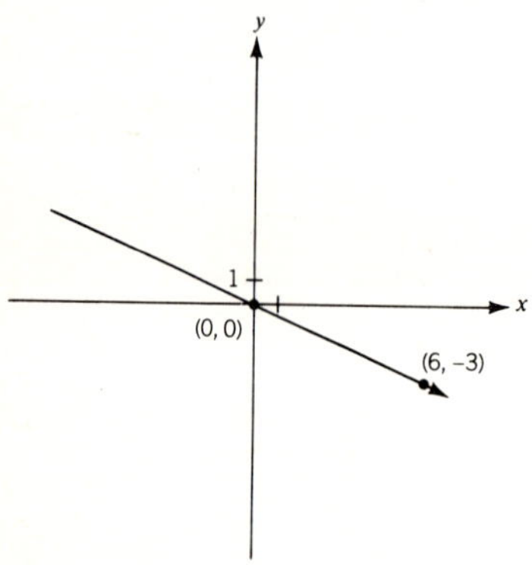

12.

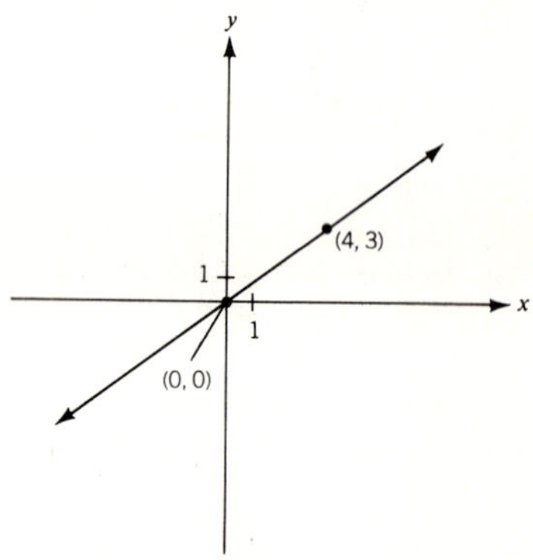

13.

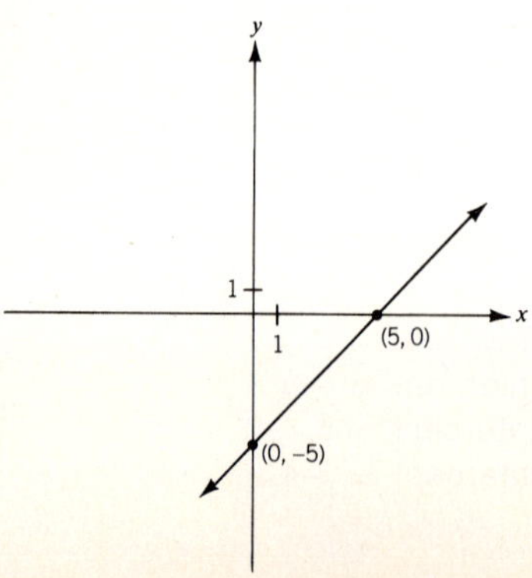

14.

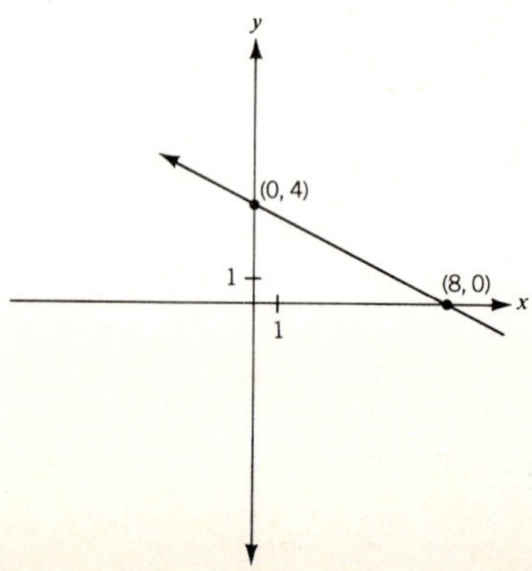

15.

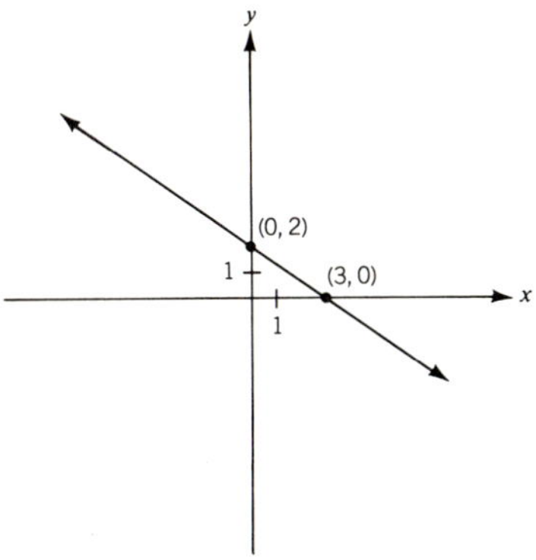

16.

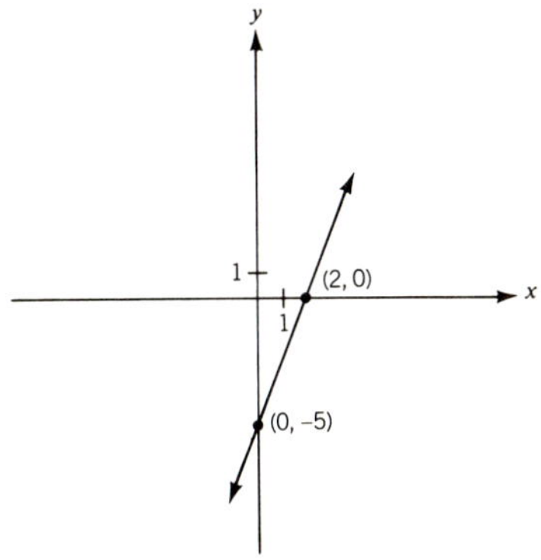

17.

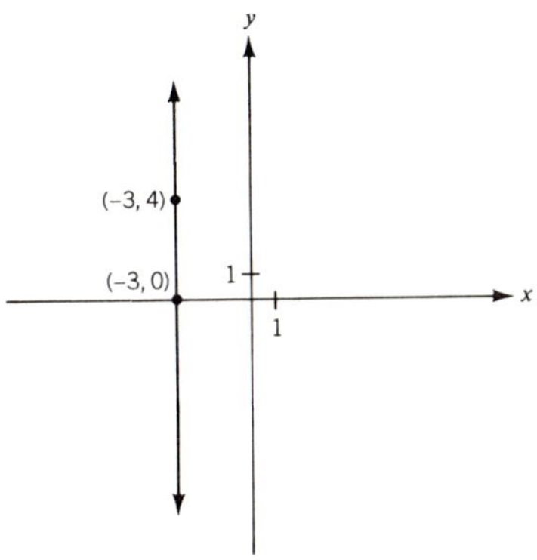

18.

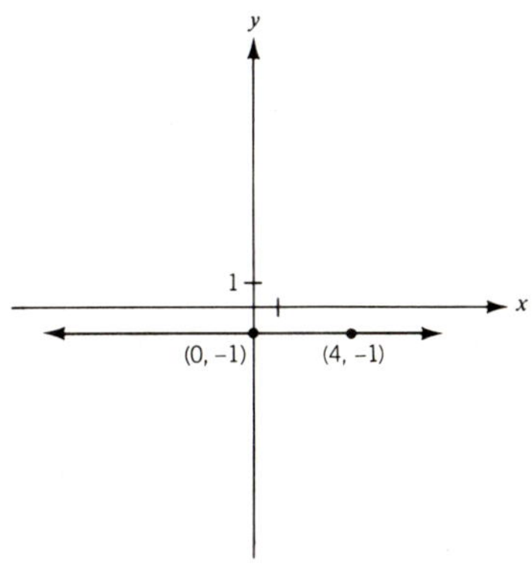

19.

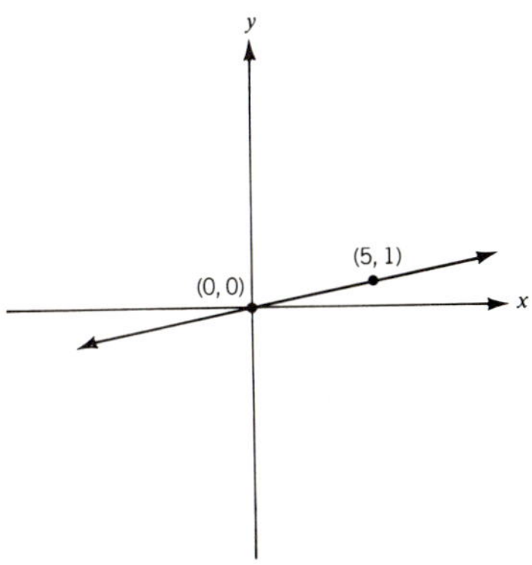

20.

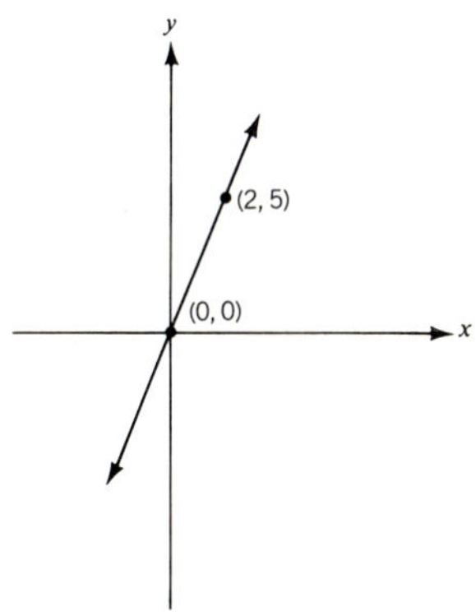

21.

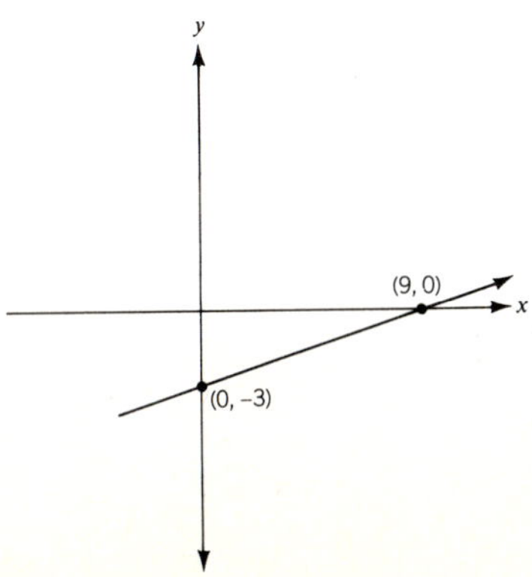

22.

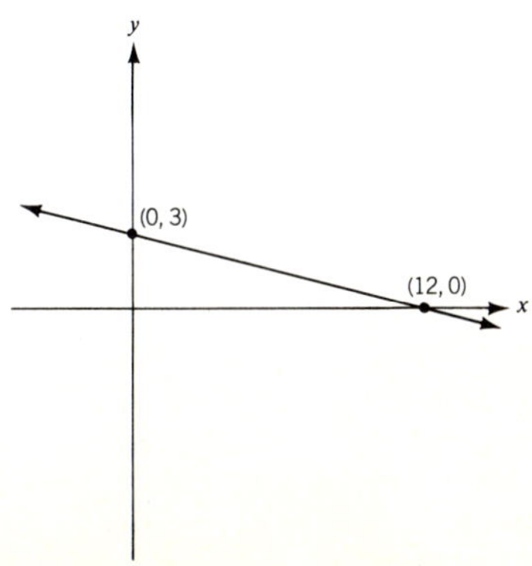

23.

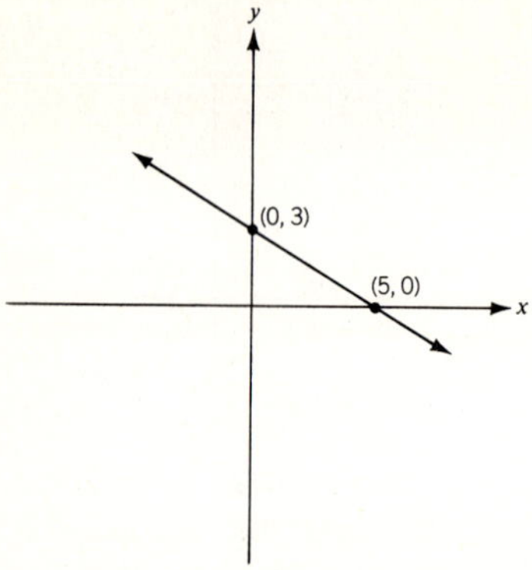

24.

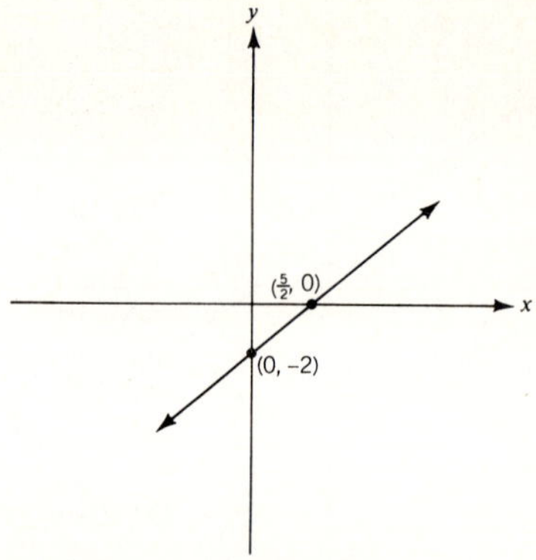

25. They graph as parallel lines, but they intersect the axes at different points.

26. All the lines pass through the point (0, 1), but their directions are different.

27. a. $k = 5$ b. $k = 9$

28.

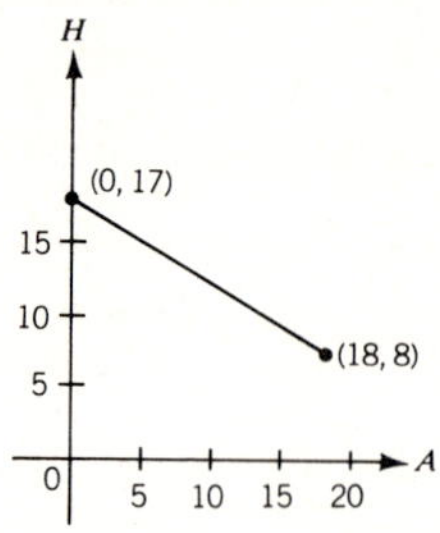

29. a. b.

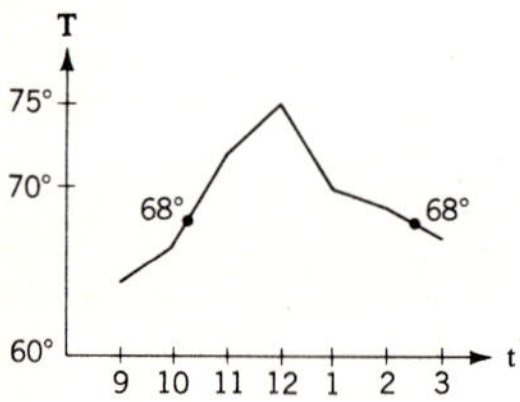

c. About 4 hours

30. b. \$130

c. 7 years

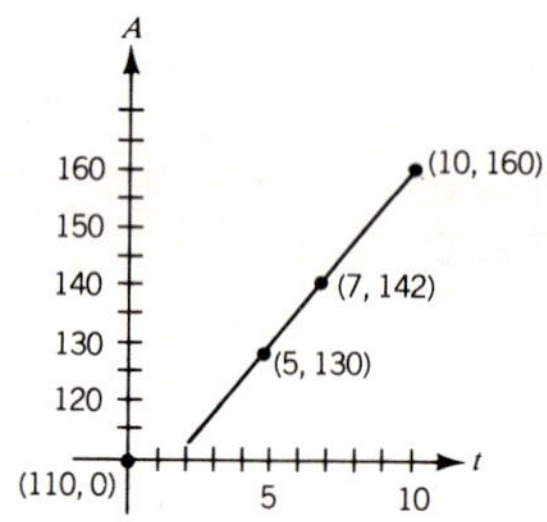

31. a.

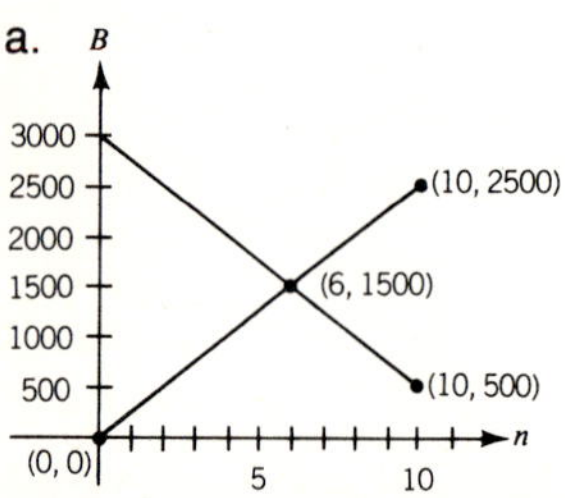

b. \$1625 c. $4\frac{1}{2}$ years

d. 6 years

Answers to Exercises 5.3

1. $\frac{1}{2}$
2. $\frac{3}{2}$
3. $-\frac{4}{3}$
4. -1
5. 1
6. -2
7. -1
8. $\frac{3}{5}$
9. $\frac{1}{2}$
10. $\frac{1}{3}$
11. $m = -2, b = 5$
12. $m = -\frac{3}{2}, b = 6$
13. $m = 2, b = -8$
14. $m = -1, b = -10$
15. $m = \frac{5}{2}, b = -5$
16. $m = \frac{2}{3}, b = -3$
17. $m = 1, b = \frac{3}{5}$
18. $m = -\frac{1}{5}, b = -1$
19. $m = 0, b = 5$
20. Slope undefined; no y-intercept
21. $m = 2, b = -4$
22. $m = 2, b = \frac{5}{2}$
23. $m = -\frac{3}{2}, b = 2$
24. $m = 2, b = 8$
25. $m = \frac{3}{2}, b = \frac{3}{2}$
26. $m = 1, b = 6$
27. Slope undefined; no y-intercept
28. $m = 0, b = -2$
29. $m = 1, b = 0$
30. $m = -1, b = 0$
31. Parallel; $m = -1$
32. Not parallel
33. Not parallel
34. Parallel; $m = 2$
35. Not parallel
36. Not parallel
37. Parallel; $m = 1$
38. Parallel; $m = \frac{5}{2}$
39. Parallel; slope undefined
40. Not parallel
41. Parallel; $m = \frac{1}{3}$
42. Parallel; $m = 0$
43. $k = 6$
44. $k = \frac{3}{4}$
45. $k = 0$
46. $k = -3$
47. $k = \frac{3}{2}$
48. $k = -3$
49. a. 132 feet
    b. 85.8 feet
50. a. and b.

Answers to Exercises 5.4

1. $2x + y - 8 = 0$
2. $3x - y + 2 = 0$
3. $6x - y + 9 = 0$
4. $x + y - 1 = 0$
5. $x - 2y - 8 = 0$
6. $x - 3y + 9 = 0$
7. $3x + 4y + 25 = 0$
8. $2x + 5y + 70 = 0$
9. $y = 7$
10. $y = 0$
11. $m = \frac{1}{2}$; $x - 2y = 0$
12. $m = \frac{3}{2}$; $3x - 2y = 0$
13. $m = \frac{1}{3}$; $x - 3y - 7 = 0$
14. $m = -1$; $x + y + 7 = 0$
15. $m = \frac{1}{4}$; $x - 4y + 3 = 0$
16. $m = -4$; $4x + y + 3 = 0$
17. $m = 0$; $y = 3$
18. Slope undefined; $x = 4$
19. Slope undefined; $x = -2$
20. $m = 0$; $y = -7$
21. $3x + 2y - 7 = 0$
22. $x + 2y - 6 = 0$
23. $x + 2y - 3 = 0$
24. $3x - 4y = 0$
25. $y - 2 = 0$
26. $x - 1 = 0$
27. $2x - y + 1 = 0$
28. $4x - 3y - 6 = 0$
29. $4x + 3y + 8 = 0$
30. $5x + 3y = 0$
31. No, slope of line $AB \neq$ slope of line BC

32. Yes, slope of line PQ = slope of line QR

33. $k = 4$

34. $k = 12$

35. a. and b.

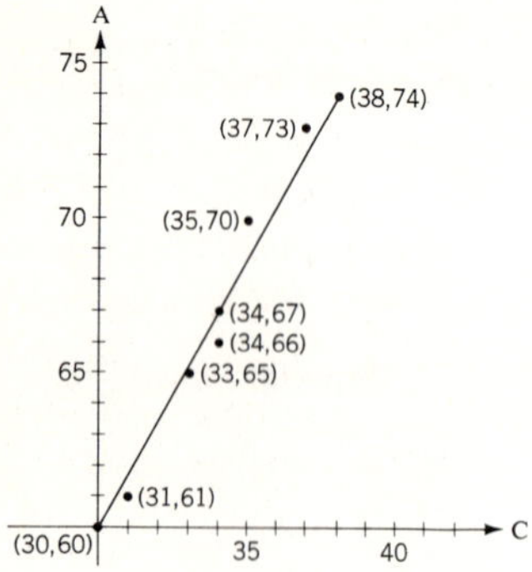

c. $7C - 4A + 30 = 0$

d. $63\frac{1}{2}$ inches, $70\frac{1}{2}$ inches, $75\frac{3}{4}$ inches

36. a. and b.

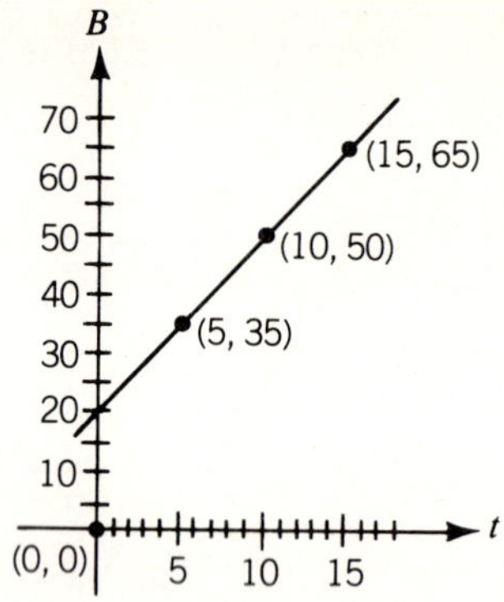

c. $B = 20$

d. $3t - B + 20 = 0$

Answers to Exercises 5.5

1. a. $m_1 = -1, m_2 = 1$
b. $b_1 = \frac{4}{3}, b_2 = -\frac{4}{3}$
c. Intersect
d. One solution

2. a. $m_1 = -1, m_2 = -1$
b. $b_1 = \frac{4}{3}, b_2 = \frac{5}{2}$
c. Parallel
d. No solution

3. a. $m_1 = -1, m_2 = -1$
b. $b_1 = \frac{4}{3}, b_2 = \frac{4}{3}$
c. Coincide
d. Line of solutions

4. a. m_1 undefined, $m_2 = 0$
b. No y-intercept, $b_2 = 4$
c. Intersect
d. One solution

5. a. $m_1 = 1, m_2 = 1$
b. $b_1 = 3, b_2 = 0$
c. Parallel
d. No solution

6. a. $m_1 = \frac{3}{4}, m_2 = \frac{3}{4}$
b. $b_1 = -3, b_2 = -3$
c. Coincide
d. Line of solutions

7. a. $m_1 = \frac{3}{2}, m_2 = -\frac{5}{2}$
b. $b_1 = -\frac{3}{2}, b_2 = \frac{5}{2}$
c. Intersect
d. One solution

8. a. $m_1 = \frac{3}{2}, m_2 = 1$
b. $b_1 = -\frac{3}{2}, b_2 = -\frac{2}{3}$
c. Intersect
d. One solution

9. a. $m_1 = 2, m_2 = 2$
b. $b_1 = -1, b_2 = -1$
c. Coincide
d. Line of solutions

10. a. $m_1 = \frac{1}{2}, m_2 = \frac{1}{2}$
b. $b_1 = -\frac{1}{2}, b_2 = -\frac{1}{2}$
c. Coincide
d. Line of solutions

11. a. $m_1 = \frac{1}{2}, m_2 = \frac{1}{2}$
b. $b_1 = -\frac{5}{2}, b_2 = \frac{5}{2}$
c. Parallel
d. No solution

12. a. $m_1 = 1, m_2 = 1$
b. $b_1 = 6, b_2 = -6$
c. Parallel
d. No solution

13. (4, 3)

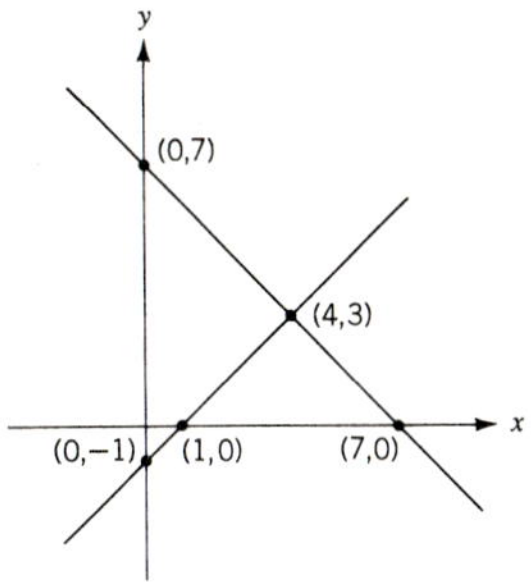

14. (−3, −4)

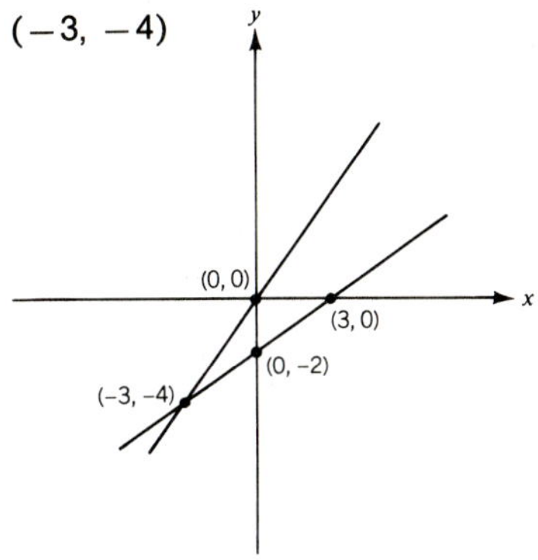

15. (2, −1)

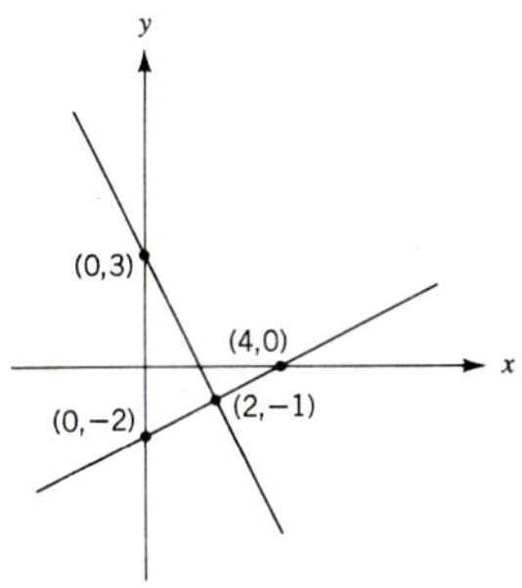

16. (3, 2)

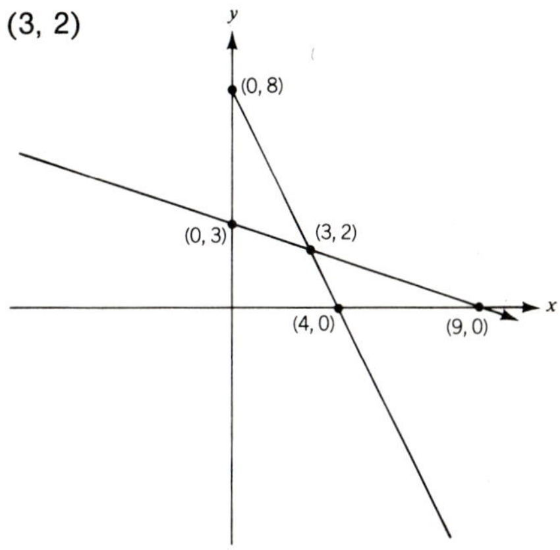

17. (8, 1)

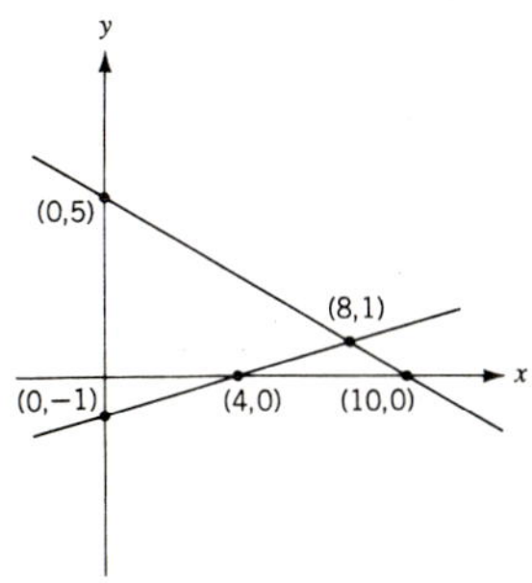

18. (2, 1)

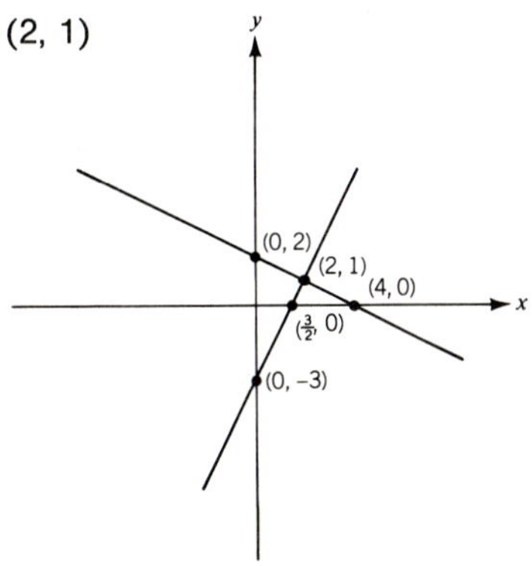

19. $(-6, 6)$

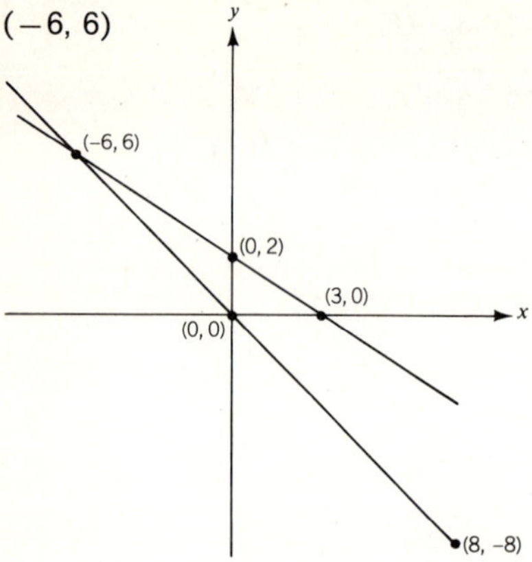

20. $(-1, 6)$

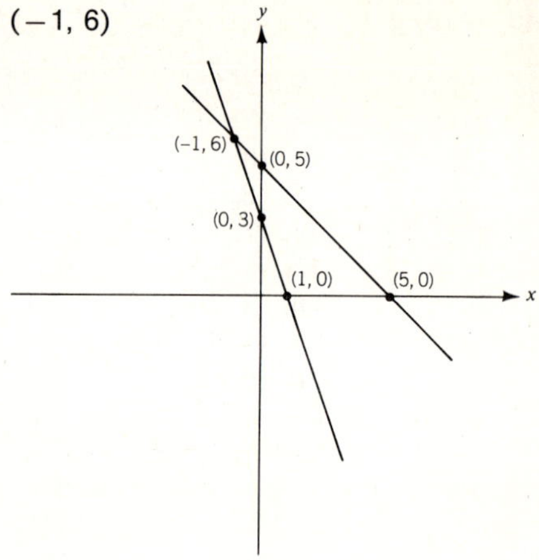

21. $(-2, -6)$

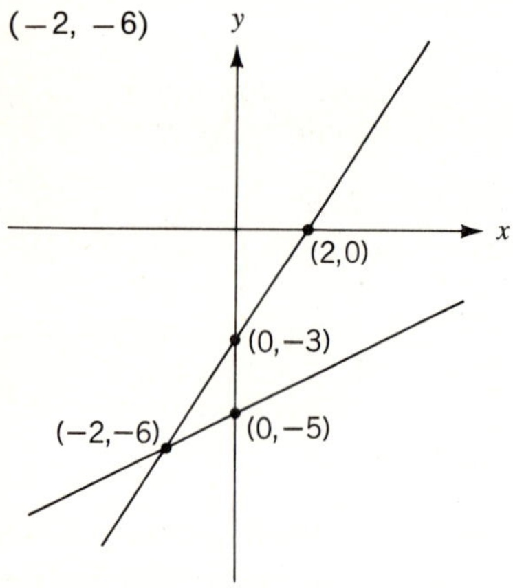

22. $(4, -5)$

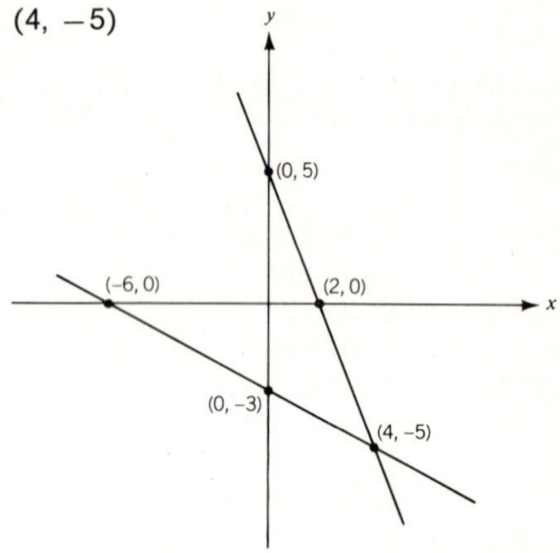

23. 30 units of Food *A*
20 units of Food *B*

24. 6 years, \$2400

Answers to Exercises 5.6

1. $(2, 4)$
2. $(4, 12)$
3. $(4, 5)$
4. $(3, 5)$
5. $(3, -1)$
6. $(-2, 8)$
7. $(-2, 5)$
8. $(3, -5)$
9. $(6, 4)$
10. $(5, 6)$
11. $(4, -2)$
12. $(2, -3)$
13. $(-7, -2)$
14. $(-2, 6)$
15. $(10, 3)$
16. $(-4, -6)$
17. $(3, 2)$
18. $(-1, 1)$
19. $(0, 0)$
20. $(0, 0)$
21. $\{(x, y) \mid y = 3x - 1\}$
22. $\varnothing$
23. $\varnothing$
24. $\{(x, y) \mid x = 6 - y\}$
25. $\left(\frac{1}{2}, \frac{7}{2}\right)$
26. $\left(\frac{13}{5}, 6\right)$
27. $(2, 1)$
28. $(-1, 2)$
29. $(3, 4)$
30. $(7, -1)$

Answers to Exercises 5.7

1. (5, 2)
2. (7, 3)
3. (3, −2)
4. (2, −5)
5. (−1, 5)
6. (−3, 7)
7. (4, 2)
8. (6, 3)
9. (2, 1)
10. (1, 3)
11. (3, 2)
12. (4, 2)
13. (7, 5)
14. (8, 10)
15. (8, −10)
16. (7, −3)
17. (−3, 6)
18. (−1, 4)
19. (4, 1)
20. (5, 10)
21. ∅
22. $\{(x, y) \mid 2x - 3y = 2\}$
23. $\{(x, y) \mid x - 3y = 4\}$
24. ∅
25. (−1, −2)
26. (−10, −4)
27. (9, 8)
28. (5, −5)
29. (4, 9)
30. (6, 10)
31. $(a + b, 2a - b)$
32. $(2a + b, a - 2b)$
33. $(a + b, a - b)$
34. $(7b - a, a - 5b)$
35. $\left(\frac{2}{a}, \frac{3}{b}\right)$
36. $\left(\frac{5}{a}, \frac{1}{b}\right)$
37. $(a + b, a - b)$
38. (a, b)

Answers to Exercises 5.8

1. 14 and −5
2. 13 and 7
3. 15 and 11
4. 6 and 3
5. Bread, $1.00 per loaf; Butter, $1.50 per pound
6. Shirts, $5 each; Sweaters, $12 each
7. Milk, 50¢ a quart; Eggs, 90¢ a dozen
8. Coats, $23.95 each; Slacks, $8.50 per pair
9. 12 miles
10. 130 miles
11. Rate of current, 1.5 mph; rate of crew in still water, 6 mph
12. Speed of current, 5 mph; still water speed of boat, 25 mph
13. 25 mph
14. Rate of wind, 20 mph; still air rate of planes, 500 mph
15. A: 65% silver B: 45% silver
16. 80% in A; 60% in B
17. 10 items of product *A* 9 items of product *B*
18. Fixed fee, 89¢; added fee, 8¢

Answers to Exercises 5.9

1. $x \geq 2$

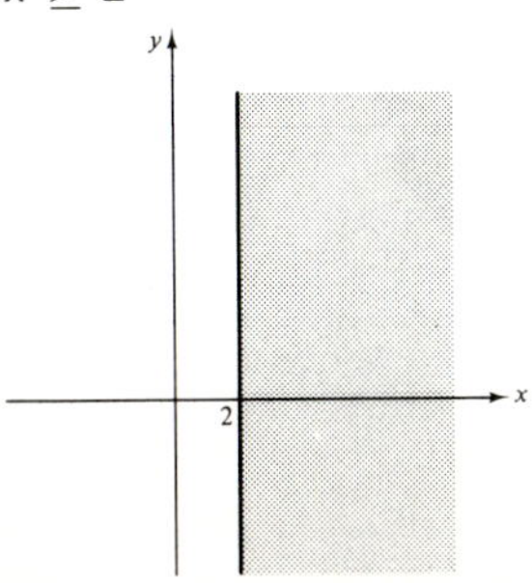

2. $x < -1$

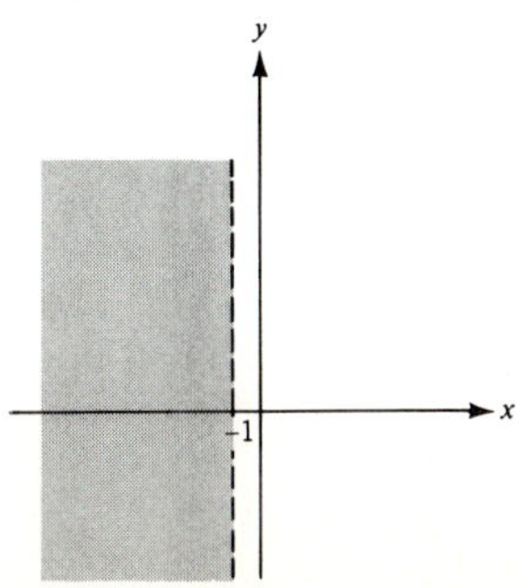

3. $y > -1$

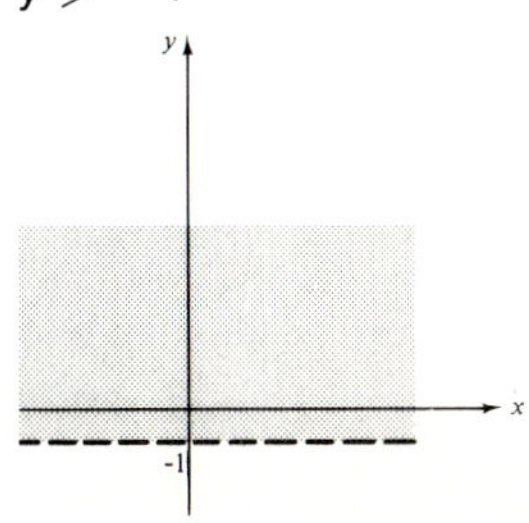

4. $y \leq 4$

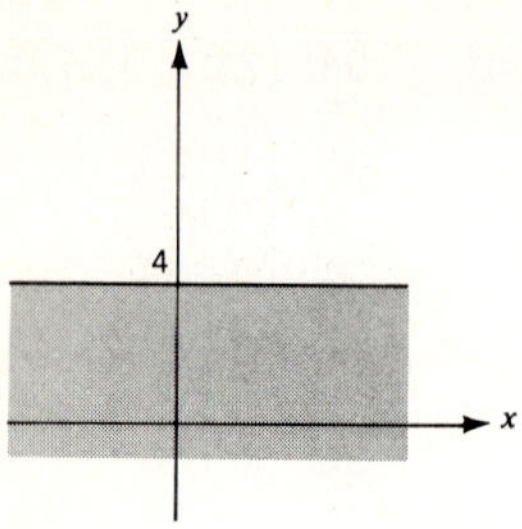

5. $x > 4$

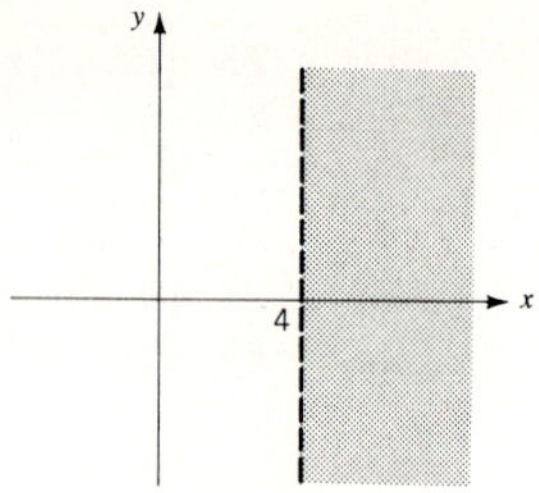

6. $x \leq -2$

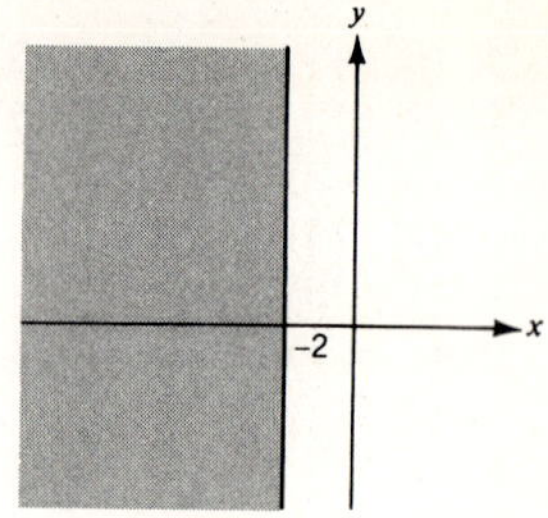

7. $y \geq 4$

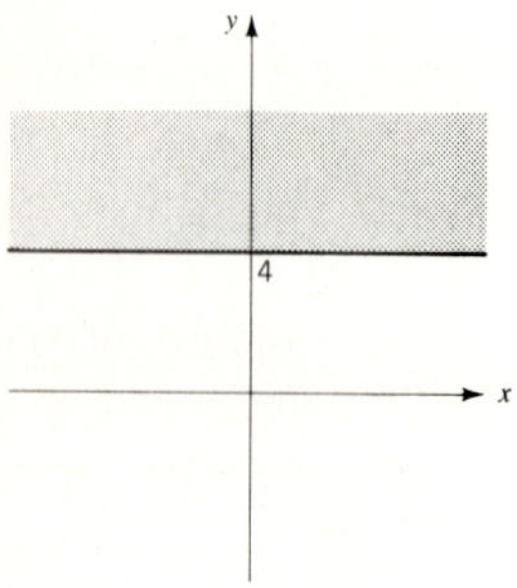

8. $y < 3$

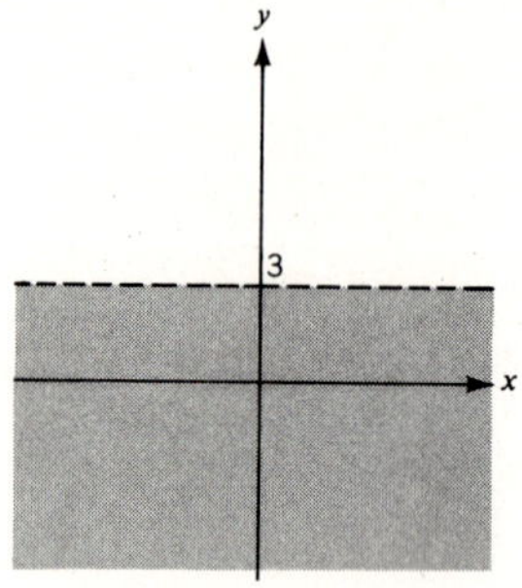

9. $1 \leq x \leq 3$

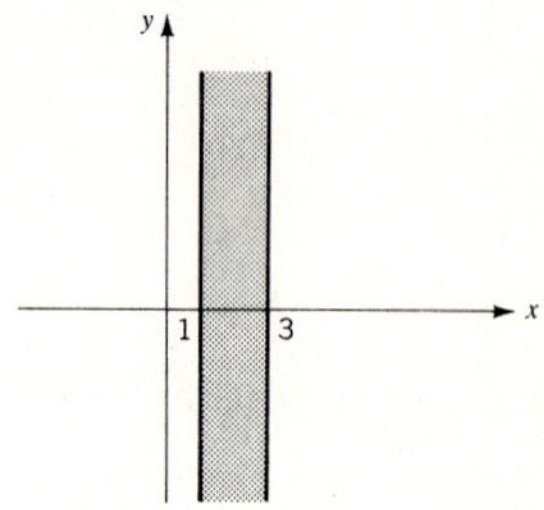

10. $2 < y \leq 5$

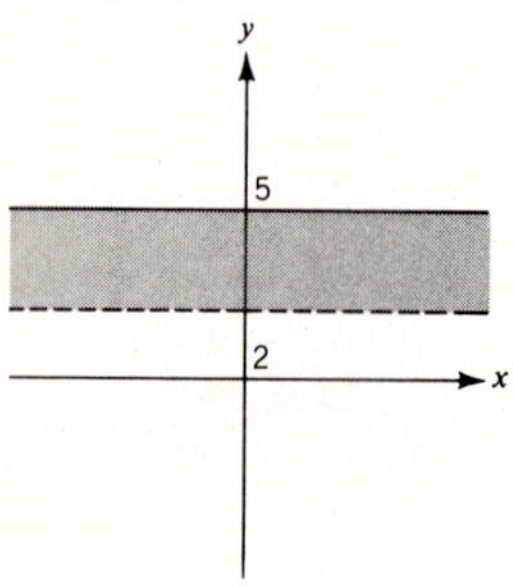

11. $x < 0$

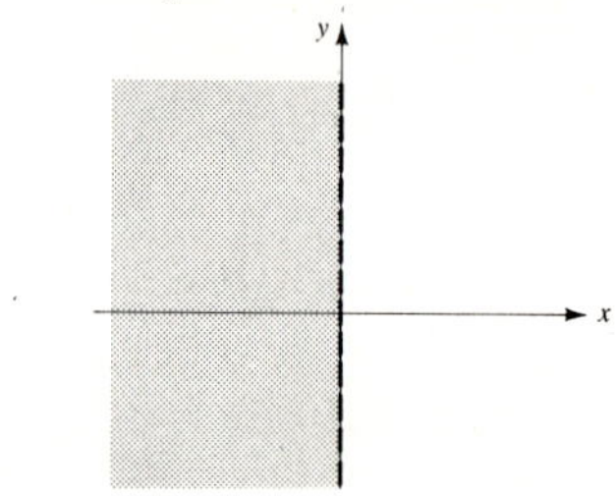

12. $y \geq 0$

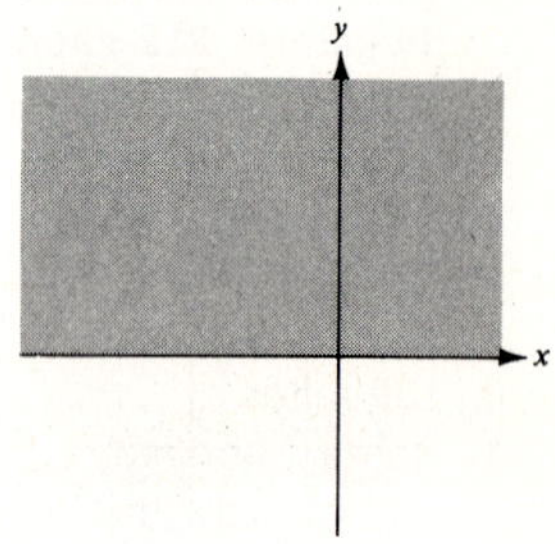

13. $-2 < x \leq 1$

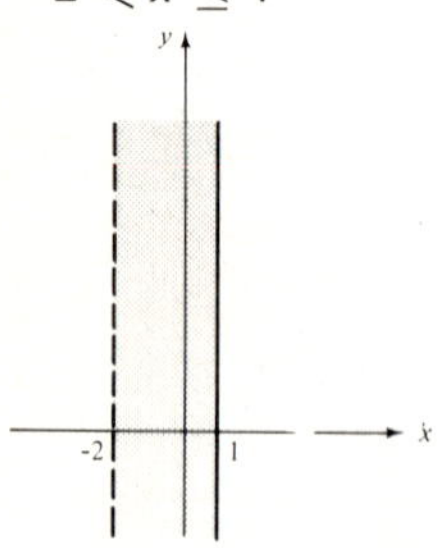

14. $1 \leq x < 3$

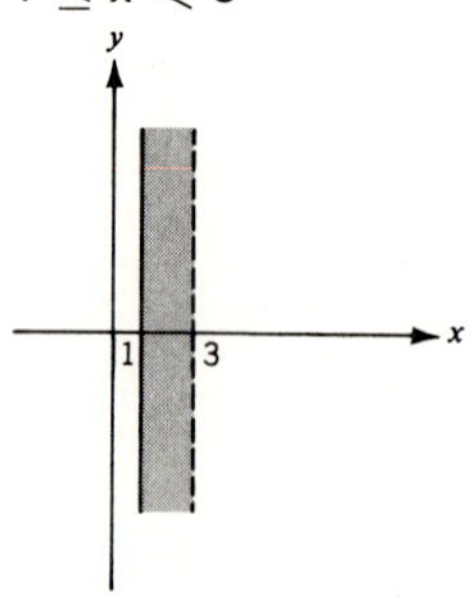

15. $y < 3x + 2$

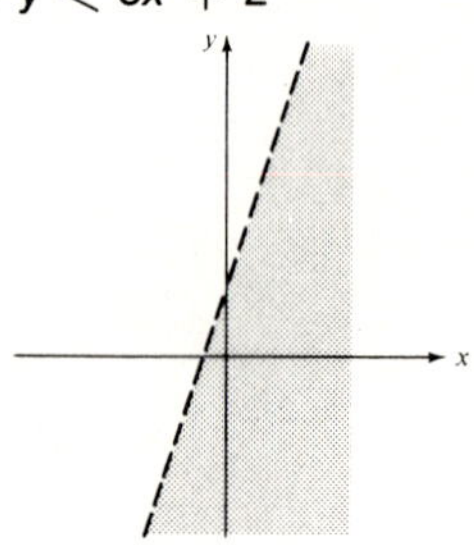

16. $y > x - 5$

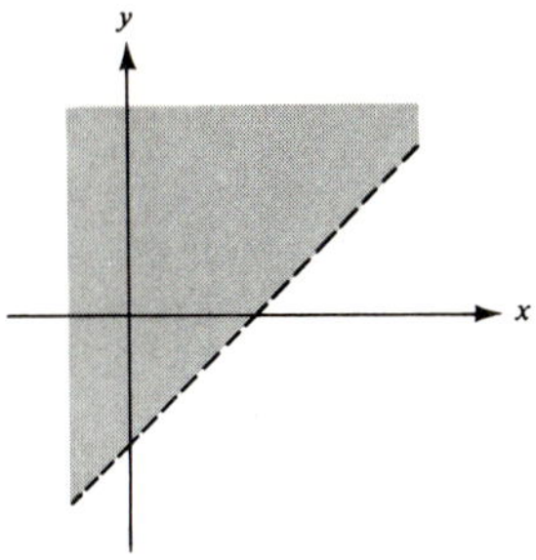

17. $y \geq 2x - 3$

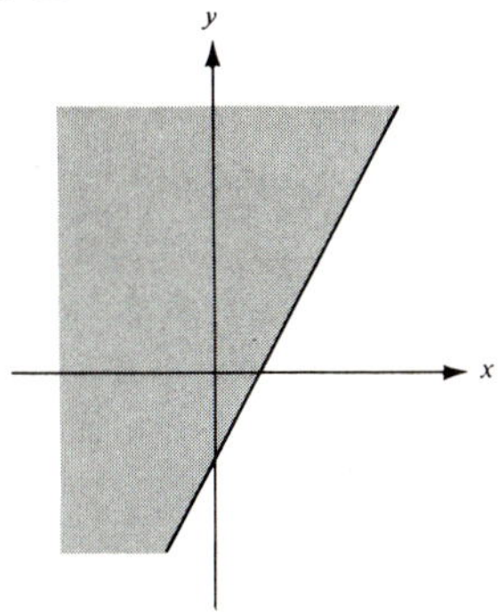

18. $y > x$

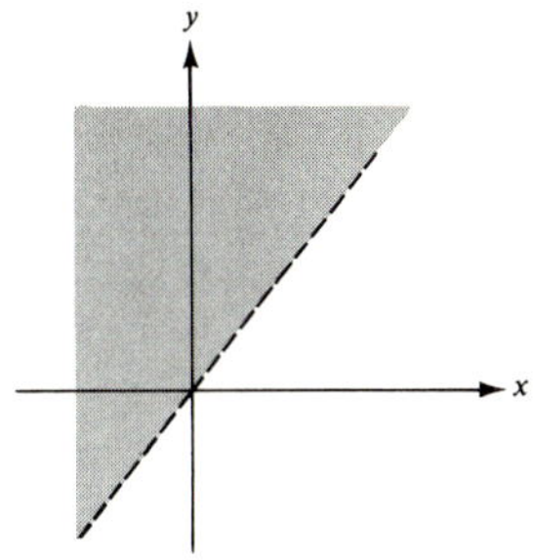

19. $y \leq 2 - x$

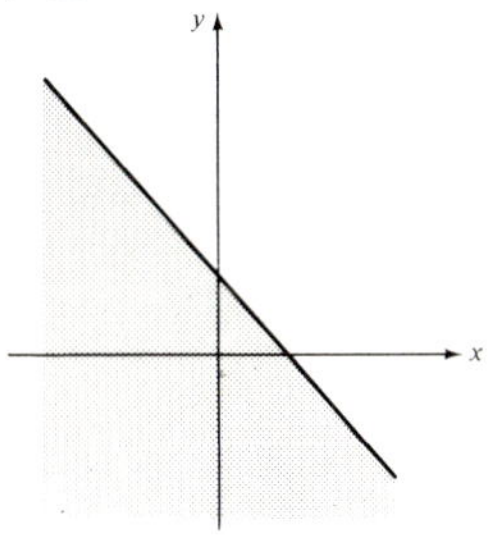

20. $y \geq x + 2$

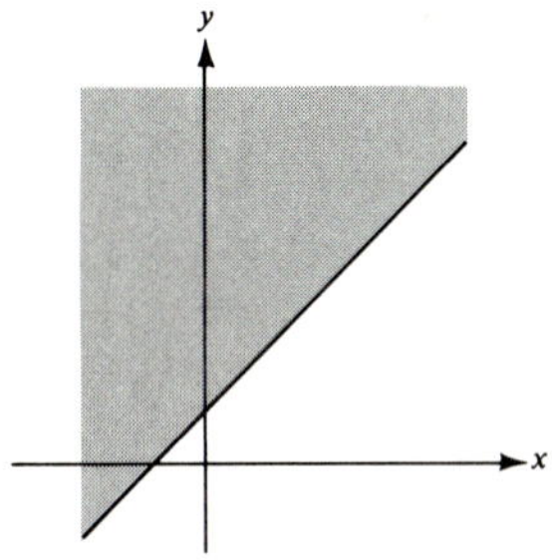

21. $y \leq x$

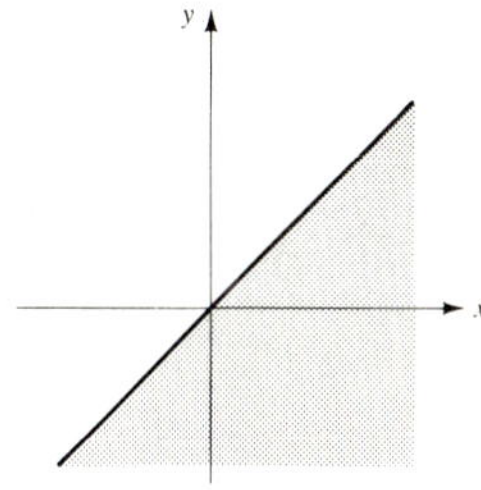

22. $x + 2y < 2$

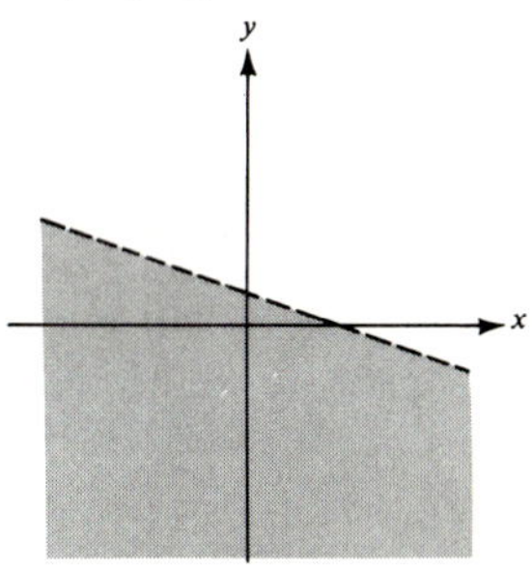

23. $2y \leq x + 4$

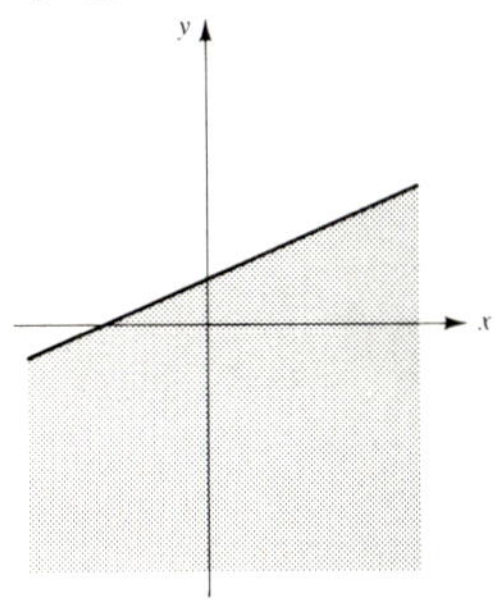

24. $3x \leq y + 2$

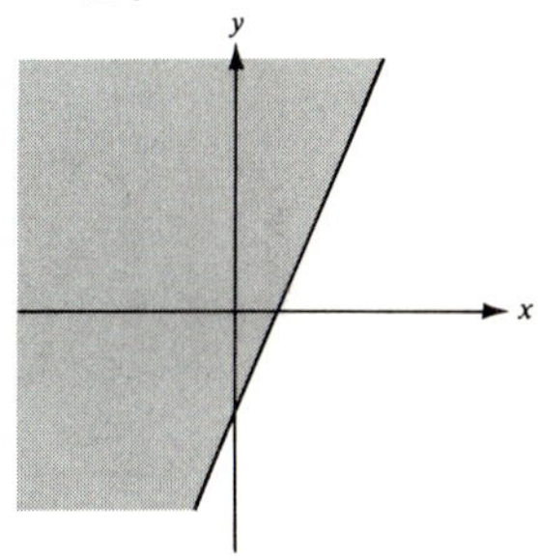

25. $y > x - 2$

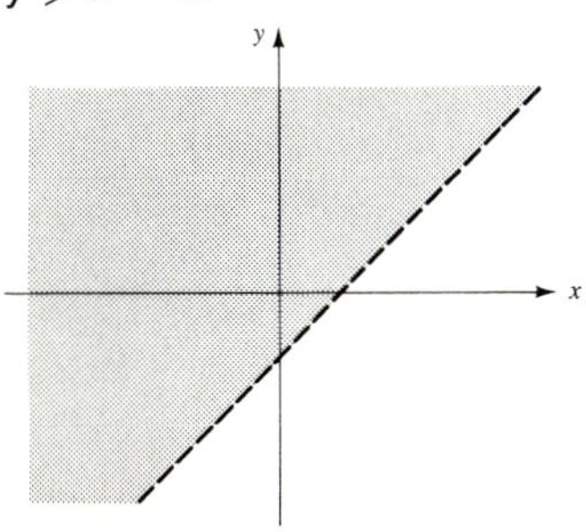

26. $y \leq x + 2$

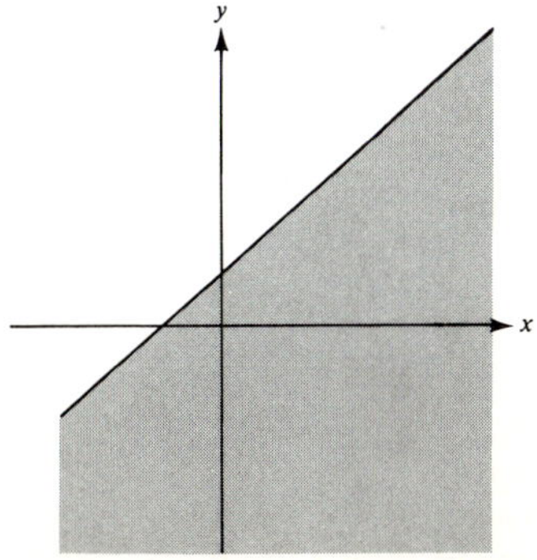

Review Exercises for Chapter 5

1. Refer to text.
2. b. and e.
3. a. $m = -\frac{3}{2}, b = 6$

b. $m = \frac{-5}{2}, b = 5$

c. $m = 2, b = -10$

d. $m = \frac{1}{3}, b = -2$

4. a. $y = \frac{-3}{2}x + 6$

b. $y = \frac{-5}{2}x + 5$

c. $y = 2x - 10$

d. $y = \frac{1}{3}x - 2$

5. a. $x - 2y + 6 = 0$
b. $3x + y - 8 = 0$
c. $y = -1$
d. $x + 3y = 0$
e. $x - 4y - 6 = 0$
f. $x = 5$
6. a. $2x - 9y + 6 = 0$
b. $7x + 3y + 3 = 0$
c. $x - y - 1 = 0$
d. $y = 5$
7. b. and c.
8. $2x - 3y + 14 = 0$
9. a. $3x - y - 11 = 0$
b. $y = 2$
c. $3x + 5y = 0$
d. $x = 5$
10. a. 3, b. 5, c. 1, d. 2, e. 4
11. $(2, -3)$
12. a. $(6, 2)$ b. $(3, 0)$
13. a. $(-4, 3)$ b. $(3.5, -3)$
14. \$20, \$10
15. 26 mph
16. 20 hours
17. \$3.50 adult and \$1.75 child
18. bonds, 12%; stocks, 21%

19. a.

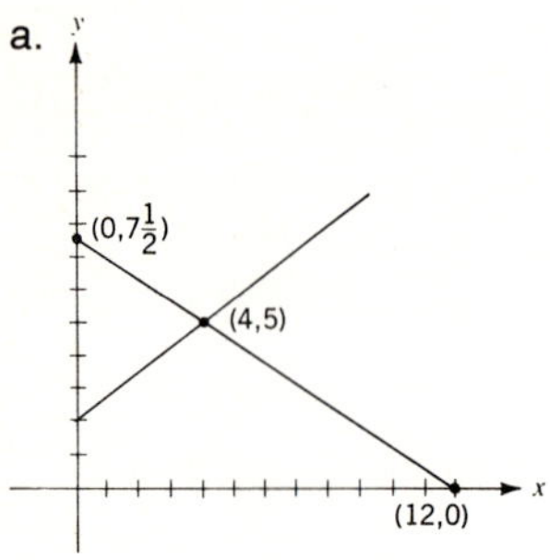

b. $\frac{15}{2}$ or \$7.50 c. 12 units

20. a. See figure for Exercise 19
b. 2 c. (4, 5)

21. a.

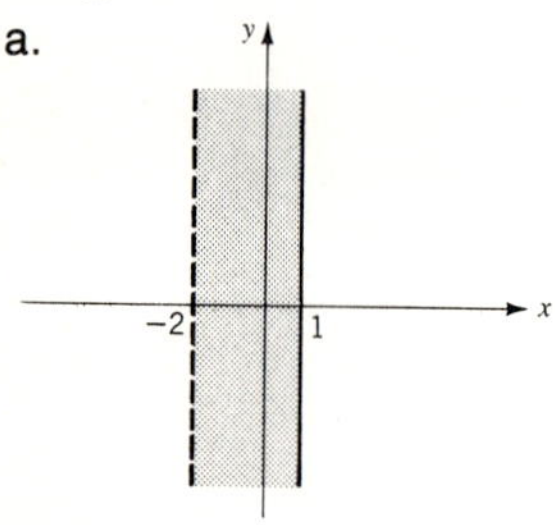

21. b.

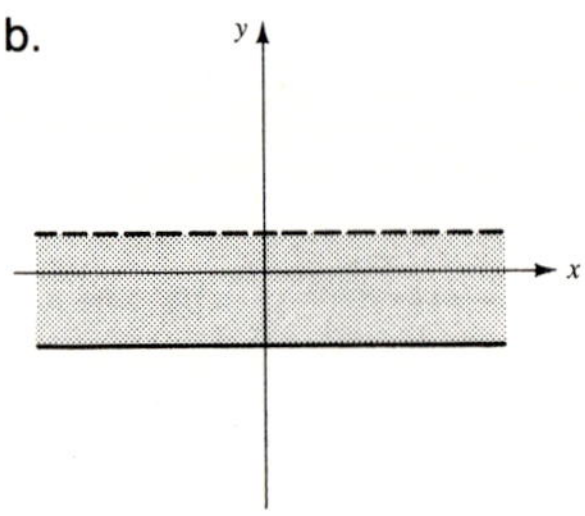

21. c.

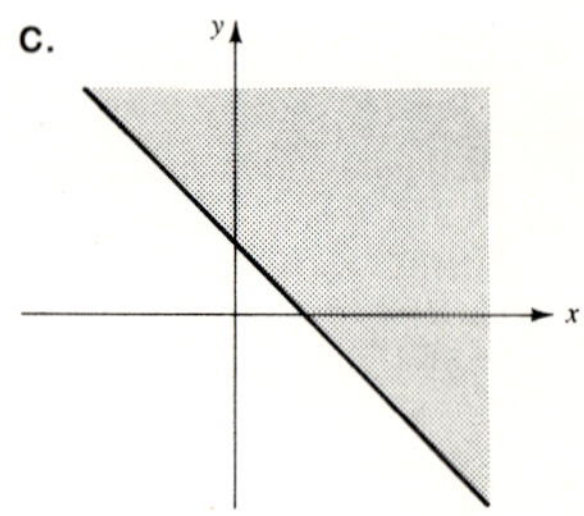

21. d.

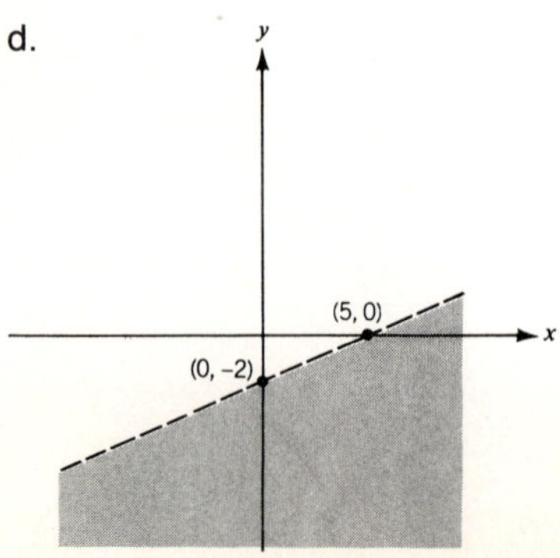

Answers to Exercises 6.1

Note: Square roots are approximated to nearest hundredth.

1. 4
2. 13
3. 3
4. 1
5. -2
6. -2
7. 20
8. 12
9. 3
10. 16
11. 4
12. 5
13. -4
14. -5
15. 1
16. 18
17. 17
18. 23
19. 13
20. 17
21. $\frac{5}{8}$
22. $\frac{3}{4}$
23. 0.1
24. 0.01
25. $\frac{5}{7}$
26. 3
27. 6.71
28. 3.16
29. 9.27
30. 1.73
31. 4.47
32. 7.94
33. 3.27
34. 0.76
35. 14.14
36. 3.46
37. 3.41
38. 3.86
39. 0.50
40. 5.96
41. 13
42. 12
43. $\sqrt{200} = 14.14$
44. $\sqrt{1369} = 37$
45. $\sqrt{2} = 1.41$
46. $\sqrt{21} = 4.58$
47. $\sqrt{64} = 8$
48. $\sqrt{26} = 5.10$
49. $\sqrt{8} = 2.83$
50. $\sqrt{20} = 4.47$
51. $\sqrt{50} = 7.07$
52. $\sqrt{5} = 2.24$
53. $a = 1, c = 2$
54. $b = \sqrt{12} = 3.46$; $c = 6.92$
55. $b = \sqrt{20} = 4.47$; $a = 8.94$
56. $a = \sqrt{2} = 1.41$; $b = 2.82$

Answers to Exercises 6.2

1. 10
2. 6
3. 10
4. 12
5. $5\sqrt{3}$
6. $20\sqrt{5}$
7. $5\sqrt{5}$
8. 6
9. $5\sqrt{5}$
10. $3\sqrt{3}$
11. $4\sqrt{2}$
12. $2\sqrt{6}$
13. $2\sqrt{2}$
14. $2\sqrt{3}$
15. $2\sqrt{10}$
16. $3\sqrt{6}$
17. $9\sqrt{2}$
18. $2\sqrt{7}$
19. $7\sqrt{2}$
20. 15
21. $5\sqrt{6}$
22. $9\sqrt{5}$
23. $7\sqrt{10}$
24. $6\sqrt{5}$
25. $6\sqrt{5}$
26. $4\sqrt{15}$
27. $6\sqrt{3}$
28. $11\sqrt{3}$
29. $20\sqrt{3}$
30. $10\sqrt{2}$
31. $10\sqrt{3}$
32. $10\sqrt{5}$
33. $22\sqrt{3}$
34. 43
35. 15
36. 14
37. 48
38. $14\sqrt{10}$
39. $2\sqrt{494}$
40. $4\sqrt{111}$
41. $\frac{\sqrt{3}}{3}$
42. $\frac{\sqrt{2}}{2}$
43. $\frac{\sqrt{7}}{7}$
44. $\frac{\sqrt{10}}{10}$
45. $\frac{3\sqrt{5}}{5}$
46. $\frac{5\sqrt{6}}{6}$
47. $2\sqrt{5}$
48. $\frac{\sqrt{6}}{3}$
49. $\frac{\sqrt{15}}{5}$
50. $2\sqrt{6}$
51. $\frac{\sqrt{3}}{6}$
52. $\frac{\sqrt{2}}{6}$
53. $\frac{\sqrt{2}}{10}$
54. $\frac{\sqrt{2}}{8}$
55. $\frac{5\sqrt{2}}{14}$
56. $\frac{4\sqrt{3}}{15}$
57. $\frac{\sqrt{3}}{9}$
58. $\frac{\sqrt{5}}{25}$
59. $\frac{\sqrt{2}}{4}$
60. $\frac{\sqrt{10}}{100}$
61. $\frac{1}{2}$
62. $\frac{1}{3}$
63. $2\sqrt{6}$
64. $3\sqrt{5}$
65. $\frac{\sqrt{3}}{9}$
66. $\frac{\sqrt{14}}{14}$
67. $\frac{\sqrt{14}}{2}$
68. $\frac{\sqrt{6}}{3}$
69. $\sqrt{15}$
70. $\sqrt{14}$
71. $\frac{\sqrt{a}}{a}$
72. $\frac{\sqrt{b}}{b}$
73. $\frac{1}{2c}$

74. $\frac{1}{3d}$
75. $\frac{5\sqrt{2b}}{b}$
76. $\frac{3\sqrt{5a}}{a}$
77. $\frac{d\sqrt{6}}{6}$
78. $\frac{c\sqrt{10}}{10}$
79. $\frac{\sqrt{b^2 - 4ac}}{2a}$
80. $\frac{\sqrt{d^2 - 9}}{3d}$

Answers to Exercises 6.3

1. $8\sqrt{5}$
2. $3\sqrt{2}$
3. $6\sqrt{6}$
4. $\sqrt{10}$
5. $3\sqrt{5} + 5\sqrt{3}$
6. $3\sqrt{3}$
7. $2\sqrt{3}$
8. $9\sqrt{7}$
9. $\sqrt{6}$
10. $3\sqrt{15}$
11. $3\sqrt{30}$
12. $7\sqrt{5} - 5\sqrt{7}$
13. $2\sqrt{28} = 4\sqrt{7}$
14. $\sqrt{10}$
15. 0
16. $7\sqrt{5}$
17. $4\sqrt{3}$
18. $7\sqrt{2}$
19. $12\sqrt{6}$
20. $8\sqrt{2}$
21. $-\sqrt{2} - 2\sqrt{3}$
22. $5\sqrt{5} + 2\sqrt{3}$
23. $\sqrt{7} + 2\sqrt{2}$
24. $\sqrt{6} + 4\sqrt{2}$
25. $31 + 10\sqrt{6}$
26. $11 - 4\sqrt{7}$
27. $21 - 8\sqrt{5}$
28. $46 + 12\sqrt{10}$
29. $4 - 2\sqrt{3}$
30. $9 - 4\sqrt{5}$
31. $59 + 14\sqrt{10}$
32. $11 + 6\sqrt{2}$
33. $15 + 3\sqrt{6}$
34. $8 - 4\sqrt{7}$
35. $-8 + 2\sqrt{5}$
36. $-30 - 5\sqrt{10}$
37. $110 - 60\sqrt{2}$
38. $98 + 40\sqrt{6}$
39. $-64 + 32\sqrt{3}$
40. $-72 - 24\sqrt{5}$
41. $13 + 7\sqrt{3}$
42. -1
43. $19 + 11\sqrt{3}$
44. $24 + 13\sqrt{3}$
45. $15 + 7\sqrt{3}$
46. $23 + 13\sqrt{3}$
47. $2 - 3\sqrt{2}$
48. $17 - 8\sqrt{2}$
49. $39 - 20\sqrt{2}$
50. $31 - 15\sqrt{2}$
51. 0
52. $9 - 3\sqrt{17}$
53. 0
54. $1 - \sqrt{5}$
55. 0
56. 0
57. $100 - 20\sqrt{2}$
58. 0
59. $\frac{4 + \sqrt{3}}{3}$
60. $\frac{1 + \sqrt{3}}{2}$
61. $\frac{2 - \sqrt{2}}{2}$
62. $\frac{6 - \sqrt{2}}{3}$
63. $-2 + \sqrt{5}$
64. $\frac{-1 - \sqrt{11}}{5}$
65. $7 + 2\sqrt{5}$
66. $5 - \sqrt{6}$
67. 2
68. $2\sqrt{5}$
69. $2\sqrt{6}$
70. $2\sqrt{3}$
71. $6\sqrt{2}$
72. 5
73. 8
74. 9
75. $2\sqrt{15}$
76. $4\sqrt{6}$

Answers to Exercises 6.4

1. $\{5, -2\}$
2. $\{4, -3\}$
3. $\{0, 4\}$
4. $\{0, -3\}$
5. $\{2, 3\}$
6. $\{-3, 2\}$
7. $\{-7, 5\}$
8. $\{-3, 6\}$
9. $\{-5, 4\}$
10. $\{-3, 10\}$
11. $\left\{-\frac{5}{2}, 2\right\}$
12. $\left\{\frac{4}{3}, -1\right\}$
13. $\left\{1, \frac{1}{6}\right\}$
14. $\left\{4, \frac{1}{5}\right\}$
15. $\left\{-\frac{5}{2}, \frac{5}{2}\right\}$
16. $\left\{-\frac{8}{3}, \frac{8}{3}\right\}$
17. $\left\{-\frac{3\sqrt{3}}{2}, \frac{3\sqrt{3}}{2}\right\}$
18. $\left\{\frac{-4\sqrt{2}}{3}, \frac{4\sqrt{2}}{3}\right\}$
19. $\left\{\frac{-\sqrt{30}}{5}, \frac{\sqrt{30}}{5}\right\}$
20. $\left\{\frac{-\sqrt{14}}{2}, \frac{\sqrt{14}}{2}\right\}$
21. $\left\{0, \frac{6}{5}\right\}$
22. $\left\{0, \frac{7}{2}\right\}$
23. $\left\{0, \frac{1}{2}\right\}$
24. $\{0, 4\}$
25. $\left\{1, -\frac{1}{2}\right\}$
26. $\{3, 1\}$
27. $\{6\}$, double root
28. $\{5\}$, double root
29. $\left\{\frac{3}{2}\right\}$, double root
30. $\left\{\frac{10}{3}\right\}$, double root
31. $\{0, 4\}$
32. $\{0, -9\}$
33. $\{-2, 2\}$
34. $\{-3, 3\}$
35. $\{4\}$, double root
36. $\{-10\}$, double root
37. $\{1, 3\}$
38. $\{-1, -5\}$
39. $\{-9, -1\}$
40. $\{1, 13\}$
41. $\{-1, 2\}$
42. $\left\{-\frac{7}{3}, 1\right\}$

43. $\left\{\dfrac{-3-\sqrt{7}}{5}, \dfrac{-3+\sqrt{7}}{5}\right\}$

44. $\left\{\dfrac{5-\sqrt{5}}{4}, \dfrac{5+\sqrt{5}}{4}\right\}$

Answers to Exercises 6.5

1. $5 \pm \sqrt{2}$
2. $-3 \pm \sqrt{7}$
3. $-4 \pm 3\sqrt{5}$
4. $2 \pm 2\sqrt{2}$
5. $2 \pm \sqrt{3}$
6. $1 \pm \sqrt{6}$
7. $4 \pm 2\sqrt{5}$
8. $3 \pm \sqrt{5}$
9. $1 \pm 2\sqrt{5}$
10. $5 \pm 3\sqrt{2}$
11. $-5 \pm 3\sqrt{3}$
12. $-4 \pm 2\sqrt{7}$
13. $-7 \pm 6\sqrt{2}$
14. $9 \pm 3\sqrt{3}$
15. $-6 \pm 3\sqrt{2}$
16. $-8 \pm 4\sqrt{2}$
17. $10 \pm 2\sqrt{10}$
18. $6 \pm 2\sqrt{6}$
19. $-6, \dfrac{5}{2}$
20. $3, -\dfrac{4}{3}$
21. $\dfrac{5 \pm 2\sqrt{2}}{2}$
22. $\dfrac{-7 \pm 2\sqrt{6}}{3}$
23. $\dfrac{5 \pm \sqrt{85}}{5}$
24. $\dfrac{-12 \pm \sqrt{138}}{6}$
25. $-\dfrac{1}{2}, \dfrac{1}{6}$
26. $\dfrac{2 \pm \sqrt{13}}{5}$
27. $\dfrac{5 \pm \sqrt{7}}{6}$
28. $\dfrac{8 \pm 3\sqrt{6}}{3}$
29. $8, \dfrac{2}{3}$
30. $2, -\dfrac{1}{3}$

Answers to Exercises 6.6

1. $\dfrac{3 \pm \sqrt{5}}{2}$
2. $\dfrac{-1 \pm \sqrt{5}}{2}$
3. $\dfrac{-3 \pm \sqrt{5}}{2}$
4. $\dfrac{1 \pm \sqrt{5}}{2}$
5. $\dfrac{-1}{2}, -2$
6. $\dfrac{-5 \pm \sqrt{13}}{6}$
7. $\dfrac{5 \pm \sqrt{41}}{4}$
8. $-1, \dfrac{-2}{3}$
9. $\dfrac{5 \pm 5\sqrt{5}}{2}$
10. $\dfrac{3 \pm 3\sqrt{21}}{10}$
11. $\dfrac{1 \pm \sqrt{6}}{5}$
12. $\dfrac{3 \pm \sqrt{3}}{3}$
13. $\dfrac{-3}{2}, \dfrac{-1}{2}$
14. $\dfrac{-7 \pm \sqrt{73}}{4}$
15. $\dfrac{2 \pm \sqrt{7}}{2}$
16. $3, \dfrac{1}{2}$
17. $\dfrac{9 \pm \sqrt{21}}{10}$
18. $\dfrac{6 \pm 2\sqrt{6}}{3}$
19. $\dfrac{2 \pm \sqrt{10}}{3}$
20. $\dfrac{3 \pm \sqrt{14}}{5}$
21. 1; rational and unequal; $(x + 2)(x + 1)$
22. 17; irrational and unequal
23. 41; irrational and unequal
24. 81; rational and unequal; $(5x - 4)(x + 1)$
25. 0; rational and equal; $(3x - 10)^2$
26. 25; rational and unequal; $(2x + 1)(x - 2)$
27. 85; irrational and unequal
28. 0; rational and equal; $(3x - 5)^2$
29. 21; irrational and unequal
30. 121; rational and unequal; $(5x + 1)(x - 2)$
31. 16; rational and unequal; $(x + 2)(x + 6)$
32. 500; irrational and unequal
33. 0; rational and equal; $(2x + 1)^2$
34. 0; rational and equal; $(x - 5)^2$
35. 0; rational and equal; $(3x - 2)^2$
36. 200; irrational and unequal
37. 89; irrational and unequal
38. 100; rational and unequal; $(3x + 4)(x - 2)$
39. 144; rational and unequal; $(3x - 2)(3x + 2)$
40. 0; rational and equal; $(5x - 2)^2$
41. $x = \dfrac{-p \pm \sqrt{p^2 - 4q}}{2}$

42. $x = \dfrac{-2p \pm \sqrt{4p^2 - 4q}}{2}$

$= -p \pm \sqrt{p^2 - q}$

43. $x = \dfrac{-8y \pm \sqrt{64y^2 + 36y^2}}{6} = \dfrac{y}{3}$ or $-3y$

44. $x = \dfrac{9y \pm \sqrt{81y^2 - 32y^2}}{8}$

$x = 2y$ or $x = \dfrac{y}{4}$

45. $W = \dfrac{-L \pm \sqrt{L^2 + 4L^2}}{2} = \dfrac{-L \pm L\sqrt{5}}{2}$

46. $L = \dfrac{W \pm \sqrt{W^2 + 4W^2}}{2}$

$= \dfrac{W \pm W\sqrt{5}}{2} = \dfrac{1 \pm \sqrt{5}}{2}W$

47. $I = \dfrac{-E \pm \sqrt{E^2 + 4RP}}{2R}$

48. $s = \dfrac{(2k + 1) \pm \sqrt{4k + 1}}{2}$

49. $x = \dfrac{-q \pm \sqrt{q^2 - 4pr}}{2p}$

50. $m = \dfrac{-b \pm \sqrt{b^2 - 4ac}}{2a}$

Answers to Exercises 6.7

1. $x^2 = 91 + 6x$; -7 or 13
2. $20x - 36 = x^2$; 2 or 18
3. $x + \dfrac{1}{x} = 2$; 1
4. $x - \dfrac{1}{x} = \dfrac{15}{4}$; 4 or $\dfrac{-1}{4}$
5. $5x^2 + 10x = 315$; 7 or -9
6. $3x^2 + 4x = 95$; 5 or $\dfrac{-19}{3}$
7. $x^2 + (x + 7)^2 = 169$; 5 and 12
8. $x^2 + x^2 = 64$; $4\sqrt{2}$
9. a. 0 or 1200
   b. 400 or 800
10. 10 or 115
11. $\dfrac{2}{3}$
12. 4 or 18
13. $6400 = 400h - 4h^2$; 20
14. $4\left(L - \dfrac{1}{2}\right)^2 = 529$; 12
15. $W(3W + 1) = 52$; 4 inches by 13 inches
16. $W(2W - 7) = 660$; 20 meters by 33 meters
17. $x^2 = (16)^2 + (30)^2$; 34 feet
18. $x^2 + (21 - x)^2 = 225$; 9 inches and 12 inches
19. $(2x + 20)(2x + 80) = 1600 + 864$; 4 feet
20. $2(x - 4)(2x - 4) = 320$; 12 centimeters by 24 centimeters
21. $\dfrac{4}{x} + \dfrac{3}{x + 3} = 1$; 6 hours
22. $\dfrac{6}{x} + \dfrac{6}{x + 5} = 1$; 10 minutes
23. $\dfrac{x}{6} + \dfrac{x}{x + 2} = 1$; $-1 + \sqrt{13} \approx$ 2 hours and 36 minutes
24. $\dfrac{15}{x} + \dfrac{15}{x + 12} - \dfrac{15}{45} = 1$; 18 minutes

25. $\frac{8}{x+2} + \frac{8}{x-2} = 3;$
6 mph

26. $\frac{2610}{x-20} = \frac{1}{3} + \frac{2610}{x+20};$
560 mph

27. $\frac{720}{210+x} + \frac{720}{210-x} = 7;$
30 kph

28. $\frac{120}{x-5} = 1 + \frac{120}{x+5};$
35 knots

29. $x^2 + 8x - 20 = 0;$
2 units

30. $x^2 + 24x - 256 = 0;$
8 units

Review Exercises for Chapter 6

1. 5.20
2. 6.24
3. 14
4. 14
5. 9.33
6. −6.63
7. −8
8. 8
9. 10
10. $2\sqrt{5}$
11. $4\sqrt{10}$
12. $10\sqrt{5}$
13. $-\sqrt{5}$
14. $8\sqrt{7}$
15. $\frac{12 - 10\sqrt{2}}{3}$
16. $\frac{7 + \sqrt{15}}{2}$
17. $\frac{-4 - \sqrt{14}}{4}$
18. $29 + 12\sqrt{5}$
19. $67 - 12\sqrt{7}$
20. $-20 + 5\sqrt{2}$
21. $-1 \pm \sqrt{2}$
22. $6 \pm 3\sqrt{5}$
23. $2 \pm \sqrt{5}$
24. $-5 \pm 2\sqrt{6}$
25. $\frac{-3 \pm \sqrt{17}}{2}$
26. $\frac{6 \pm 5\sqrt{2}}{2}$
27. $\frac{-3 \pm \sqrt{3}}{3}$
28. $\frac{3 \pm \sqrt{3}}{3}$
29. $\frac{3}{2}, -2$
30. $-\frac{4}{3}, \frac{1}{2}$
31. $1 \pm \sqrt{13}$
32. $\frac{1 \pm \sqrt{7}}{3}$
33. $\frac{3}{2}, \frac{7}{2}$
34. $\frac{-2 \pm \sqrt{6}}{3}$
35. $\frac{4 \pm 3\sqrt{2}}{2}$
36. $\frac{2 \pm \sqrt{13}}{3}$
37. 0, 4
38. −9 (double root)
39. $\frac{5 \pm \sqrt{13}}{6}$
40. $\frac{1 \pm \sqrt{3}}{3}$
41. 30 minutes
42. 0 or 1
43. 5 and 6
44. 3 feet by 15 feet
45. 180 mph
46. 27 inches, 36 inches
47. $\frac{cv \pm \sqrt{c^2v^2 - 2gH}}{g}$
48. $\frac{-\pi h \pm \sqrt{\pi^2h^2 + 2\pi T}}{2\pi}$

Index